Lecture Notes in Artificial Intel

Edited by J. G. Carbonell and J. Siekmann

Subseries of Lecture Notes in Computer Science

Rosa Meo Pier Luca Lanzi
Mika Klemettinen (Eds.)

Database Support for Data Mining Applications

Discovering Knowledge with Inductive Queries

Series Editors

Jaime G. Carbonell, Carnegie Mellon University, Pittsburgh, PA, USA
Jörg Siekmann, University of Saarland, Saarbrücken, Germany

Volume Editors

Rosa Meo
Università degli Studi di Torino
Dipartimento di Informatica
Corso Svizzera, 185, 10149 Torino, Italy
E-mail: meo@di.unito.it

Pier Luca Lanzi
Politecnico di Milano
Dipartimento di Elettronica e Informazione
Piazza Leonardo da Vinci 32, 20133 Milano, Italy
E-mail: lanzi@elet.polimi.it

Mika Klemettinen
Nokia Research Center
P.O.Box 407, Itämerenkatu 11-13, 00045 Nokia Group, Finland
E-mail: mika.klemettinen@nokia.com

Library of Congress Control Number: 2004095564

CR Subject Classification (1998): I.2, H.2, H.3

ISSN 0302-9743
ISBN 3-540-22479-3 Springer-Verlag Berlin Heidelberg New York

Springer-Verlag is a part of Springer Science+Business Media

springeronline.com

Printed in Germany

Typesetting: Camera-ready by author, data conversion by Scientific Publishing Services, Chennai, India
Printed on acid-free paper SPIN: 11301172 06/3142 5 4 3 2 1 0

Preface

Data mining from traditional relational databases as well as from non-traditional ones such as semi-structured data, Web data and scientific databases such as biological, linguistic and sensor data has recently become a popular way of discovering hidden knowledge. In the context of relational and traditional data, methods such as association rules, chi square rules, ratio rules, implication rules, etc. have been proposed in multiple, varied contexts. In the context of non-traditional data, newer, more experimental yet novel techniques are being proposed. There is an agreement among the researchers across communities that data mining is a key ingredient for success in their respective areas of research and development. Consequently, interest in developing new techniques for data mining has peaked and a tremendous stride is being made to answer interesting and fundamental questions in various disciplines using data mining.

In the past, researchers mainly focused on algorithmic issues in data mining and placed much emphasis on scalability. Recently, the focus has shifted towards a more declarative way of answering questions using data mining that has given rise to the concept of mining queries.

Data mining has recently been applied with success to discovering hidden knowledge from relational databases. Methods such as association rules, chi square rules, ratio rules, implication rules, etc. have been proposed in several and very different contexts. To cite just the most frequent and famous ones: the market basket analysis, failures in telecommunication networks, text analysis for information retrieval, Web content mining, Web usage, log analysis, graph mining, information security and privacy, and finally analysis of objects traversal by queries in distributed information systems.

From these widespread and various application domains it results that data mining rules constitute a successful and intuitive descriptive paradigm able to offer complementary choices in rule induction. Other than inductive and abductive logic programming, research into data mining from knowledge bases has been almost non-existent, because contemporary methods place the emphasis on the scalability and efficiency of algorithmic solutions, whose inherent procedurality is difficult to cast into the declarativity of knowledge base systems.

In particular, researchers convincingly argue that the ability to declaratively mine and analyze relational databases for decision support is a critical requirement for the success of the acclaimed data mining technology. Indeed, DBMSs constitute today one of the most advanced and sophisticated achievements that applied computer science has made in the past years. Unfortunately, almost all the most powerful DBMSs we have today have been developed with a focus on On-Line Transaction-Processing tasks. Instead, database technology for On-Line Analytical-Processing tasks, such as data mining, is more recent and in need of further research.

Although there have been several encouraging attempts at developing methods for data mining using SQL, simplicity and efficiency still remain significant prerequisites for further development. It is well known that today database technology is mature enough: popular DBMSs, such as Oracle, DB2 and SQL-Server, provide interfaces, services, packages and APIs that embed data mining algorithms for classification, clustering, association rules extraction and temporal sequences, such that they are directly available to programmers and ready to be called by applications.

Therefore, it is envisioned that we should be able now to mine relational databases for interesting rules directly from database query languages, without any data restructuring or preprocessing steps. Hence no additional machineries with respect to database languages would be necessary. This vision entails that the optimization issues should be addressed at the system level for which we have now a significant body of research, while the analyst could concentrate better on the declarative and conceptual level, in which the difficult task of interpretation of the extracted knowledge occurs. Therefore, it is now time to develop declarative paradigms for data mining so that these developments can be exploited at the lower and system level, for query optimization.

With this aim we planned this book on "Data Mining" with an emphasis on approaches that exploit the available database technology, declarative data mining, intelligent querying and associated issues such as optimization, indexing, query processing, languages and constraints. Attention is also paid to solution of data preprocessing problems, such as data cleaning, discretization and sampling, developed using database tools and declarative approaches, etc.

Most of this book resulted also as a consequence of the work we conducted during the development of the *cInQ* project (**c**onsortium on discovering knowledge with **In**ductive **Q**ueries) an EU funded project (IST 2000-26469) aiming at developing database technology for leveraging decision support systems by means of query languages and inductive approaches to knowledge extraction from databases. It presents new and invited contributions, plus the best papers, extensively revised and enlarged, presented during workshops on the topics of database technology, data mining and inductive databases at international conferences such as EDBT and PKDD/ECML, in 2002.

May 2004

Rosa Meo
Pier Luca Lanzi
Mika Klemettinen

Volume Organization

This volume is organized in two main sections. The former focuses on *Database Languages and Query Execution*, while the latter focuses on methodologies, techniques and new approaches that provide *Support for Knowledge Discovery Process*. Here, we briefly overview each contribution.

Database Languages and Query Execution

The first contribution is *Inductive Databases and Multiple Uses of Frequent Itemsets: The cInQ Approach* which presents the main contributions of theoretical and applied nature, in the field of inductive databases obtained in the `cInQ` project.

In *Query Languages Supporting Descriptive Rule Mining: A Comparative Study* we provide a comparison of features of available relational query languages for data mining, such as `DMQL`, `MSQL`, `MINE RULE`, and standardization efforts for coupling database technology and data mining systems, such as `OLEDB-DM` and `PMML`.

Declarative Data Mining Using SQL-3 shows a new approach, compared to existing SQL approaches, to mine association rules from an object-relational database: it uses a recursive join in SQL-3 that allows no restructuring or preprocessing of the data. It proposes a new `mine by` SQL-3 operator for capturing the functionality of the proposed approach.

Towards a Logic Query Language for Data Mining presents a logic database language with elementary data mining mechanisms, such as user-defined aggregates that provide a model, powerful and general as well, of the relevant aspects and tasks of knowledge discovery.

Data Mining Query Language for Knowledge Discovery in a Geographical Information System presents *SDMOQL* a spatial data mining query language for knowledge extraction from GIS. The language supports the extraction of classification rules and association rules, the use of background models, various interestingness measures and the visualization.

Towards Query Evaluation in Inductive Databases Using Version Spaces studies inductive queries. These ones specify constraints that should be satisfied by the data mining patterns in which the user is interested. This work investigates the properties of solution spaces of queries with monotonic and anti-monotonic constraints and their boolean combinations.

The GUHA Method, Data Preprocessing and Mining surveys the basic principles and foundations of the *GUHA* method, the available systems and related works. This method originated in the Czechoslovak Academy of Sciences of Prague in the mid 1960s with strong logical and statistical foundations. Its main principle is to let the computer generate and evaluate all the hypotheses that may be interesting given the available data and the domain problem. This work discusses also the relationships between the GUHA method and relational data mining and discovery science.

Constraint Based Mining of First Order Sequences in SeqLog presents a logical language, *SeqLog*, for mining and querying sequential data and databases. This language is used as a representation language for an inductive database system. In this system, variants of level-wise algorithms for computing the version space of the solutions are proposed and experimented in the user-modeling domain.

Support for Knowledge Discovery Process

Interactivity, Scalability and Resource Control for Efficient KDD Support in DBMS proposes a new approach for combining preprocessing and data mining operators in a KDD-aware implementation algebra. In this way data mining operators can be integrated smoothly into a database system, thus allowing interactivity, scalability and resource control. This framework is based on the extensive use of pipelining and is built upon an extended version of a specialized database index.

Frequent Itemset Discovery with SQL Using Universal Quantification investigates the integration of data analysis functionalities into two basic components of a database management system: query execution and optimization. It employs universal and existential quantifications in queries and a vertical layout to ease the set containment operations needed for frequent itemsets discovery.

Deducing Bounds on the Support of Itemsets provides a complete set of rules for deducing tight bounds on the support of an itemset if the support of all its subsets are known. These bounds can be used by the data mining system to choose the best access path to data and provide a better representation of the collection of frequent itemsets.

Model-Independent Bounding of the Supports of Boolean Formulae in Binary Data considers frequencies of arbitrary boolean formulas, a new class of aggregates: the summaries. These ones are computed for descriptive purposes on a sparse binary data set. This work considers the problem of finding tight upper bounds on these frequencies and gives a general formulation of the problem with a linear programming solution.

Condensed Representations for Sets of Mining Queries proposes a general framework for condensed representations of sets of mining queries, defined by monotonic and anti-monotonic selection predicates. This work proves important for inductive and database systems for data mining since it deals with *sets* of queries, whereas previous work in maximal, closed and condensed representations treated so far the representation of a *single* query only.

One-Sided Instance-Based Boundary Sets introduces a family of version-space representations that are important for their applicability to inductive databases. They correspond to the task of concept learning from a database of examples when this database is updated. One-sided instance-based boundary sets are shown to be correctly and efficiently computable.

Domain Structures in Filtering Irrelevant Frequent Patterns introduces a notion of domain constraints, based on distance measures and in terms of domain structure and concept taxonomies. Domain structures are useful in the analysis of communications networks and complex systems. Indeed they allow irrelevant combinations of events that reflect the simultaneous information of independent processes in the same database to be pruned.

Integrity Constraints over Association Rules investigates the notion of integrity constraints in inductive databases. This concept is useful in detecting inconsistencies in the results of common data mining tasks. This work proposes a form of integrity constraints called *association map constraints* that specifies the allowed variations in confidence and support of association rules.

Acknowledgments

For reviewing the contributions included in this volume we relied on a group of internationally well-known experts who we wish to thank here.

Roberto Bayardo	IBM Almaden Research Center, USA
Elena Baralis	Politecnico di Torino, Italy
Christian Böhm	University for Health Informatics and Technology, Austria
Saso Dzeroski	Jozef Stefan Institute, Slovenia
Kris Koperski	Insightful Corporation, USA
Stefan Kramer	Technische Universität München, Germany
Dominique Laurent	Université F. Rabelais, Tours, France
Giuseppe Manco	ICAR-CNR, Italy
Stefano Paraboschi	Università di Bergamo, Italy
Jian Pei	The State University of New York, USA
Giuseppe Psaila	Università di Bergamo, Italy
Lorenza Saitta	Università del Piemonte Orientale, Italy
Maria Luisa Sapino	Università di Torino, Italy
Hannu Toivonen	University of Helsinki, Finland
Jiong Yang	University of Illinois at Urbana Champaign, USA
Mohammed Zaki	Rensselaer Polytechnic Institute, USA

Table of Contents

I Database Languages and Query Execution

II Support for KDD-Process

Inductive Databases and Multiple Uses of Frequent Itemsets: The cINQ Approach

Jean-François Boulicaut

Institut National des Sciences Appliquées de Lyon,
LIRIS CNRS FRE 2672, Bâtiment Blaise Pascal
F-69621 Villeurbanne cedex, France

Abstract. Inductive databases (IDBs) have been proposed to afford the problem of knowledge discovery from huge databases. With an IDB the user/analyst performs a set of very different operations on data using a query language, powerful enough to perform all the required elaborations, such as data preprocessing, pattern discovery and pattern post-processing. We present a synthetic view on important concepts that have been studied within the cINQ European project when considering the pattern domain of itemsets. Mining itemsets has been proved useful not only for association rule mining but also feature construction, classification, clustering, etc. We introduce the concepts of pattern domain, evaluation functions, primitive constraints, inductive queries and solvers for itemsets. We focus on simple high-level definitions that enable to forget about technical details that the interested reader will find, among others, in cINQ publications.

1 Introduction

Knowledge Discovery in Databases (KDD) is a complex interactive process which involves many steps that must be done sequentially. In the cINQ project[1], we want to develop a new generation of databases, called "*inductive databases*" (IDBs), suggested by Imielinski and Mannila in [42] and for which a simple formalization has been proposed in [20]. This kind of databases integrate *raw data* with *knowledge* extracted from *raw data*, materialized under the form of patterns into a common framework that supports the knowledge discovery process within a database framework. In this way, the process of KDD consists essentially in a querying process, enabled by a query language that can deal either with raw data or patterns and that can be used throughout the whole KDD process across many different applications. A few query languages can be considered as candidates for inductive databases. For instance, considering the prototypical case of assoc-

[1] This research is partially funded by the Future and Emerging Technologies arm of the IST Programme FET-Open scheme (cINQ project IST-2000-26469). The author warmly acknowledges all the contributors to the cINQ project and more particularly Luc De Raedt, Baptiste Jeudy, Mika Klemettinen, and Rosa Meo.

R. Meo et al.(Eds.): Database Support for Data Mining Applications,LNAI 2682, pp. 1–23, 2004.

iation rule mining, [10] is a comparative evaluation of three proposals (`MSQL` [43], `DMQL` [38], and `MINE RULE` [59]) in the light of the IDBs' requirements.

In this paper, we focus on mining queries, the so-called *inductive queries*, i.e., queries that return patterns from a given database. More precisely, we consider the pattern domain of itemsets and databases that are transactional databases. Doing so, we can provide examples of concepts that have emerged as important within the cINQ project after 18 months of work.

It is useful to abstract the meaning of mining queries. A simple model has been introduced in [55] that considers a data mining process as a sequence of queries over the data but also the so-called *theory* of the data. Given a language $\mathcal{L}$ of patterns (e.g., itemsets, sequences, association rules), the theory of a database r with respect to $\mathcal{L}$ and a selection predicate q is the set $Th(r, \mathcal{L}, q) = \{\phi \in \mathcal{L} \mid q(r, \phi) \text{ is true}\}$. The predicate q indicates whether a pattern ϕ is considered interesting (e.g., ϕ denotes a property that is "frequent" in r). The selection predicate can be defined as a combination (boolean expression) of primitive constraints that have to be satisfied by the patterns. Some of them refer to the "behavior" of a pattern in the data, e.g., its "frequency" in a given data set is above or below a user-given threshold, some others define syntactical restrictions on desired patterns, e.g., its "length" is below a user-given threshold. Preprocessing concerns the definition of the database r, the mining phase is often the computation of the specified theory while post-processing can be considered as a querying activity on a materialized theory or the computation of a new theory.

This formalization however does not reflect the context of many classical data mining processes. Quite often, the user is interested not only in a collection of patterns that satisfy some constraints (e.g., frequent patterns, strong rules, approximate inclusion or functional dependencies) but also to some properties of these patterns in the selected database (e.g., their frequencies, the error for approximate dependencies). In that case, we will consider the so-called *extended theories*. For instance, when mining frequent itemsets or frequent and valid association rules [2], the user needs for the frequency of the specified patterns or rules. Indeed, during the needed post-processing phase, the user/analyst often uses various objective interestingness measures like the confidence [2], the conviction [23] or the J-mesure [72] that are computed efficiently provided that the frequency of each frequent itemset is available. Otherwise, it might be extremely expensive to look at the data again.

Designing *solvers* for more or less primitive constraints concerns the core of data mining algorithmic research. We must have solvers that can compute the (extended) theories and that have good properties in practice (e.g., scalability w.r.t. the size of the database or the size of the search space). A "generate and test" approach that would enumerate the sentences of $\mathcal{L}$ and then test the selection predicate q is generally impossible. A huge effort has concerned a clever use of the constraints occurring in q to have a tractable evaluation of useful inductive queries. This is the research area of *constraint-based data mining*. Most of the algorithmic research in pattern discovery tackles the design of complete algorithms for computing (extended) theories given more or less specific conjunctions of primitive constraints. Typically, many researchers have considered the computation of frequent patterns, i.e., patterns that satisfy a minimal frequency constraint. An important paper on a generic algorithm for such a typical mining task is [55]. However, if the active use of the so-called *anti-monotonic constraints* (e.g., the minimal frequency) is now well-understood, the situation is far less clear for non anti-monotonic constraints [64,51,18].

A second major issue is the possibility to approximate the results of (extended) inductive queries. This approximation can concern a collection of patterns that is a superset or a subset of the desired collection. This is the typical case when the theories are computed from a sample of the data (see, e.g., [74]) or when a relaxed constraint is used. Another important case of approximation for extended theories is the exact computation of the underlying theory while the evaluation functions are only approximated. This has lead to an important research area, the computation of the so-called *condensed representations* [54], a domain in which we have been playing a major role since the study of frequent closed itemsets as an ϵ-adequate representation for frequency queries [12].

This paper is organized as follows. Section 2 introduces notations and definitions that are needed for discussing inductive queries that return itemsets. It contains an instance of the definition of a pattern domain. Section 3 identifies several important open problems. Section 4 provides elements of solution that are currently studied within the cINQ project. Section 5 is a short conclusion.

2 A Pattern Domain for Itemsets

The definition of a *pattern domain* is made of the definition of a language of patterns $\mathcal{L}$, evaluation functions that assign a semantics to each pattern in a given database $\mathbf{r}$, languages for primitive constraints that specify the desired patterns, and inductive query languages that provide a language for combining the primitive constraints.

We do not claim that this paper is an exhaustive description of the itemset pattern domain. Even though we selected representative examples of evaluation functions and primitive constraints, many others have been or might be defined and used.

2.1 Language of Patterns and Terminology

We introduce some notations that are used for defining the pattern domain of itemsets. In that context, we consider that:

- A so-called *transactional database* contains the data,
- Patterns are the so-called *itemsets* and one kind of descriptive rule that can be derived from them, i.e., the *association rules*.

Definition 1 (Transactional Databases). *Assume that* $\mathtt{Items}$ *is a finite set of symbols denoted by capital letters, e.g.,* $\mathtt{Items}=\{\mathtt{A},\mathtt{B},\mathtt{C},\ldots\}$. *A* transaction *t is a subset of* `Items`. *A* transactional database $\mathbf{r}$ *is a finite and non empty multiset* $\mathbf{r} = \{t_1, t_2, \ldots, t_n\}$ *of transactions.*

Typical examples of transactional databases concern basket data (transactions are sets of products that are bought by customers), textual data (transactions are sets of keywords or descriptors that characterize documents), or gene expression data (transactions are sets of genes that are over-expressed in given biological conditions).

Definition 2 (Itemsets). *An* itemset *is a subset of* `Items`. *The language of patterns for itemsets is* $\mathcal{L} = 2^{\texttt{Items}}$.

We often use a string notation for sets, e.g., `AB` for $\{\texttt{A}, \texttt{B}\}$. Figure 1 provides an example of a transactional database and some information about itemsets within this database.

Association rules are not only a classical kind of pattern derived from itemsets [1,2] but are also used for some important definitions.

Definition 3 (Association Rules). *An association rule is denoted* $X \Rightarrow Y$ *where* $X \cap Y = \emptyset$ *and* $X \subseteq \texttt{Items}$ *is the* body *of the rule and* $Y \subseteq \texttt{Items}$ *is the* head *of the rule.*

Let us now define constraints on itemsets.

Definition 4 (Constraint). *If* $\mathcal{T}$ *denotes the set of all transactional databases and* $2^{\texttt{Items}}$ *the set of all itemsets, an itemset constraint* $\mathcal{C}$ *is a predicate over* $2^{\texttt{Items}} \times \mathcal{T}$. *An itemset* $S \in 2^{\texttt{Items}}$ satisfies *a constraint* $\mathcal{C}$ *in the database* $\mathbf{r} \in \mathcal{T}$ *iff* $\mathcal{C}(S, \mathbf{r}) = \mathit{true}$. *When it is clear from the context, we write* $\mathcal{C}(S)$. *Given a subset* I *of* `Items`, *we define* $\mathrm{SAT}_{\mathcal{C}}(I) = \{S \in I,\ S \text{ satisfies } \mathcal{C}\}$. $\mathrm{SAT}_{\mathcal{C}}$ *denotes* $\mathrm{SAT}_{\mathcal{C}}(2^{\texttt{Items}})$. *The same definitions can be easily extended to rules.*

2.2 Evaluation Functions

Evaluation functions return information about the properties of a given pattern in a given database. Notice that using these evaluation functions can be considered as a useful task for the user/analyst. It corresponds to *hypothesis testing* when hypothesis can be expressed as itemsets or association rules, e.g., what are the transactions that support the H hypothesis? How many transactions support H? Do I have less than n counter-examples for hypothesis H?

Several evaluation functions are related to the "satisfiability" of a pattern in a given data set, i.e., deciding whether a pattern hold or not in a given database.

Definition 5 (Support for Itemsets and Association Rules). *A transaction* t supports *an itemset* X *if every item in* X *belongs to* t, *i.e.,* $S \subseteq t$. *It is then possible to define a boolean evaluation function* e_1 *such that* $e_1(X, \mathbf{r})$ *is true if all the transactions in* $\mathbf{r}$ *support* X *and false elsewhere. The same definition can be adapted for association rules:* $e_1(X \Rightarrow Y, \mathbf{r})$ *returns true iff when* $\mathbf{r}$ *supports* X, *it supports* Y *as well. The* support *(denoted* $\mathrm{support}(S, \mathbf{r})$*) of an itemset* S *is* *the multiset of all transactions of* $\mathbf{r}$ *that supports* S *(e.g.,* $\text{support}(\emptyset) = \mathbf{r}$*). The* support *of a rule is defined as the support of the itemset* $X \cup Y$. *A transaction* t supports *a rule* $X \Rightarrow Y$ *if it supports* $X \cup Y$.

Definition 6 (Exceptions to Rules). *A transaction* t *is an exception for a rule* $X \Rightarrow Y$ *if it supports* X *and it does not support* Y. *It is then possible to define a new boolean evaluation function* $e_2(X \Rightarrow Y, \mathbf{r})$ *that returns true if none of the transactions* t *in* $\mathbf{r}$ *is an exception to the rule* $X \Rightarrow Y$. *A rule with no exception is called a* logical *rule.*

These evaluation functions that return sets of transactions or boolean values are useful when crossing over the patterns and the transactional data. Also, the size of the supporting set is often used.

Definition 7 (Frequency). *The* absolute frequency *of an itemset* S *in* $\mathbf{r}$ *is defined by* $\mathcal{F}_a(S, \mathbf{r}) = |\text{support}(S)|$ *where* $|.|$ *denote the cardinality of the multiset (each transaction is counted with its multiplicity). The relative frequency of* S *in* $\mathbf{r}$ *is* $\mathcal{F}(S, \mathbf{r}) = |\text{support}(S)|/|\text{support}(\emptyset)|$. *When there is no ambiguity from the context, parameter* $\mathbf{r}$ *is omitted and the frequency denotes the relative frequency (i.e., a number in [0,1]).*

Figure 1 provides an example of a transactional database and the supports and the frequencies of some itemsets.

$T =$

t_1	ABCD
t_2	BC
t_3	AC
t_4	AC
t_5	ABCD
t_6	ABC

Itemset	Support	Frequency
A	$\{t_1, t_3, t_4, t_5, t_6\}$	0.83
B	$\{t_1, t_2, t_5, t_6\}$	0.67
AB	$\{t_1, t_5, t_6\}$	0.5
AC	$\{t_1, t_3, t_4, t_5, t_6\}$	0.83
CD	$\{t_1, t_5\}$	0.33
ACD	$\{t_1, t_5\}$	0.33

Fig. 1. Supports and frequencies of some itemsets in a transactional database

Other measures might be introduced for itemsets that, e.g., returns the degree of correlation between the attributes it contains. We must then provide evaluation functions that compute these measures.

It is straightforward to define the frequency evaluation function of an association rule $X \Rightarrow Y$ in $\mathbf{r}$ as $\mathcal{F}(X \Rightarrow Y, \mathbf{r}) = \mathcal{F}(X \cup Y, \mathbf{r})$ [1,2]. When mining association rules, we often use objective interestingness measures like confidence [2], conviction [23], J-mesure [72], etc. These can be considered as new evaluation functions. Most of these measures can be computed from the frequencies of rule components. For instance, the *confidence* of a rule $X \Rightarrow Y$ in $\mathbf{r}$ is $\text{conf}(X \Rightarrow Y) = \mathcal{F}(X \Rightarrow Y, \mathbf{r})/\mathcal{F}(Y, \mathbf{r})$. It gives the conditional probability that a transaction from $\mathbf{r}$ supports $X \cup Y$ when it supports X. The confidence of a logical rule is thus equal to 1.

We consider now several evaluation functions that have been less studied in the data mining context but have been proved quite useful in the last 3 years (see, e.g., [65,12,75,15,16]). Notice however that these concepts have been used for a quite a long time in other contexts, e.g., in concept lattices.

Definition 8 (Closures of Itemsets). *The closure of an itemset S in* $\mathbf{r}$ *(denoted by* $\texttt{closure}(S, \mathbf{r})$*) is the maximal (for set inclusion) superset of S which has the same support as S. In other terms, the closure of S is the set of items that are common to all the transactions which support S.*

Notice that when the closure of an itemset X is a proper superset of X, say Y, it means that an association rule $X \Rightarrow Y \setminus X$ holds in $\mathbf{r}$ with confidence 1.

Example 1. In the database of Figure 1, let us compute $\texttt{closure}(\texttt{AB})$. Items A and B occur in transactions 1, 5 and 6. Item C is the only other item that is also present in these transactions, thus $\texttt{closure}(\texttt{AB}) = \texttt{ABC}$. Also, $\texttt{closure}(\texttt{A}) = \texttt{AC}$, $\texttt{closure}(\texttt{B}) = \texttt{BC}$, and $\texttt{closure}(\texttt{BC}) = \texttt{BC}$.

We now introduce an extension of this evaluation function [15,16].

Definition 9 (δ-closure). *Let δ be an integer and S an itemset. The δ-closure of S, $\texttt{closure}_\delta(S)$ is the maximal (w.r.t. the set inclusion) superset Y of S such that for every item $A \in Y - S$, $|\texttt{Support}(S \cup \{\texttt{A}\})|$ is at least $|\texttt{Support}(S)| - \delta$. In other terms, $\mathcal{F}_a(\texttt{closure}_\delta(S))$ has almost the same value than $\mathcal{F}_a(S)$ when δ is small w.r.t. the number of transactions.*

Example 2. In the database of Figure 1, $\texttt{closure}_2(\texttt{B}) = \texttt{BCD}$ while $\texttt{closure}_0(\texttt{B}) = \texttt{BC}$.

Notice that $\texttt{closure}_0 = \texttt{closure}$. Also, the δ-closure of a set X provides an association rule with high confidence between X and $\texttt{closure}_\delta(X) \setminus X$ when δ is a positive integer that is small w.r.t. the number of transactions.

It is of course possible to define many other evaluation functions. We gave representative examples of such functions and we now consider examples of primitive constraints that can be built from them.

2.3 Primitive Constraints

Many primitive constraints can be defined. We consider some examples that have been proved useful. These examples are representative of two important kinds of constraints: constraints based on evaluation functions and syntactic constraints. These later can be checked without any access to the data and are related to the well known machine learning concept of linguistic bias (see, e.g., [63]).

Let us consider primitive constraints based on frequency. First, we can enforce that a given pattern is frequent enough ($\mathcal{C}_{\text{minfreq}}(S)$) and then we specify that a given pattern has to be infrequent or not too frequent ($\mathcal{C}_{\text{maxfreq}}(S)$).

Definition 10 (Minimal Frequency). *Given an itemset S and a frequency threshold $\gamma \in [0, 1]$, $\mathcal{C}_{\text{minfreq}}(S) \equiv \mathcal{F}(S) \geq \gamma$. Itemsets that satisfy $\mathcal{C}_{\text{minfreq}}$ are said* γ-frequent *or* frequent *in* $\mathbf{r}$. *Indeed, this constraint can be defined also on association rules: $\mathcal{C}_{\text{minfreq}}(X \Rightarrow Y) \equiv \mathcal{F}(X \Rightarrow Y) \geq \gamma$.*

Definition 11 (Maximal Frequency). *Given an itemset S and a frequency threshold $\gamma \in [0,1]$, $\mathcal{C}_{\text{maxfreq}}(S) \equiv \mathcal{F}(S) \leq \gamma$. Indeed, this constraint can be defined also on association rules.*

Definition 12 (Minimal Confidence on Rules). *Given a rule $X \Rightarrow Y$ and a confidence threshold $\theta \in [0,1]$, $\mathcal{C}_{\text{minconf}}(X \Rightarrow Y) \equiv \text{conf}(S) \geq \theta$. Rules that satisfy $\mathcal{C}_{\text{minconf}}$ are called valid rules. A dual constraint for maximal confidence might be introduced as well.*

Example 3. Considering the database of Figure 1, if $\mathcal{C}_{\text{minfreq}}$ specifies that an itemset (or a rule) must be 0.6-frequent, then $\text{SAT}_{\mathcal{C}_{\text{minfreq}}} = \{\mathtt{A}, \mathtt{B}, \mathtt{C}, \mathtt{AC}, \mathtt{BC}\}$. For rules, if the confidence threshold is 0.7, then the frequent and valid rules are $\text{SAT}_{\mathcal{C}_{\text{minfreq}} \wedge \mathcal{C}_{\text{minconf}}}(\emptyset \Rightarrow A, \emptyset \Rightarrow C, \emptyset \Rightarrow AC, A \Rightarrow C, C \Rightarrow A, B \Rightarrow C)$.

It is straightforward to generalize these definitions to all the other insterestingness measures that we can use for itemsets and rules. However, let us notice that not all the interestingness measures have bounded domain values. It motivates the introduction of the optimal constraints.

Definition 13 (Optimality). *Given an evaluation function $\mathcal{E}$ that returns an ordinal value, let us denote by $\mathcal{C}_{opt}(\mathcal{E}, \phi, n)$ the constraint that is satisfied if ϕ belongs to the n best patterns according to $\mathcal{E}$ values (the n patterns with the highest values).*

For instance, such a constraint can be used to specify that only the n most frequent patterns are desired (see [7,70,71] for other examples).

Another kind of primitive constraint concerns the *syntactical restrictions* that can be defined on one pattern. By syntactical, we mean constraints that can be checked without any access to the data and/or the background knowledge, just by looking at the pattern. A systematic study of syntactical constraints for itemsets and rules has been described in [64,51].

Definition 14 (Syntactic Constraints). *It is of the form $S \in \mathcal{L}_C$, where $\mathcal{L}_C \subseteq \mathcal{L} = 2^{\mathtt{Items}}$. Various means can be used to specify $\mathcal{L}_C$, e.g., regular expressions.*

Some other interesting constraints can use additional information about the items, i.e., some background knowledge encoded in, e.g., relational tables. In [64], the concept of *aggregate constraint* is introduced.

Definition 15 (Aggregate Constraint). *It is of the form $agg(S)\theta v$, where agg is one of the aggregate functions min, max, sum, count, avg, and θ is one of the boolean operators $=, \neq, <, \leq, >, \geq$. It says the aggregate of the set of numeric values in S stands in relationship θ to v.*

Example 4. Consider the database of Figure 1, assume that $\mathcal{C}_{size}(S) \equiv |S| \leq 2$ (it is equivalent to $count(S) \leq 2$) and $\mathcal{C}_{miss}(S) \equiv \mathtt{B} \notin S$, then $\text{SAT}_{\mathcal{C}_{size}} = \{\emptyset, \mathtt{A}, \mathtt{B}, \mathtt{C}, \mathtt{D}, \mathtt{AB}, \mathtt{AC}, \mathtt{AD}, \mathtt{BC}, \mathtt{BD}, \mathtt{CD}\}$ and $\text{SAT}_{\mathcal{C}_{miss}} = \{\emptyset, \mathtt{A}, \mathtt{C}, \mathtt{D}, \mathtt{AC}, \mathtt{AD}, \mathtt{ACD}\}$.

The same kind of syntactical constraint can be expressed on association rules, including the possibility to express constraints on the body and/or the head of the rule.

We now consider primitive constraints based on closures.

Definition 16 (Closed Itemsets and Constraint $\mathcal{C}_{close}$). *A closed itemset is an itemset that is equal to its closure in* **r**. *Let us assume that $\mathcal{C}_{close}(S) \equiv$ $\texttt{closure}(S) = S$. In other terms, closed itemsets are maximal sets of items that are supported by a multiset of transactions.*

Example 5. In the database of Figure 1, the closed itemsets are `C`, `AC`, `BC`, `ABC`, and `ABCD`.

Free itemsets are sets of items that are not "strongly" correlated [15]. They have been designed as a useful intermediate representation for computing closed sets since the closed sets are the closures of the free sets.

Definition 17 (Free Itemsets and Constraint $\mathcal{C}_{free}$). *An itemset S is free if no logical rule holds between its items, i.e., it does not exist two distinct itemsets X, Y such that $S = X \cup Y$, $Y \neq \emptyset$ and $X \Rightarrow Y$ is a logical rule.*

Example 6. In the database of Figure 1, the free sets are $\emptyset$, `A`, `B`, `D`, and `AB`.

An alternative definition is that all the proper subsets of a free set S have a different frequency than S. Notice that free itemsets have been formalized independently as the co-called *key* patterns [5]. Furthermore, the concept of free itemset formalizes the concept of generator [65] in an extended framework since free itemsets are a special case of δ-free itemsets [15,16].

Definition 18 (δ-free Itemsets and Constraint $\mathcal{C}_{\delta-free}$). *Let δ be an integer and S an itemset, an itemset S is δ-free if no association rule with at most δ exceptions holds between its subsets. δ-free sets satisfy the constraint $\mathcal{C}_{\delta-free}(S) \equiv (\forall S' \subset S) \Rightarrow S \not\subseteq \texttt{closure}_\delta(S')$.*

Example 7. In the database of Figure 1, the 1-free sets are $\emptyset$, `A`, `B`, and `D`.

2.4 Example of Inductive Queries

Now, it is interesting to consider boolean combinations of primitive constraints. Notice that in this section, we consider neither the problem of query evaluation nor the availability of concrete query languages.

The Standard Association Rule Mining Task. Mining the frequent itemsets means the computation of $\text{SAT}_{\mathcal{C}_{\text{minfreq}}}$ for a given frequency threshold. The standard association rule mining problem introduced in [1] is to find all the association rules that verify the minimal frequency and minimal confidence constraints for some user-defined thresholds. In other terms, we are looking for each pattern ϕ (rules) such that $\mathcal{C}_{\text{minfreq}}(\phi) \wedge \mathcal{C}_{\text{minconf}}(\phi)$ is true. Filtering rules according to syntactical criteria can also be expressed by further constraints.

Example 8. Provided the dataset of Figure 1 and the constraints from Example 3 and 4, $\text{SAT}_{\mathcal{C}_{\text{minfreq}} \wedge \mathcal{C}_{size} \wedge \mathcal{C}_{miss}} = \{\texttt{A}, \texttt{C}, \texttt{AC}\}$ is returned when the query specifies that the desired itemsets must be 0.6-frequent, with size less than 3 and without the attribute B. It is straightforward to consider queries on rules. Let us consider an example where the user/analyst wants all the frequent and valid association rules but also quite restricted rules with high confidence (but without any minimal frequency constraint). Such a query could be based, e.g., on the constraint $(\mathcal{C}_{\text{minfreq}}(\phi) \wedge \mathcal{C}_{\text{minconf}}(\phi)) \vee (\mathcal{C}_s(\phi) \wedge \mathcal{C}_{\text{minconf}}(\phi))$ where $\mathcal{C}_s(X \Rightarrow Y) \equiv |X| = |Y| = 1 \wedge Y = \texttt{A}$.

Mining Discriminant Patterns. An interesting application of frequency constraints concerns the search for patterns that are frequent in one data set and infrequent in another one. This has been studied in [33] as the emerging pattern mining task. More recently, it has been studied within an inductive logic programming setting in [32] and applied to molecular fragment discovery.

Assume a transactional database for which one of the item defined a class value (e.g., item A is present when the transaction has the class value "interesting" and false when the transaction has the class value "irrelevant"). It is then possible to split the database $\mathbf{r}$ into two databases, the one of interesting transactions $\mathbf{r}_1$ and the one of irrelevant transactions (say $\mathbf{r}_2$). Now, a useful mining task concerns the computation of every itemset such that $\mathcal{C}_{\text{minfreq}}(S, \mathbf{r}_1) \wedge \mathcal{C}_{\text{maxfreq}}(S, \mathbf{r}_2)$. Indeed, these itemsets are supported by interesting transactions and not supported by irrelevant ones. Thresholds can be assigned thanks to a statistical analysis and such patterns can be used for predictive mining tasks.

Mining Association Rules with Negations. Let $\texttt{Items}^+ = \{\texttt{A}, \texttt{B}, ...\}$ be a finite set of symbols called the positive items and a set $\texttt{Items}^-$ of same cardinality as $\texttt{Items}^+$ whose elements are denoted $\overline{\texttt{A}}, \overline{\texttt{B}}, \ldots$ and called the negative items. Given a transaction database $\mathbf{r}$ over $\texttt{Items}^+$, let us define a complemented transaction database over $\texttt{Items} = \texttt{Items}^+ \cup \texttt{Items}^-$ as follows: for a given transaction $t \in \mathbf{r}$, we add to t negative items corresponding to positive items not present in t. Generalized itemsets are subsets of $\texttt{Items}$ and can contain positive and negative items. In other terms, we want to have a symmetrical impact for the presence or the absence of items in transactions [14]. It leads to extremely dense transactional databases, i.e., extremely difficult extraction processes.

In [13], the authors studied the extraction of frequent itemsets ($\mathcal{C}_{\text{minfreq}}$) that do not involve only negative items ($\mathcal{C}_{alpp}$). $\mathcal{C}_{alpp}(S)$ is true when S involves at least p positive items. Also, this constraint has been relaxed into $\mathcal{C}_{alppoam1n} = \mathcal{C}_{alpp} \vee \mathcal{C}_{am1n}$ (at least p positive attributes or at most 1 negative attribute). On different real data sets, it has been possible to get interesting results when it was combined with condensed representations (see Section 4).

Mining Condensed Representation of Frequent Itemsets. Condensed representation is a general concept (see, e.g., [54]) that can be extremely useful for the concise representation of the collection of frequent itemsets and their frequencies. In Section 3, we define more precisely this approach. Let us notice at that stage that several algorithms exist to compute various condensed representations of the frequent itemsets: CLOSE [65], CLOSET[69], CHARM [75], MIN-EX [12,15,16], or PASCAL [5]. These algorithms compute different condensed representations: the frequent closed itemsets (CLOSE, CLOSET, CHARM), the frequent free itemsets (MIN-EX, PASCAL), or the frequent δ-free itemsets for MIN-EX. From an abstract point of view, these algorithms are respectively looking for itemsets that satisfy $\mathcal{C}_{\text{minfreq}} \wedge \mathcal{C}_{close}$, $\mathcal{C}_{\text{minfreq}} \wedge \mathcal{C}_{free}$, and $\mathcal{C}_{\text{minfreq}} \wedge \mathcal{C}_{\delta-free}$. Furthermore, it can be interesting to provide association rules whose components satisfy some constraints based on closures. For instance, association rules that are based on free itemsets on their left-hand side ($\mathcal{C}_{free}(BODY)$) and their closures on the right-hand side ($\mathcal{C}_{close}(BODY \cup HEAD)$) are of a particular interest: they constitute a kind of cover for the whole collection of frequent and valid association rules [4,8]. Notice also that the use of classification rules based on δ-free sets ($\mathcal{C}_{\delta-free}$) for the body and a class value in the head has been studied in [17,28].

Postprocessing Queries. Post-processing queries can be understood as queries on materialized collections of itemsets or association rules: the user selects the itemsets or the rules that fulfill some new criteria (while these itemsets or rules have been mined, e.g., they are all frequent and valid). However, from the specification point of view, they are not different from data mining queries even though the evaluation does not need an extraction phase and can be performed on materialized collections of itemsets or rules.

One important post-processing use is to cross over the patterns and the data, e.g., when looking at transactions that are exceptions to some rules. For instance, given an association rule $A \Rightarrow B$, one wants all the transactions t from $\mathbf{r}$ (say the transactional database $\mathbf{r}_1$ for which $e_2(A \Rightarrow B, t)$ is true. Notice that $\mathbf{r} \setminus \mathbf{r}_1$ is the collection of exceptions to the rule. A rule mining query language like `MSQL` [43] offers a few built-in primitives for rule post-processing, including primitives that cross-over the rules and the transactions (see also [10] in this volume for examples of post-processing queries).

3 A Selection on Some Open Problems

We consider several important open problems that are related to itemset and rule queries.

3.1 Tractability of Frequent Itemset and Association Rule Mining

Computing the result of the classical association rule mining problem is generally done in two steps [2]: first the computation of all the frequent itemsets and their frequency and then the computation of every valid association rule that can be made from disjoint subsets of each frequent itemset. This second step is far less expensive than the first one because no access to the database is needed: only the collection of the frequent itemsets and their frequencies are needed. Furthermore, the frequent itemsets can be used for many other applications, far beyond the classical association rule mining task. Notice among others, clustering (see, e.g., [61]), classification (see, e.g., [53,28]), generalized rule mining (see, e.g., [54,14]).

Computing the frequent itemsets is an important data mining task that has been studied by many researchers since 1994. The famous APRIORI algorithm [2] has inspired many research and efficient implementations of APRIORI-like algorithms can be used provided that the collection of the frequent itemsets is not too large. In other terms, for the desired frequency threshold, the size of the maximal frequent itemsets must not be too long (around 15). Indeed, this kind of algorithm must count the frequencies of at least every frequent itemset and useful tasks, according to the user/analyst, become intractable as soon as the size of $SAT_{\mathcal{C}_{\text{minfreq}}}$ is too large for the chosen frequency threshold. It is the case of dense and correlated datasets and many real datasets fall in this category. In Section 4, we consider solutions thanks to the design of condensed representations for frequent itemsets.

The efficiency of the extraction of the answer to an itemset query relies on the possibility to use constraints during the itemset computation. A classical result is that effective safe pruning can be achieved when considering anti-monotonic constraints [55,64], e.g., the minimal frequency constraint. It relies on the fact that if an itemset violates an anti-monotonic constraint then all its supersets violate it as well and therefore this itemset and its supersets can be pruned and thus not considered for further evaluation.

Definition 19 (Anti-monotonicity). *An* anti-monotonic *constraint is a constraint $\mathcal{C}$ such that for all itemsets S, S': $(S' \subseteq S \wedge \mathcal{C}(S)) \Rightarrow \mathcal{C}(S')$.*

Example 9. Examples of anti-monotonic constraints are: $\mathcal{C}_{\text{minfreq}}(S)$, $\mathcal{C}(S) \equiv \mathtt{A} \notin S$, $\mathcal{C}(S) \equiv S \subseteq \{\mathtt{A},\mathtt{B},\mathtt{C}\}$, $\mathcal{C}(S) \equiv S \cap \{A,B,C\} = \emptyset$, $\mathcal{C}_{free}(S)$, $\mathcal{C}_{am1n}(S)$, $\mathcal{C}_{\delta-free}(S)$, $\mathcal{C}(S) \equiv S.price > 50$ and $\mathcal{C}(S) \equiv Sum(S.price) < 500$. The two last constraints mean respectively that the price of all items must be lower than fifty and that the sum of the prices of the items must be lower than five hundred.

Notice that the conjunction or disjunction of anti-monotonic constraints is anti-monotonic.

Even though the anti-monotonic constraints, when used actively, can drastically reduce the search space, it is not possible to ensure the tractability of an inductive query evaluation. In that case, the user/analyst has to use more selective constraints, e.g., a higher frequency threshold. Indeed, a side-effect can be that the extracted patterns become not enough interesting, e.g., they are so frequent that they correspond to trivial statements.

Furthermore, itemset queries do not involve only anti-monotonic constraints. For instance, $\mathcal{C}_{close}$ is not anti-monotonic. Sometimes, it is possible to post-process the collection of itemsets that satisfy the anti-monotonic part of the selection predicate to check the remaining constraints afterwards.

3.2 Tractability of Constraint-Based Itemset Mining

Pushing constraints is useful for anti-monotonic ones. Other constraints can be pushed like the *monotonic* constraints or the *succinct* constraints [64].

Definition 20 (Monotonicity). *A* monotonic *constraint is a constraint* $\mathcal{C}$ *such that for all itemsets* S, S': *(*$S \subseteq S' \wedge S$ *satisfies* $\mathcal{C}$*)* $\Rightarrow S'$ *satisfies* $\mathcal{C}$.

The negation of an anti-monotonic constraint is a monotonic constraint and the conjunction or disjunction of monotonic constraints is still monotonic.

Example 10. $\mathcal{C}(S) \equiv \{A, B, C, D\} \subseteq S$, $\mathcal{C}_{alpp}(S)$, $\mathcal{C}(S) \equiv Sum(S.price) > 100$ (the sum of the prices of items from S is greater than 100) and $\mathcal{C}(S) \equiv S \cap \{A, B, C\} \neq \emptyset$ are examples of monotonic constraints.

Indeed, monotonic constraints can also be used to improve the efficiency of itemset extraction (optimization of the candidate generation phase that prevents to consider candidates that do not satisfy the monotonic constraint (see, e.g., [18]).

The succinctness property that has been introduced in [64] are syntactic constraints that can be put under the form of a conjunction of monotonic and anti-monotonic constraints. Clearly, it is possible to use such a property for the optimization of the constraint-based extraction (optimization of candidate generation and pruning).

Pushing non anti-monotonic constraints sometimes increases the computation times since it prevents effective pruning based on anti-monotonic constraints [73,18,34]. For instance, as described in [13], experiments have shown that it has been needed to relax the monotonic constraint $\mathcal{C}_{alpp}$ ("pushing" it gave rise to a lack of pruning) by $\mathcal{C}_{alppoam1n} = \mathcal{C}_{alpp} \vee \mathcal{C}_{am1n}$ where $\mathcal{C}_{am1n}$ is anti-monotonic. The identification of a good strategy for pushing constraints needs for an a priori knowledge of constraint selectivity. However, this is in general not available at extraction time. Designing adaptative strategies for pushing constraints during itemset mining is still an open problem. Notice however that some algorithms have been already proposed for specific strategies on itemset mining under conjunctions of constraints that are monotonic and anti-monotonic [64,18]. This has been explored further within the cINQ project (see Section 4).

Notice also that the constraints defined by a user on the desired association rules have to be transformed into suitable itemset constraints. So far, this has to be done by an ad-hoc processing and designing semi-automatic strategies for that goal is still an open problem.

3.3 Interactive Itemset Mining

From the user point of view, pattern discovery is an interactive and iterative process. The user defines a query by specifying various constraints on the patterns he/she wants. When a discovery process starts, it is difficult to figure out the collection of constraints that leads to an interesting result. The result of a data mining query is often unpredictable and the users have to produce sequences of queries until he/she gets an actionable collection of patterns. So, we have not only to optimize single inductive query evaluations but also the evaluation of sequences of queries. This has been studied for itemsets and association rules in, e.g., [37,3,36,62].

One classical challenge is the design of incremental algorithms for computing theories (e.g., itemsets that satisfy a complex constraint) or extended theories (e.g., itemsets and their frequencies) when the data changes. More generally, reusing previously computed theories to answer more efficiently to new inductive queries is important. It means that results about, equivalence and containment of inductive queries are needed. Furthermore, the concept of dominance has emerged [3]. In that case, only data scans are needed to update the value of the evaluation functions.

However, here again, a trade-off has to be found between the optimization of single queries by the best strategy for pushing its associated constraints and the optimization of the whole sequence. Indeed, the more we push the constraint and materialize only the constrained collection, the less it will be possible to reuse it for further evaluations [37,36]. In Section 4, we refer to recent advances in that area.

3.4 Concrete Query Languages

There is no dedicated query languages for itemsets but several proposal exist for association rule mining, e.g., (`MSQL` [43], `DMQL` [38], and `MINE RULE` [59]). Among them, the `MINE RULE` query language is one of the few proposals for which a formal operational semantics has been published [59]. Ideally, these query languages must support not only the selection of the mining context and its pre-processing (e.g., sampling, selection and grouping, discretization), the specification of a mining task (i.e., the expression of various constraints on the desired rules), and the post-processing of these rules (e.g., the support of subjective interestingness evaluation, redundancy elimination and grouping strategies, etc.).

A comparative study of the available concrete query languages is published in this volume [10]. It illustrates that we are still lacking from an "ideal" query language for supporting KDD processes based on association rules. From our perspective, we are still looking for a good set of primitives that might be supported by such languages. Furthermore, a language like `MINE RULE` enables to use various kinds of constraints on the desired rules (i.e., the relevant constraints on the itemsets are not explicit) and optimizing the evaluation of the mining queries (or sequences of queries) still need further research. This challenge is also considered within the cINQ project.

4 Elements of Solution

We now provide pointers to elements of solution that have been studied by the cINQ partners these last 18 months. It concerns each of the issues we have been discussing in Section 3. Even though the consortium has not studied only itemsets but also molecular fragments, sequential patterns and strings, the main results can be illustrated on itemsets.

Let us recall that, again, we do not claim that this section considers all the solutions studied so far. Many other projects and/or research groups are interested in the same open problems and study other solutions. This is the typical case for depth-first algorithms like [39] which opens new possibilities for efficient constraint-based itemset computation [67].

Let us first formalize that inductive queries that return itemsets might also provide the results of the frequencies for further use.

Definition 21 (Itemset Query). *A itemset query is a pair* $(\mathcal{C}, \mathbf{r})$ *where* $\mathbf{r}$ *is a transactional database and* $\mathcal{C}$ *is an itemset constraint. The result of a query* $Q = (\mathcal{C}, \mathbf{r})$ *is defined as the set* $Res(Q) = \{(S, \mathcal{F}(S)) \mid S \in SAT_{\mathcal{C}}\}$.

There are two main approaches for the approximation of $Res(Q)$:

- The result is $Approx(Q) = \{(S, \mathcal{F}(S)) \mid S \in SAT_{\mathcal{C}'}\}$ where $\mathcal{C}' \neq \mathcal{C}$. In that case, $Approx(Q)$ and $Res(Q)$ are different. When $\mathcal{C}$ is more selective in $\mathbf{r}$ than $\mathcal{C}'$, we have $Approx(Q) \subseteq Res(Q)$. A post-processing on $Approx(Q)$ might be used to eliminate itemsets that do not verify $\mathcal{C}$. When $\mathcal{C}$ is less selective than $\mathcal{C}'$ then $Approx(Q)$ is said incomplete.
- The result is $Approx(Q) = \{(S, \mathcal{F}'(S)) \mid S \in SAT_{\mathcal{C}}\}$ where $\mathcal{F}'$ provides an approximation of the frequency of each itemset in $Approx(Q)$.

Indeed, it can be so that the two situations occur simultaneously. A typical case is the use of sampling on the database: one can sample the database ($\mathbf{r}' \subset \mathbf{r}$ is the mining context) and compute $Res(Q)$ not in $\mathbf{r}$ but in $\mathbf{r}'$. In that case, both the collection of the frequent itemsets and their frequencies are approximated. Notice however that clever strategies can be used to avoid, in practice, an incomplete answer [74].

A classical result is that it is possible to represent the collection of the frequent itemsets by its maximal elements, the so-called positive border in [55] or the S set in the machine learning terminology [60]. Also, it is possible to compute these maximal itemsets and their frequencies without computing every frequency of every frequent itemsets (see, e.g., [6]). This can be generalized to any anti-monotonic constraint: the collection of the most specific sentences $Approx(Q)$ (e.g., the maximal itemsets) is a compact representation of $Res(Q)$ from which (a) it is easy to derive the exact collections of patterns (every sentence that is more general belongs to the solution, e.g., every subset of the maximal frequent itemsets) but, (b) the evaluation functions (e.g., the frequency) are only approximated. Thus, the maximal itemsets can be considered as an example of an approximative condensed representation of the frequent itemsets.

First, we have been studying algorithms that compute itemsets under more general constraints, e.g., conjunctions of anti-monotonic and monotonic constraints. Next, we have designed other approximative condensed representations and exact ones as well.

4.1 Algorithms for Constraint-Based Mining

cInQ partners have studied the extraction of itemsets (and rather similar pattern domains like strings, sequences or molecular fragments) under a conjunction of monotonic and anti-monotonic constraints. Notice also that since disjunctions of anti-monotonic (resp. monotonic) constraints are anti-monotonic (resp. monotonic), it enables to consider rather general forms of inductive queries.

[46] provides a generic algorithm that generalizes previous work for constraint-based itemset mining in a levelwise approach (e.g., [73,64]). The idea is that, given a conjunction of an anti-monotonic constraint and a monotonic constraint ($\mathcal{C}_{am} \wedge \mathcal{C}_m$), it is possible to start a levelwise search from the minimal (w.r.t. set inclusion) itemsets that satisfy $\mathcal{C}_m$ and completes this collection until the maximal itemsets that satisfy the $\mathcal{C}_{am}$ constraint are reached. Such a levelwise algorithm provides the complete collection $Res(Q)$ when Q can be expressed by means of a conjunction $\mathcal{C}_{am} \wedge \mathcal{C}_m$. [46] introduces strategies (e.g., for computing the minimal itemsets that satisfy $\mathcal{C}_m$ by using the duality between monotonic and anti-monotonic constraints). Details are available in [44].

Mining itemsets under $\mathcal{C}_{am} \wedge \mathcal{C}_m$ can also be considered as a special case of the general algorithm introduced in [30]. This paper considers queries that are boolean expressions over monotonic and anti-monotonic primitives on a single pattern variable ϕ. This is a quite general form of inductive query and it is shown that the solution space corresponds to the union of various version spaces [60,57,40,41]. Because each version space can be represented in a concise way using its border sets S and G, [30] shows that the solution space of a query can be represented using the border sets of several version spaces. When a query enforces a conjunction $\mathcal{C}_{am} \wedge \mathcal{C}_m$, [30] proposes to compute $S(\mathcal{C}_{am} \wedge \mathcal{C}_m)$ as $\{s \in S(\mathcal{C}_{am}) \mid \exists g \in G(\mathcal{C}_m) : g \subseteq s\}$ and dually for $G(\mathcal{C}_{am} \wedge \mathcal{C}_m)$. Thus, the borders for $\mathcal{C}_{am} \wedge \mathcal{C}_m$ can be computed from $S(\mathcal{C}_{am})$ and from $G(\mathcal{C}_m)$ as usual for the classical version space approach. Sets such as $S(\mathcal{C}_{am})$ can be computed using classical algorithms such as the levelwise algorithm [55] and the dual set $G(\mathcal{C}_m)$ can be computed using the dual algorithms [32]. These border sets are an approximative condensed representation of the solution. For the MolFea specific inductive database, it has been proved quite effective for molecular fragment finding [49,32,50,48].

Sequential pattern mining has been studied as well. Notice that molecular fragments can be considered as a special case of sequences or strings. [27] studies sequential pattern mining under a specific conjunction of constraint that ask for minimal frequency and similarity w.r.t. a reference pattern. In this work, the main contribution has been to relax the similarity constraint into an anti-monotonic one to improve pruning efficiency. It is an application of the framework for convertible constraints [68]. Also, logical sequence mining under constraints has been studied, in a restricted framework [56] (regular expressions on the sequence of predicate symbols and minimal frequency) and in a more general setting [52] (conjunction of anti-monotonic and monotonic constraints).

4.2 Condensed Representations for Frequent Itemsets

cInQ partners have studied the condensed approximations in two complementary directions: the use of border sets in a very general setting (i.e., version spaces) but also several condensed representations of the frequent itemsets.

- Border sets represent the maximally general and/or maximally specific solutions to an inductive query. They can be used to bound the set of all

solutions [30]. This can be used in many different pattern domains provided that the search space is structured by a specialization relation and that the solution space is a version space. They are useful in case only membership of the solution set is important.
- Closed sets [65,12], δ-free sets [15,16], and disjoint-free sets [25] are condensed representations that have been designed as ϵ-adequate representations w.r.t. frequency queries, i.e., representations from which the frequency of any itemset can be inferred or approximated within a bounded error.

The collection of the γ-frequent itemsets and their frequencies can be considered as an $\gamma/2$-adequate representation w.r.t. frequency queries [12]. It means that the error on the inference of a frequency for a given itemset is bounded by $\gamma/2$. Indeed, the frequency of an infrequent itemset can be set to $\gamma/2$ while the frequency of a frequent one is known exactly. Given a set $\mathcal{S}$ of pairs $(X, \mathcal{F}(X))$, e.g., the collection of all the frequent itemsets and their frequencies, we are interested in condensed representations of $\mathcal{S}$ that are subsets of $\mathcal{S}$ with two properties: (1) They are much smaller than $\mathcal{S}$ and faster to compute, and (2), the whole set $\mathcal{S}$ can be generated from the condensed representation with no access to the database, i.e., efficiently.

We have introduced in Section 2.3 the concepts of closed sets, free sets and δ-free sets. Disjoint-free itemsets are a generalization of free itemsets [25]. They are all condensed representations of the frequent itemsets that are exact representations (no loss of information w.r.t. the frequent itemsets and their frequencies), except for the δ-free itemsets (with $\delta \neq 0$) which is an approximative one. Let us now give the principle of regeneration from the frequent closed itemsets:

- Given an itemset S and the set of frequent closed itemsets,
  - If S is not included in a frequent closed itemset then S is not frequent.
  - Else S is frequent and $\mathcal{F}(S) = \text{Max}\{\mathcal{F}(X),\ S \subseteq X \wedge \mathcal{C}_{close}(X)\}$.

As a result, γ-frequent closed itemsets are like the γ-frequent itemsets a $\gamma/2$-adequate representation for frequency queries.

Example 11. In the database of Figure 1, if the frequency threshold is 0.2, every itemset is frequent and the frequent closed sets are `C`, `AC`, `BC`, `ABC`, and `ABCD`. $\mathcal{F}(\texttt{AB}) = \mathcal{F}(\texttt{ABC})$ since `ABC` is the smallest closed superset of `AB`.

The regeneration from δ-free itemsets is provided later. By construction, $|SAT_{\mathcal{C}_{close}}| \leq |SAT_{\mathcal{C}_{free}}|$ and $|SAT_{\mathcal{C}_{\delta-free}}| \leq |SAT_{\mathcal{C}_{free}}|$ when $\delta > 0$. Also, in practice, the size of these condensed representations are several orders of magnitude lower than the size of the frequent itemsets for dense data sets [24].

Several algorithms exist to compute various condensed representations of frequent itemsets [65,69,75,12,15,5,25]. These algorithms compute different condensed representations: the frequent closed itemsets (CLOSE, CLOSET, CHARM), the frequent free itemsets (MIN-EX, PASCAL), the frequent δ-free itemsets (MIN-EX), or the disjoint-free itemsets (H/VLINEX). Tractable extractions from dense and highly-correlated data have become possible for frequency thresholds on which previous algorithms are intractable.

Representations based on δ-free itemsets are quite interesting when it is not possible to mine the closed sets or even the disjoint-free sets, i.e., when the computation is intractable given the user-defined frequency threshold. Indeed, algorithms like CLOSE [65] or PASCAL [5] or H/VLIN-EX [25] use special kinds of logical rules to prune candidate itemsets because their frequencies can be inferred from the frequencies of others. However, to be efficient, these algorithms need that such logical rules hold in the data.

Let us now consider the δ-free itemsets and how they can be used to answer frequency queries. The output of the MIN-EX algorithm [16] is formally given by the three following sets: $FF(\mathbf{r}, \gamma, \delta)$ is the set of the γ-frequent δ-free itemsets, $IF(\mathbf{r}, \gamma, \delta)$ is the set of the minimal (w.r.t. the set inclusion) infrequent δ-free itemsets (i.e., the infrequent δ-free itemsets whose all subsets are γ-frequent). $FN(\mathbf{r}, \gamma, \delta)$ is the set of the minimal γ-frequent non-δ-free itemsets (i.e., the γ-frequent non-δ-free itemsets whose all subsets are δ-free). The two pairs (FF, IF) and (FF, FN) are two condensed representations based on δ-free itemsets.

It is possible to compute an approximation of the frequency of an itemset using one of these two condensed representations:

- Let S be an itemset. If there exists $X \in IF(\mathbf{r}, \gamma, \delta)$ such that $X \subseteq S$ then S is infrequent. If $S \notin FF(\mathbf{r}, \gamma, \delta)$ and there does not exist $X \in FN(\mathbf{r}, \gamma, \delta)$ such that $X \subseteq S$ then S is infrequent. In these two cases, the frequency of S can be approximated by $\gamma/2$ Else, let F be the δ-free itemset such that: $\mathcal{F}(F) = \text{Min}\{\mathcal{F}(X),\ X \subseteq S \text{ and } X \text{ is } \delta\text{-free}\}$. Assuming that $n_S = |\text{support}(S)|$ and $n_F = |\text{support}(F)|$, then $n_F \geq n_S \geq n_F - \delta(|S| - |F|)$, or, dividing this by n, the number of rows in $\mathbf{r}$, $\mathcal{F}(F) \geq \mathcal{F}(S) \geq \mathcal{F}(F) - \frac{\delta}{n}(|S| - |F|)$.

It is thus possible to regenerate an approximation of the answer to a frequent itemset query from one of the condensed representation (FF, IF) or (FF, FN). Typical δ values range from zero to a few hundreds. With a database size of several tens of thousands of rows, the error made is below few percents [16]. If $\delta = 0$, then the two condensed representations enable to regenerate exactly the answer to a frequent itemset query.

This line of work has inspired other researchers. For instance, [26] proposed a new exact condensed representation of the frequent itemsets that generalizes the previous ones. It is, to the best of our knowledge, the most interesting exact representation identified so far. In [66], new approximative condensed representations are proposed that are built from the maximal frequent itemsets for various frequency values.

The condensed representations can be used also for constraint-based mining of itemsets and the optimization of sequence of queries. In [46,45], constraint-based mining under conjunctions of anti-monotonic and monotonic constraints is combined with condensed representations. Some technical problems have to be solved and it has lead to the concept of *contextual δ-free itemsets* w.r.t. a monotonic constraint [19]. The use of condensed representations is not limited to the optimization of single queries. [47] describes the use of a cache that contains free itemsets to optimize the evaluation of sequences of itemset queries.

Notice that other researchers also consider the optimization of sequences based on condensed representations like the free itemsets [35].

4.3 Optimizing Association Rule Mining Queries

MINE RULE [59] has been designed by researchers who belong to the cInQ consortium. This is one of the query languages dedicated to association rule mining [9,10]. New extensions to the MINE RULE operator have been studied [11,58]. Two important and challenging new notions include: *pattern views* and relations among inductive queries. Both of these notions have also been included (and were actually inspired on MINE RULE) in the logical inductive database theory [29]. Pattern views intensionally specify a set of patterns using an inductive query in MINE RULE. This is similar in spirit to a traditional relation view in a traditional database. The view relation is defined by a query and can be queried like any other relation later on. It is the task of the (inductive) database management to take care (using, e.g., query materialization or query transformation) that the right answers are generated to such views. Pattern views raise many new challenges to data mining. The other notion that is nicely elaborated in MINE RULE concerns the dominance and subsumption relation between consecutive inductive queries. [11] studies the properties that the sets of patterns generated by two MINE RULE queries present in interesting situations. For instance, given two similar queries that are identical apart from one or more clauses in which they differ for an attribute, the result-sets of the two queries exhibit an inclusion relationship when a functional dependency is present between the differing attributes. [11] studies also the equivalence properties that two MINE RULE queries present when they have two clauses with constraints on attributes that are functionally dependent. Finally, it studies the properties that the queries have when multiple keys of a relation are involved. All these notions, if elaborated in the context of the inductive databases, will help the system to speed-up the query answering procedures. Again these ideas have been carried over to the logical theory of inductive databases [29].

Partners of the consortium have been inspired by the MINE RULE query language to study information discovery from XML data by means of association rules [22,21].

4.4 Towards a Theory of Inductive Databases

The final goal of the cInQ project is to propose a theory of inductive databases. As a first valuable step, a logical and set-oriented theory of inductive databases has been proposed [29,30,31], where the key idea is that a database consists of sets of data sets and sets of pattern sets, Furthermore there is an inductive query language, where each query either generates a data or a pattern set. Queries generating patterns sets are – in their most general form – arbitrary boolean expression over monotonic and anti-monotonic primitives. This corresponds to a logical view of inductive databases because the queries are boolean expressions as well as a set oriented one because the answers to inductive queries are sets of patterns.

Issues concerned with the evaluation and optimization of such inductive queries based on the border set representations can be found in [30]. Furthermore, various other issues concerned with inductive pattern views and the memory organization of such logical inductive databases are explored in [29]. Finally, various formal properties of arbitrary boolean inductive queries (e.g., normal forms, minimal number of version spaces needed) have been studied [31].

Interestingly, these theoretical results have emerged from an abstraction of useful KDD processes, e.g., for molecular fragment discovery with the domain specific inductive database MolFea [32,50,49,48] or for association rule mining processes with, e.g., the `MINE RULE` operator.

5 Conclusions

We provided a presentation of the itemset pattern domain. Any progress on constraint-based mining for itemsets can influence the research on the multiple uses of frequent itemsets (feature construction, similarity measures and clustering, classification rule mining or bayesian network construction, etc). It means that, not only (more or less generalized) association rule mining in difficult contexts like dense data sets can become tractable but also many other data mining processes can benefit from this outcome.

We introduced most of the results obtained by the cInQ consortium after 18 months of work. A few concepts have emerged that are now studied in depth, e.g., approximative and exact condensed representations, relationships between inductive query solutions and versions spaces, strategies for the active use of constraints during inductive query evaluation, containment and dominance between mining queries.

A lot has yet to be done, e.g., towards the use of these concepts for predictive data mining tasks. Also, we have to study the robustness of these concepts in various application domains and thus different pattern domains. It is a key issue to identify a set of data mining primitives and thus figure out what could be a good query language for inductive databases. Indeed, the design of dedicated inductive databases, e.g., inductive databases for molecular fragment discovery, is an invaluable step. Not only it solves interesting applicative problems but also it gives the material for abstraction and thus the foundations of the inductive database framework.

References

1. R. Agrawal, T. Imielinski, and A. Swami. Mining association rules between sets of items in large databases. In *Proceedings SIGMOD'93*, pages 207–216, Washington, USA, May 1993. ACM Press.
2. R. Agrawal, H. Mannila, R. Srikant, H. Toivonen, and A. I. Verkamo. Fast discovery of association rules. In U. M. Fayyad, G. Piatetsky-Shapiro, P. Smyth, and R. Uthurusamy, editors, *Advances in Knowledge Discovery and Data Mining*, pages 307–328. AAAI Press, 1996.

3. E. Baralis and G. Psaila. Incremental refinement of mining queries. In *Proceedings DaWaK'99*, volume 1676 of *LNCS*, pages 173–182, Firenze, I, Sept. 1999. Springer-Verlag.
4. Y. Bastide, N. Pasquier, R. Taouil, G. Stumme, and L. Lakhal. Mining minimal non-redundant association rules using frequent closed itemsets. In *Proceedings CL 2000*, volume 1861 of *LNCS*, pages 972–986, London, UK, 2000. Springer-Verlag.
5. Y. Bastide, R. Taouil, N. Pasquier, G. Stumme, and L. Lakhal. Mining frequent patterns with counting inference. *SIGKDD Explorations*, 2(2):66–75, Dec. 2000.
6. R. J. Bayardo. Efficiently mining long patterns from databases. In *Proceedings SIGMOD'98*, pages 85–93, Seattle, USA, May 1998. ACM Press.
7. R. J. Bayardo and R. Agrawal. Mining the most interesting rules. In *Proceedings SIGKDD'99*, pages 145–154, San Diego, USA, Aug. 1999. ACM Press.
8. C. Becquet, S. Blachon, B. Jeudy, J.-F. Boulicaut, and O. Gandrillon. Strong association rule mining for large gene expression data analysis: a case study on human SAGE data. *Genome Biology*, 3(12), Dec. 2002.
9. M. Botta, J.-F. Boulicaut, C. Masson, and R. Meo. A comparison between query languages for the extraction of association rules. In *Proceedings DaWaK'02*, volume 2454 of *LNCS*, pages 1–10, Aix-en-Provence, F, Sept. 2002. Springer-Verlag.
10. M. Botta, J.-F. Boulicaut, C. Masson, and R. Meo. Query languages supporting descriptive rule mining: a comparative study. In P. L. Lanzi and R. Meo, editors, *Database Support for Data Mining Applications*, number 2682 in LNCS. Springer-Verlag, 2003. This volume.
11. M. Botta, R. Meo, and M.-L. Sapino. Incremental execution of the MINE RULE operator. Technical Report RT 66/2002, Dipartimento di Informatica, Università degli Studi di Torino, Corso Svizzera 185, I-10149 Torino, Italy, May 2002.
12. J.-F. Boulicaut and A. Bykowski. Frequent closures as a concise representation for binary data mining. In *Proceedings PAKDD'00*, volume 1805 of *LNAI*, pages 62–73, Kyoto, JP, Apr. 2000. Springer-Verlag.
13. J.-F. Boulicaut, A. Bykowski, and B. Jeudy. Mining association rules with negations. Technical Report 2000-14, INSA Lyon, LISI, Batiment Blaise Pascal, F-69621 Villeurbanne, France, Nov. 2000.
14. J.-F. Boulicaut, A. Bykowski, and B. Jeudy. Towards the tractable discovery of association rules with negations. In *Proceedings FQAS'00*, Advances in Soft Computing series, pages 425–434, Warsaw, PL, Oct. 2000. Springer-Verlag.
15. J.-F. Boulicaut, A. Bykowski, and C. Rigotti. Approximation of frequency queries by mean of free-sets. In *Proceedings PKDD'00*, volume 1910 of *LNAI*, pages 75–85, Lyon, F, Sept. 2000. Springer-Verlag.
16. J.-F. Boulicaut, A. Bykowski, and C. Rigotti. Free-sets: a condensed representation of boolean data for the approximation of frequency queries. *Data Mining and Knowledge Discovery journal*, 7(1):5–22, 2003.
17. J.-F. Boulicaut and B. Crémilleux. Delta-strong classification rules for predicting collagen diseases. In *Proceedings of the ECML-PKDD'01 Discovery Challenge on Thrombosis Data*, pages 29–38, Freiburg, D, Sept. 2001. Available on line.
18. J.-F. Boulicaut and B. Jeudy. Using constraint for itemset mining: should we prune or not? In *Proceedings BDA'00*, pages 221–237, Blois, F, Oct. 2000.
19. J.-F. Boulicaut and B. Jeudy. Mining free-sets under constraints. In *Proceedings IDEAS'01*, pages 322–329, Grenoble, F, July 2001. IEEE Computer Society.
20. J.-F. Boulicaut, M. Klemettinen, and H. Mannila. Modeling KDD processes within the inductive database framework. In *Proceedings DaWaK'99*, volume 1676 of *LNCS*, pages 293–302, Firenze, I, Sept. 1999. Springer-Verlag.

21. D. Braga, A. Campi, S. Ceri, M. Klemettinen, and P. L. Lanzi. Discovering interesting information in XML data with association rules. In *Proceedings SAC 2003 Data Mining track*, Melbourne, USA, Mar. 2003. ACM Press.
22. D. Braga, A. Campi, M. Klemettinen, and P. L. Lanzi. Mining association rules from XML data. In *Proceedings DaWaK'02*, volume 2454 of *LNCS*, pages 21–30, Aix-en-Provence, F, Sept. 2002. Springer-Verlag.
23. S. Brin, R. Motwani, and C. Silverstein. Beyond market baskets: Generalizing association rules to correlations. In *Proceedings SIGMOD'97*, pages 265–276, Tucson, USA, May 1997. ACM Press.
24. A. Bykowski. *Condensed representations of frequent sets: application to descriptive pattern discovery.* PhD thesis, Institut National des Sciences Appliquées de Lyon, LISI, F-69621 Villeurbanne cedex, France, Oct. 2002.
25. A. Bykowski and C. Rigotti. A condensed representation to find frequent patterns. In *Proceedings PODS'01*, pages 267–273. ACM Press, May 2001.
26. T. Calders and B. Goethals. Mining all non derivable frequent itemsets. In *Proceedings PKDD'02*, volume 2431 of *LNAI*, pages 74–83, Helsinki, FIN, Aug. 2002. Springer-Verlag.
27. M. Capelle, C. Masson, and J.-F. Boulicaut. Mining frequent sequential patterns under a similarity constraint. In *Proceedings IDEAL'02*, volume 2412 of *LNCS*, pages 1–6, Manchester, UK, Aug. 2002. Springer-Verlag.
28. B. Crémilleux and J.-F. Boulicaut. Simplest rules characterizing classes generated by delta-free sets. In *Proceedings ES 2002*, pages 33–46, Cambridge, UK, Dec. 2002. Springer-Verlag.
29. L. de Raedt. A logical view of inductive databases. Technical report, Institut fur Informatik, Albert-Ludwigs-Universitat, Georges-Kohler-Allee, Gebaude 079, D-79110 Freiburg, Germany, May 2002. 13 pages.
30. L. de Raedt. Query evaluation and optimization for inductive database using version spaces (extended abstract). In *Proceedings DTDM'02 co-located with EDBT'02*, pages 19–28, Praha, CZ, Mar. 2002. An extended version appears in this volume.
31. L. de Raedt, M. Jaeger, S. D. Lee, and H. Mannila. A theory of inductive query answering (extended abstract). In *Proceedings ICDM'02*, pages 123–130, Maebashi City, Japan, December 2002. IEEE Computer Press.
32. L. de Raedt and S. Kramer. The levelwise version space algorithm and its application to molecular fragment finding. In *Proceedings IJCAI'01*, pages 853–862, Seattle, USA, Aug. 2001. Morgan Kaufmann.
33. G. Dong and J. Li. Efficient mining of emerging patterns: Discovering trends and differences. In *Proceedings SIGKDD'99*, pages 43–52, San Diego, USA, Aug. 1999. ACM Press.
34. M. N. Garofalakis, R. Rastogi, and K. Shim. SPIRIT: Sequential pattern mining with regular expression constraints. In *Proceedings VLDB'99*, pages 223–234, Edinburgh, UK, September 1999. Morgan Kaufmann.
35. A. Giacommetti, D. Laurent, and C. T. Diop. Condensed representations for sets of mining queries. In *Proceedings KDID'02 co-located with ECML-PKDD'02*, Helinski, FIN, Aug. 2002. An extended version appears in this volume.
36. B. Goethals and J. V. den Bussche. On supporting interactive association rule mining. In *Proceedings DaWaK'00*, volume 1874 of *LNCS*, pages 307–316, London, UK, Sept. 2000. Springer-Verlag.
37. B. Goethals and J. van den Bussche. A priori versus a posteriori filtering of association rules. In *Proceedings SIGMOD Workshop DMKD'99*, Philadelphia, USA, May 1999.

38. J. Han and M. Kamber. *Data Mining: Concepts and techniques.* Morgan Kaufmann Publishers, San Francisco, USA, 2000. 533 pages.
39. J. Han, J. Pei, and Y. Yin. Mining frequent patterns without candidate generation. In *Proceedings ACM SIGMOD'00*, pages 1–12, Dallas, Texas, USA, May 2000. ACM Press.
40. H. Hirsh. Theoretical underpinnings of version spaces. In *Proceedings IJCAI'91*, pages 665–670, Sydney, Australia, Aug. 1991. Morgan Kaufmann.
41. H. Hirsh. Generalizing version spaces. *Machine Learning*, 17(1):5–46, 1994.
42. T. Imielinski and H. Mannila. A database perspective on knowledge discovery. *Communications of the ACM*, 39(11):58–64, Nov. 1996.
43. T. Imielinski and A. Virmani. MSQL: A query language for database mining. *Data Mining and Knowledge Discovery*, 3(4):373–408, 1999.
44. B. Jeudy. *Extraction de motifs sous contraintes: application à l'évaluation de requêtes inductives.* PhD thesis, Institut National des Sciences Appliquées de Lyon, LISI, F-69621 Villeurbanne cedex, France, Dec. 2002. In French.
45. B. Jeudy and J.-F. Boulicaut. Constraint-based discovery and inductive queries: application to association rule mining. In *Proceedings ESF Exploratory Workshop on Pattern Detection and Discovery*, volume 2447 of *LNAI*, pages 110–124, London, UK, Sept. 2002. Springer-Verlag.
46. B. Jeudy and J.-F. Boulicaut. Optimization of association rule mining queries. *Intelligent Data Analysis journal*, 6:341–357, 2002.
47. B. Jeudy and J.-F. Boulicaut. Using condensed representations for interactive association rule mining. In *Proceedings PKDD'02*, volume 2431 of *LNAI*, pages 225–236, Helsinki, FIN, Aug. 2002. Springer-Verlag.
48. S. Kramer. Demand-driven construction of structural features in ILP. In *Proceedings ILP'01*, volume 2157 of *LNCS*, pages 132–141, Strasbourg, F, Sept. 2001. Springer-Verlag.
49. S. Kramer and L. de Raedt. Feature construction with version spaces for biochemical applications. In *Proceedings ICML'01*, pages 258–265, William College, USA, July 2001. Morgan Kaufmann.
50. S. Kramer, L. de Raedt, and C. Helma. Molecular feature mining in HIV data. In *Proceedings SIGKDD'01*, pages 136–143, San Francisco, USA, Aug. 2001. ACM Press.
51. L. V. Lakshmanan, R. Ng, J. Han, and A. Pang. Optimization of constrained frequent set queries with 2-variable constraints. In *Proceedings SIGMOD'99*, pages 157–168, Philadelphia, USA, 1999. ACM Press.
52. S. D. Lee and L. de Raedt. Constraint-based mining of first order sequences in SEQLOG. In *Proceedings KDID'02 co-located with ECML-PKDD'02*, Helsinki, FIN, Aug. 2002. An extended version appears in this volume.
53. B. Liu, W. Hsu, and Y. Ma. Integrating classification and association rule mining. In *Proceedings KDD'98*, pages 80–86, New York, USA, 1998. AAAI Press.
54. H. Mannila and H. Toivonen. Multiple uses of frequent sets and condensed representations. In *Proceedings KDD'96*, pages 189–194, Portland, USA, Aug. 1996. AAAI Press.
55. H. Mannila and H. Toivonen. Levelwise search and borders of theories in knowledge discovery. *Data Mining and Knowledge Discovery*, 1(3):241–258, 1997.
56. C. Masson and F. Jacquenet. Mining frequent logical sequences with SPIRIT-LoG. In *Proceedings ILP'02*, volume 2583 of *LNAI*, pages 166–182, Sydney, Australia, July 2002. Springer-Verlag.
57. C. Mellish. The description identification problem. *Artificial Intelligence*, 52(2):151–168, 1992.

58. R. Meo. Optimization of a language for data mining. In *Proceedings of the 18th Symposium on Applied Computing SAC 2003 Data Mining track*, Melbourne, USA, Mar. 2003. ACM Press. To appear.
59. R. Meo, G. Psaila, and S. Ceri. An extension to SQL for mining association rules. *Data Mining and Knowledge Discovery*, 2(2):195–224, 1998.
60. T. Mitchell. Generalization as search. *Artificial Intelligence*, 18(2):203–226, 1980.
61. P. Moen. *Attribute, Event Sequence, and Event Type Similarity Notions for Data Mining.* PhD thesis, Department of Computer Science, P.O. Box 26, FIN-00014 University of Helsinki, Jan. 2000.
62. B. Nag, P. M. Deshpande, and D. J. DeWitt. Using a knowledge cache for interactive discovery of association rules. In *Proceedings SIGKDD'99*, pages 244–253, San Diego, USA, Aug. 1999. ACM Press.
63. C. Nedellec, C. Rouveirol, H. Ade, and F. Bergadano. Declarative bias in inductive logic programming. In L. de Raedt, editor, *Advances in Logic Programming*, pages 82–103. IOS Press, 1996.
64. R. Ng, L. V. Lakshmanan, J. Han, and A. Pang. Exploratory mining and pruning optimizations of constrained associations rules. In *Proceedings SIGMOD'98*, pages 13–24, Seattle, USA, 1998. ACM Press.
65. N. Pasquier, Y. Bastide, R. Taouil, and L. Lakhal. Efficient mining of association rules using closed itemset lattices. *Information Systems*, 24(1):25–46, Jan. 1999.
66. J. Pei, G. Dong, W. Zou, and J. Han. On computing condensed frequent pattern bases. In *Proceedings ICDM'02*, pages 378–385, Maebashi City, JP, Dec. 2002. IEEE Computer Press.
67. J. Pei and J. Han. Constrained frequent pattern mining:a pattern-growth view. *SIGKDD Explorations*, 4(1):31–39, June 2002.
68. J. Pei, J. Han, and L. V. S. Lakshmanan. Mining frequent itemsets with convertible constraints. In *Proceedings ICDE'01*, pages 433–442, Heidelberg, D, Apr. 2001. IEEE Computer Press.
69. J. Pei, J. Han, and R. Mao. CLOSET an efficient algorithm for mining frequent closed itemsets. In *Proceedings SIGMOD Workshop DMKD'00*, Dallas, USA, May 2000.
70. T. Scheffer. Finding association rules that trade support optimally against confidence. In *Proceedings PKDD'01*, volume 2168 of *LNCS*, pages 424–435, Freiburg, D, Sept. 2001. Springer-Verlag.
71. J. Sese and S. Morishita. Answering the most correlated N association rules efficiently. In *Proceedings PKDD'02*, volume 2431 of *LNAI*, pages 410–422, Helsinki, FIN, Aug. 2002. Springer-Verlag.
72. P. Smyth and R. M. Goodman. An information theoretic approach to rule induction from databases. *IEEE Transactions on Knowledge and Data Engineering*, 4(4):301–316, Aug. 1992.
73. R. Srikant, Q. Vu, and R. Agrawal. Mining association rules with item constraints. In *Proceedings KDD'97*, pages 67–73, Newport Beach, USA, 1997. AAAI Press.
74. H. Toivonen. Sampling large databases for association rules. In *Proceedings VLDB'96*, pages 134–145, Mumbay, India, Sept. 1996. Morgan Kaufmann.
75. M. J. Zaki. Generating non-redundant association rules. In *Proceedings SIGKDD'00*, pages 34–43, Boston, USA, Aug. 2000. ACM Press.

Query Languages Supporting Descriptive Rule Mining: A Comparative Study

Marco Botta[1], Jean-François Boulicaut[2], Cyrille Masson[2], and Rosa Meo[1]

[1] Universitá di Torino, Dipartimento di Informatica,
corso Svizzera 185, I-10149, Torino, Italy
[2] Institut National des Sciences Appliquées de Lyon, LISI/LIRIS,
Bat. Blaise Pascal, F-69621 Villeurbanne cedex, France

Abstract. Recently, inductive databases (IDBs) have been proposed to tackle the problem of knowledge discovery from huge databases. With an IDB, the user/analyst performs a set of very different operations on data using a query language, powerful enough to support all the required manipulations, such as data preprocessing, pattern discovery and pattern post-processing. We provide a comparison between three query languages (`MSQL`, `DMQL` and `MINE RULE`) that have been proposed for descriptive rule mining and discuss their common features and differences. These query languages look like extensions of SQL. We present them using a set of examples, taken from the real practice of rule mining. In the paper we discuss also OLE DB for Data Mining and Predictive Model Markup Language, two recent proposals that like the first three query languages respectively provide native support to data mining primitives and provide a description in a standard language of statistical and data mining models.

1 Introduction

Knowledge Discovery in Databases (KDD) is a complex process which involves many steps that must be done sequentially. When considering the whole KDD process, the proposed approaches and querying tools are still unsatisfactory. The relation among the various proposals is also sometimes unclear because, at the moment, a general understanding of the fundamental primitives and principles that are necessary to support the search of knowledge in databases is still lacking.

In the cInQ project[1], we want to develop a new generation of databases, called "*inductive databases*", as suggested in [5]. This kind of databases integrates *raw data* with *knowledge* extracted from *raw data*, materialized under the form of patterns, into a common framework that supports the KDD process. In this way, the KDD process consists essentially in a querying process, enabled by a powerful query language that can deal with both raw data and patterns. A few query languages can be considered as candidates for inductive databases. Most proposals emphasize one of the different phases of the KDD process. This paper

[1] Project (IST 2000-26469) partially funded by the EC IST Programme - FET.

R. Meo et al. (Eds.): Database Support for Data Mining Applications, LNAI 2682, pp. 24–51, 2004.

is a critical evaluation of three proposals in the light of the IDBs' requirements: MSQL [6,7], DMQL [10,11] and MINE RULE [12,13]. In the paper we discuss also OLE DB for Data Mining (OLE DB DM) by Microsoft and Predictive Model Markup Language (PMML) by Data Mining Group [18]. OLE DB DM is an Application Programming Interface whose aim is to ease the task of developing data mining applications over databases. It is related to the other query languages because like them it provides native support for data mining primitives. PMML, instead, is a standard markup language, based on XML, and describes statistical and data mining models.

The paper is organized as follows. Section 2 summarizes the desired properties of a language for mining inside an inductive database. Section 3 introduces the main features of the analyzed languages, whereas in Section 4 some real examples of queries are discussed, so that the comparison between the languages is straightforward. Finally Section 5 draws some conclusions.

2 Desired Properties of a Data Mining Query Language

A *query language* for IDBs, is an extension of a database query language that includes primitives for supporting the steps of a KDD process, that are:

- The selection of data to be mined. The language must offer the possibility to select (e.g., via standard queries but also by means of sampling), to manipulate and to query data and views in the database. It must also provide support for multi-dimensional data manipulation.
  DMQL, MINE RULE and OLE DB DM allow the selection of data. Neither of them has primitives for sampling. All of them allow multi-dimensional data manipulation (because this is inherent to SQL).
- The specification of the type of patterns to be mined. Clearly, real-life KDD processes need for different kinds of patterns like various types of descriptive rules, clusters or predictive models.
  DMQL considers different patterns beyond association rules.
- The specification of the needed background knowledge (e.g., the definition of a concept hierarchy).
  Even though both MINE RULE and MSQL can treat hierarchies if the relationship 'is-a' is represented in a companion relation, DMQL allows its explicit definition and use during the pattern extraction.
- The definition of constraints that the extracted patterns must satisfy. This implies that the language allows the user to define constraints that specify the interesting patterns (e.g., using measures like frequency, generality, coverage, similarity, etc).
  DMQL, MSQL and MINE RULE allow the specification of various kinds of constraints based on rule elements, rule cardinality and aggregate values. They allow the specification of primitive constraints based on support and confidence measures. DMQL allows some other measures like novelty.

- The satisfaction of the *closure property* (by storing the results in the database). **All of them satisfy this property.**
- The post-processing of results. The language must allow to browse the patterns, apply selection templates, *cross over* patterns and data, e.g., by selecting the data in which some patterns hold, or aggregating results. **MSQL is richer than the other languages in its offer of few post-processing primitives (it has a dedicated operator, SelectRules). DMQL allows some visualization options. However, all the languages are quite poor for rule post-processing.**

3 Query Languages for Rule Mining

3.1 MSQL

MSQL [7,14] has been designed at the Rutgers University, New Jersey, USA. Rules in MSQL are based on descriptors, each descriptor being an expression of the form $(A_i = a_{ij})$, where A_i is an attribute and a_{ij} is a value or a range of values in the domain of A_i. A conjunctset is the conjunction of an arbitrary number of descriptors, provided that there is no pair of descriptors built on the same attribute. In practice, MSQL extracts propositional rules like $\mathcal{A} \Rightarrow \mathcal{B}$, where $\mathcal{A}$ is a conjunctset and $\mathcal{B}$ is a descriptor (it follows that only a single proposition is allowed in the consequent). We say that a tuple t of a relation R satisfies a descriptor $(A_i = a_{ij})$ if the value of A_i in t is equal to a_{ij}. Moreover, t satisfies a conjunctset C if it satisfies all the descriptors of C. Finally, t satisfies a rule $\mathcal{A} \Rightarrow \mathcal{B}$ if it satisfies all the descriptors in $\mathcal{A}$ and $\mathcal{B}$, but it violates the rule $\mathcal{A} \Rightarrow \mathcal{B}$ if it does not satisfy $\mathcal{A}$ or $\mathcal{B}$. Notice that support of a rule is defined as the number of tuples satisfying $\mathcal{A}$ in the relation on which the mining has been performed, and the confidence is the ratio between the number of tuples satisfying both $\mathcal{A}$ and $\mathcal{B}$ and the support of the rule. An example of a rule extracted from $Emp(emp\_id, job, sex, car)$ relation containing employee data is $(job = doctor) \wedge (sex = male) \Rightarrow (car = BMW)$.

The main features of MSQL, as stated by the authors, are:

- *Ability to nest SQL expressions* such as sorting and grouping in a MSQL statement and allowing nested SQL queries by means of the WHERE clause.
- *Satisfaction of the closure property* and availability of operators to further manipulate results of previous MSQL queries.
- *Cross-over between data and rules* with operations allowing to identify subsets of data satisfying or violating a given set of rules.
- *Distinction between rule generation and rule querying.* This allows splitting rule generation, that is computationally expensive from rule post-processing, that must be as interactive as possible.

MSQL comprises four basic statements (see Section 4 for examples):

- Create Encoding that encodes continuous valued attributes into discrete values. Notice that during mining, the discretization is done "on the fly", so that it is not necessary to materialize a separate copy of the table.

– A `GetRules` query computes rules from the data and materializes them into a rule database. Its syntax is as follows:

```
[Project Body, Consequent, confidence, support]
GetRules(C) [as R1] [into <rulebase_name>]
[where (RC|PC|MC|SQ)]
[sql-group-by clause] [using-encoding-clause]
```

A `GetRules` query can deal with different conditions on the rules:
  - Rule format condition (RC), that enables to restrict the items occuring in the rules elements. RC has the following format:
```
Body { in | has | is } <descriptor-list>
Consequent { in | is } <descriptor-list>
```
  - Pruning condition (PC), that defines thresholds for support and confidence values, and constraints on the length of the rules. PC has the format:
```
confidence <relop> <float-val in [0.0,1.0]>
support <relop> <integer>
support <relop> <float-val in [0.0,1.0]>
length <relop> <integer>
relop ::= { < | <= | = | >= | > }
```
  - Mutex condition (MC), that avoids two given attributes to occur in the same rule (useful when we know some functional dependencies between attributes). Its syntax is:
```
Where <other-conditions>
    { AND | OR }    mutex(method, method [, method])
    [{ AND | OR}    mutex(method, method [, method])]
```
  - Subquery conditions (SQ), which are subqueries connected with the conventional `WHERE` keyword using `IN` and `(NOT) EXISTS`.

– A `SelectRules` query can be used for rule post-processing, i.e., querying previously extracted rules. Its syntax is as follows:

```
SelectRules(rulebase_name) [where <conditions>]
```

where `<conditions>` concerns the body, the consequent, the support and/or the confidence of the rules.

– `Satisfies` and `Violates`, that allow to cross-over data and rules. These two statements can be used together with a database selection statement, inside the `WHERE` clause of a query.

3.2 MINE RULE

`MINE RULE` has been designed at the Politecnico di Torino and the Politecnico di Milano, Italy [12,13]. This operator extracts a set of association rules from the database and stores them back in the database in a separate relation.

An association rule extracted by `MINE RULE` from a source relation is defined as follows. Let us consider a source relation over the schema $\mathcal{S}$. Let $\mathcal{R}$ and $\mathcal{G}$

be two disjoint subsets of $\mathcal{S}$ called respectively the schema of the rules and the grouping attributes. An association rule is extracted from (or satisfied by) at least a group of the source relation, where each group is a partition of the relation by the values of the grouping attributes $\mathcal{G}$. An association rule has the form $\mathcal{A} \Rightarrow \mathcal{B}$, where $\mathcal{A}$ and $\mathcal{B}$ are sets of rule elements ($\mathcal{A}$ is the body of the rule and $\mathcal{B}$ the head). The elements of a rule are taken from the tuples of one group. In particular, each rule element is a projection over (a subset of) $\mathcal{R}$. Note that however, for a given `MINE RULE` statement, the schema of the body and head elements is unique, even though they may be different.

An example of a rule extracted from the relation $Emp(emp\_id, job, sex, car)$ grouped by *emp_id* with the schema of the body (job, sex) and the schema of the head (car) is the following: $\{(doctor, male)\} \Rightarrow \{(BMW)\}$. This rule is extracted from within tuples, because each group coincides with a tuple of the relation. Instead, for the relation $Sales(transaction\_id, item, customer, payment)$, collecting data on customers purchases, grouped by *customer* and with the rule schema $(item)$ (where body and head schemas are coincident) a rule could be $\{(pasta), (oil), (tomatoes)\} \Rightarrow \{(wine)\}$.

The `MINE RULE` language is an extension of SQL. Its main features are:

- *Selection of the relevant set of data* for a data mining process. This feature is applied at different granularity levels, that is at the row level (selection of a subset of the rows of a relation) or at the group level (*group condition*). The *grouping condition* determines which data of the relation can take part to an association rule. This feature is similar to the grouping conditions that we can find in conventional SQL. The definition of groups, i.e. the partitions from which the rules are extracted, is made at run time and is not decided a priori with the key of the source relation (as in `DMQL`).
- Definition of the *structure of the rules*. This feature defines single-dimensional association rules (i.e., rule elements are the different values of the same dimension or attribute), or multi-dimensional rules (rule elements involve the value of more than one attribute). The structure of the rules can also be constrained by specifying the cardinality of the rule's body and head.
- Definition of *constraints applied at different granularity levels*. Constraints belong to two categories: constraints applied at the rule level (*mining conditions*), and constraints applied at the *cluster* level (*cluster conditions*). A mining condition is a constraint that is evaluated and satisfied by each tuple whose attributes, as rule elements, are involved in the rule. A cluster condition is a constraint evaluated for each cluster. Clusters are subgroups (or partitions) of the main groups that are created keeping together tuples of the same group that present common features (i.e., the value of the clustering attributes). In presence of clusters, rule body and head are extracted from a pair of clusters of the same group satisfying the cluster conditions. For instance, clusters and cluster condition may be exploited in order to extract association rules in which body and head are ordered and therefore constitute the elementary patterns of sequences.

– Definition of *rule evaluation measures.* Practically, the language allows to define support and confidence thresholds.

The general syntax of MINE RULE follows:

```
<MineRuleOp> := MINE RULE <TableName> AS
 SELECT DISTINCT <BodyDescr>, <HeadDescr> [,SUPPORT] [,CONFIDENCE]
 [WHERE <WhereClause>]
 FROM <FromList> [ WHERE <WhereClause> ]
 GROUP BY <AttrList> [ HAVING <HavingClause> ]
 [ CLUSTER BY <AttrList> [ HAVING <HavingClause> ]]
 EXTRACTING RULES WITH SUPPORT:<real>, CONFIDENCE:<real>

<BodyDescr>:= [ <CardSpec> ] <AttrList> AS BODY
<BodyDescr>:= [ <CardSpec> ] <AttrList> AS HEAD
<CardSpec>:=<Number> .. (<Number> | n)
<AttrList>:=<AttrName>[,<AttrList>]
```

3.3 DMQL

DMQL has been designed at the Simon Fraser University, Canada [10,11]. In DMQL, an association rule is a relation between the values of two sets of predicates that are evaluated on the relations of the database. These predicates are of the form $P(X, c)$ where P is a predicate that takes the name of an attribute of the underlying relation, X is a variable and c is a constant value belonging to the attribute's domain. The predicate is satisfied if in the relation there exists a tuple identified by the variable X whose homonymous attribute takes the value c. Notice that it is possible for the predicates to be evaluated on different relations of the database. For instance, DMQL can extract rules like $town(X,' London') \Rightarrow buys(X,' DVD')$ where town and buys may be two attributes of different relations and X is an attribute present in the both relations. Rules may belong to different categories: a single-dimensional rule contains multiple occurrences of a single predicate (e.g., *buys*) while a multi-dimensional rule involves more predicates, each of which occurs only once in the rule. However, the presence of one or more instances of a predicate in the same rule can be specified by the name of the predicate followed by +. Another important feature of DMQL is that it allows to guide the discovery process by using metapatterns. Metapatterns are a kind of templates that restricts the syntactical aspect of the association rules to be extracted. Moreover, they represent a way to push some hypotheses of the user and it is possible to incorporate some further constraints in them. An example of metapattern could be $town(X : customer, London) \wedge income(X, Y) \Rightarrow buys(X, Z)$, which restricts the discovery to rules with a body concerning town and income levels of the customers and a head concerning one item bought by those customers. Furthermore, a metapattern can allow the presence of non instantiated predicates that the mining task will take care to instantiate to the name of a valid attribute of the underlying relation. For instance, if we want to extract association rules describing the customers traits that are frequently related to the purchase of certain items by those customers we could use the following metapattern to guide the association rule mining:

$P(X : customer, W) \wedge Q(X, Y) \Rightarrow buys(X, Z)$

where P and Q are predicate variables that can be instantiated to the relevant attributes of the relations under examination, X is a key of the *customer* relation, W, Y and Z are object variables that can assume the values of the respective predicates for customer X.

`DMQL` consists of the specification of four major primitives in data mining, that are the following:

- *The set of relevant data w.r.t. a data mining process.*
  This primitive can be specified like in a conventional relational query extracting the set of relevant data from the database.
- *The kind of knowledge to be discovered.*
  This primitive may include association rules, classification rules (rules that assign data to disjoint classes according to the value of a chosen classifying attribute), characteristics (descriptions that constitute a summarization of the common properties in a given set of data), comparisons (descriptions that allow to compare the total number of tuples belonging to a class with different contrasting classes), generalized relations (obtained by generalizing a set of data corresponding to low level concepts with data corresponding to higher level concepts according to a specified concept hierarchy).
- *The background knowledge.*
  This primitive manages a set of concept hierachies or generalization operators which assist the generalization process.
- *The justification of the interestingness of the knowledge (i.e., thresholds).*
  This primitive is included as a set of different constraints depending on the kind of target rules. For association rules, e.g., besides the classical support and confidence thresholds, `DMQL` allows the specification of noise (the minimum percentage of tuples in the database that must satisfy a rule so that it is not discarded) and rule novelty, for selecting the most specific rules.

The `DMQL` grammar for extracting association rules is an extension of the conventional `SQL` grammar. Thus, we can find in it traditional relational operators like `HAVING`, `WHERE`, `ORDER BY` and `GROUP BY`, but we can also specify the database, select the relevant attributes of the database relation and the concept hierarchy, define thresholds and guide the mining process using a metapattern. The general syntax of a `DMQL` query is:

use database ⟨*database_name*⟩
{**use hierarchy** ⟨*hierarchy_name*⟩ **for** ⟨*attribute_or_dimension*⟩ }
in relevance to ⟨*attribute_or_dimension_list*⟩
mine associations [as ⟨*pattern_name*⟩] [**matching** ⟨*metapattern*⟩]
from ⟨*relation(s)/cube(s)*⟩ [**where** ⟨*condition*⟩]
[**order by** ⟨*order_list*⟩]
[**group by** ⟨*grouping_list*⟩][**having** ⟨*condition*⟩]
with ⟨*interest_measure*⟩ **threshold** = value

3.4 OLE DB DM

OLE DB DM has been designed at Microsoft Corporation [17] [16]. It is an extension of the OLE DB Application Programming Interface (API) that allows any application to easily access a relational data source under the Windows family OS. The main motivation of the design of OLE DB for DM is to ease the development of data mining projects with applications that are not stand-alone but are tightly-coupled with the DBMS. Indeed, research work in data mining focused on scaling analysis and algorithms running outside the DBMS on data exported from the databases in files. This situation generates problems in the deployment of the data mining models produced because the data management and maintenance of the model occurs outside of the DBMS and must be solved by ad-hoc solutions. OLE DB DM aims at ease the burden of making data sources communicate with data mining algorithms (also called mining model provider).

The key idea of OLE DB DM is the definition of a data mining model, i.e. a special sort of table whose rows contain an abstraction, a synthetic description of input data (called case set). The user can populate this model with predicted or summary data obtained running a data mining algorithm, specified as part of the model, over the case set. Once the mining task is done, it is possible to use the data mining model, for instance to predict some values over new cases, or browse the model for post-processing activities, such as reporting or visualization.

The representation of the data in the model depends on the format of data produced by the algorithm. This one could produce output data for instance by using PMML (Predictive Model Markup Language [18]). PMML is a standard proposed by DMG based on XML. It is a mark-up language for the description of statistical and data mining models. PMML describes the inputs of data mining models, the data transformations used for the preparation of data and the parameters used for the generation of the models themselves.

OLE DB DM provides an SQL-like language that allows client applications to perform the key operations in the OLE DB DM framework: definition of a data mining model (with the CREATE MINING MODEL statement), execution of an external mining algorithms on data provided by a relational source and population of the data mining model (INSERT INTO statement), prediction of the value of some attributes on new data (PREDICTION JOIN), browsing of the model (SELECT statement).

Thus, elaboration of an OLE DB DM model can be done using classical SQL queries. Once the mining algorithm has been executed, it is prossible to do some crossing-over between the data mining model and the data fitting the mining model using the PREDICTION JOIN statement. This is a special form of the SQL join that allows to predict the value of some attributes in the input data (test data) according to the model, provided that these attributes were specified as prediction attributes in the mining model.

The grammar for the creation of a data mining model is the following:

```
<dm_create>::=CREATE MINING MODEL <identifier> (<col_def_list>)
             USING <algorithm> [(<algo_param_list>)]

<col_def_list>::= <col_def> |<col_def_list> , <col_def>
<col_def>::= <col_def_reg> | <col_def_tbl>
<col_def_reg>::= <identifier> <col_type> [<col_distribution>]
          [<col_binary>] [<col_content>] [<col_content_qual>]
        [<col_qualif>] [<col_prediction>] [<relation_clause>]

<col_def_tbl> ::= <identifier> TABLE <col_prediction>
( <col_def_list> )

// 2 algorithms currently implemented in SQL server 2000
<algorithm> ::= MICROSOFT_DECISION_TREES | MICROSOFT_CLUSTERING

<algo_param_list>::=<algo_param> | <algo_param>,<algo_param_list>
<algo_param>::= <identifier> = <value>

<col_type>::= LONG | BOOLEAN | TEXT | DOUBLE | DATE

<col_distribution>-> NORMAL | UNIFORM

<col_binary>::= MODEL_EXISTENCE_ONLY | NOT NULL

<col_content>::= DISCRETE | CONTINUOUS
          | DISCRETIZED ( [<disc_method> [, <numeric_const>]] )
          | SEQUENCE_TIME

<disc_method>::=AUTOMATIC | EQUAL_AREAS | THRESHOLDS | CLUSTERS

<col_content_qual>::= ORDERED | CYCLICAL

<col_qualif>::= KEY | PROBABILITY | VARIANCE | STDEV | STDDEV
             | PROBABILITY_VARIANCE | PROBABILITY_STDEV
             | PROBABILITY_STDDEV | SUPPORT

<col_prediction>::= PREDICT | PREDICT_ONLY

<relation_clause>::= <related_to_clause> | <of_clause>

<related_to_clause>::=RELATED TO <identifier> | RELATED TO KEY

<of_clause>::= OF <identifier> | OF KEY
```

Notice that the grammar allows to specify many kinds of qualifiers for an attribute. For instance, it allows to specify the role of an attribute in the model (key), the type of an attribute, if the attribute domain is ordered or cyclical, if it is continuous or discrete (and in this latter case the type of discretization used), if the attribute is a measurement of time, and its range, etc. It is possible to give a probability and other statistical features associated to an attribute value. The probability specifies the degree of certainty that the value of the attribute is correct.

`PREDICT` keyword specifies that it is a prediction attribute. This means that the content of the attribute will be predicted on test data by the data mining algorithm according to the values of the other attributes of the model.

`RELATED TO` allows to associate the current attribute to other attributes, for instance for a foreign key relationship or because the attribute is used to classify the values of another attribute.

Notice that <col_def_tbl> production rule allows a data mining model to contain nested tables. Nested tables are tables stored as the single values of a column in an outer table. The input data of a mining algorithm are often obtained by gathering and joining information that is scattered in different tables of the database. For instance, customer information and sales information are generally kept in different tables. Thus, when joining the customer and the sales tables, it is possible to store in a nested table of the model all the items that have been bought by a given customer. Thus, nested tables allow to reduce redundant information in the model.

Notice that `OLE DB DM` seems particularly tailored to predictive tasks, i.e. to predict the value of an attribute in a relational table. Indeed, the current implementation of `OLE DB DM` in Microsoft SQL Server 2000, only two algorithms are provided (Microsoft Decision Trees and Microsoft Clustering) and both of them are designed for attribute prediction. Instead, algorithms that use data mining models for the discovery of association rules, therefore for tasks without a direct predictive purpose, seems not currently supported by `OLE DB DM`. However, according to the specifications [17], `OLE DB DM` should be soon extended for association rules mining.

Notice also that it is possible to directly create a mining model that conforms to the `PMML` standard using the following statement:

```
<pmml_create>::=CREATE MINING MODEL <id> FROM PMML <string>
```

We recall here the schema used by `PMML` for the definition of models based on association rules.

```
<!ENTITY  \% FIELD-USAGE-TYPE "(active |
                               predicted |
                               supplementary)" >

<!ENTITY  \% OUTLIER-TREAT-METHOD "( asIs |
                                     asMissingValues |
                                     asExtremeValues ) " >
```

```
<!ENTITY \% MISS-VALUE-TREAT-METHOD "(asIs | asMean |
                                         asMode | asMedian |
                                         asValue) " >

<!ELEMENT MiningField (Extension*)>
<!ATTLIST MiningField
  name                      \%FIELD-NAME;                #REQUIRED
  usageType                 \%FIELD-USAGE-TYPE;          "active"
  outliers                  \%OUTLIER-TREAT-METHOD;      "asIs"
  lowValue                  \%NUMBER;                    #IMPLIED
  highValue                 \%NUMBER;                    #IMPLIED
  missingValueReplacement CDATA                          #IMPLIED
  missingValueTreatment     \%MISS-VALUE-TREAT-METHOD; #IMPLIED

<!ELEMENT MiningSchema   (MiningField+) >
```

Notice that according to this specification it is possible to specify the schema of a model giving the name, type, range of values of each attribute. Furthermore, it is possible to specify the treatment method if the value of the attribute is missing, or if it is an outlier w.r.t. the predicted value for that attribute.

3.5 Feature Summary

Table 1 summarizes the different features of an ideal query language for rule mining and shows how the studied proposals satisfy them as discussed in previous Sections. Notice that the fact that OLE DB DM supports or not some of

Table 1. Summary of the main features of the different languages. [1]Depending on the algorithm. [2]Only association rules. [3]Association rules and elementary sequential patterns. [4]Concept hierarchies. [5]Selectrules, satisfies and violates. [6]Operators for visualization. [7]PREDICTION JOIN. [8]Algorithm parameters

Feature	MSQL	MINE RULE	DMQL	OLE DB DM
Satisfaction of the closure property	Yes	Yes	Yes	Yes[1]
Selection of source data	No	Yes	Yes	Yes
Specification of different types of patterns	No[2]	Some[3]	Yes	Not directly[1]
Specification of the Background Knowledge	No	No	Some[4]	No
Post-processing of the generated results	Yes[5]	No	Some[6]	Some[7]
Specification of constraints	Yes	Yes	Yes	No[8]

the features reported in Table 1 depends strictly by the data mining algorithm referenced in the data mining model. Instead, `OLE DB DM` guarantees naturally the selection of source data, since this feature is its main purpose.

When considering different languages, it is important to identify precisely the kind of descriptive rules that are extracted. All the languages can extract intra-tuple association rules, i.e. rules that associate values of attributes of a tuple. The obtained association rules describe the common properties of (a sufficient number of) tuples of the relation. Instead, only `DMQL` and `MINE RULE` can extract inter-tuple association rules, i.e. rules that associate the values of attributes of different tuples and therefore describe the properties of a set of tuples. Nested tables in the data mining model of `OLE DB DM` could ease the extraction of inter-tuple association rules by the data mining algorithm. Indeed, nested tables include in an unique row of the model the features of different tuples of the source, original tables. Thus, intra-tuple association rules seem to constitute the common "core" of the expressive capabilities of the three languages.

The language capability of dealing with inter-tuple rules affects the representation of the input for the mining engine. As already said, `MSQL` considers only intra-tuple association rules. As illustrated in the next section, this limit may be overcome by a change of representation of the input relation, i.e., by inclusion of the relevant attributes of different tuples in a unique tuple of a new relation. However, this can be a tedious and long pre-processing work. Furthermore in these cases, the `MSQL` statements that catch the same semantics of the analogous statements in `DMQL` and `MINE RULE`, can be very complex and difficult to understand.

As a last example of the different capabilities of the languages, we can mention that while `DMQL` and `MINE RULE` effectively use aggregate functions (resp. on rule elements and on clusters) for the extraction of association rules, `MSQL` provides them only as a post-processing tool over the results.

4 Comparative Examples

We describe here a complete KDD process centered around the classical basket analysis problem that will serve as a running example throughout the paper.

We are considering information of relations *Sales*, *Transactions* and *Customers* shown in Figure 1. In relation *Sales* we have stored information on sold items in the purchase transactions; in relation *Transactions* we identify the customers that have purchased in the transactions and record the method of payment; in relation *Customers* we collect information on the customers.

From the information of these tables we want to look for association rules between bought items and customer's age for payments with credit cards. The discovered association rules are meant to predict the age of customers according to their purchase habits. This data mining step requires at first some manipulations as a preprocessing step (selection of the items bought by credit card and encoding of the `age` attribute) in order to prepare data for the successive pattern extraction; then the actual pattern extraction step may take place.

transaction_id	item
1	ski_pants
1	hiking_boots
2	col_shirts
2	brown_boots
3	col_shirts
3	brown_boots
4	jackets
5	col_shirts
5	jackets
6	hiking_boots
6	brown_boots
7	ski_pants
7	hiking_boots
7	brown_boots
8	ski_pants
8	hiking_boots
8	brown_boots
8	jackets
9	hiking_boots
10	ski_pants
11	ski_pants
11	brown_boots
11	jackets

transaction_id	customer	payment
1	c1	credit_card
2	c2	credit_card
3	c3	cash
4	c4	credit_card
5	c5	credit_card
6	c6	cash
7	c7	credit_card
8	c8	credit_card
9	c9	credit_card
10	c3	credit_card
11	c2	cash

customer_id	customer_age	job
c1	26	employee
c2	35	manager
c3	48	manager
c4	39	engineer
c5	46	teacher
c6	25	student
c7	29	employee
c8	24	student
c9	28	employee

Fig. 1. `Sales` table (on the left); `Transactions` table (on the right above); `Customers` table (on the right below)

Suppose that by inspecting the result of a previous data mining extraction step, we are now interested in investigating the purchases that violate certain extracted patterns. In particular, we are interested in obtaining association rules between sets of bought items in the purchase transactions that violate the rules with `'ski_pants'` in their antecedent. To this aim, we can cross-over between extracted rules and original data, selecting tuples of the source table that violate the interesting rules, and perform a second mining step, based on the results of the previous mining step: from the selected set of tuples, we extract the association rules between two sets of items with a high confidence threshold. Finally, we allow two post-processing operations over the extracted association rules: selection of rules with 2 items in the body and selection of rules with a maximal body among the rules with the same consequent.

4.1 MSQL

The first thing to do is to represent source data in a suitable format for MSQL. Indeed, MSQL expects to receive a unique relation obtained by joining the source relations *Sales*, *Transactions* and *Customers* on attributes *transaction_id* and *customer_id*. Furthermore, the obtained relation must be encoded in a binary format such that each tuple represents a transaction with as many boolean

Table 2. `Boolean_Sales` transactional table used with `MSQL`

t_id	ski_pants	hiking_boots	col_shirts	brown_boots	jackets	customer_age	payment
t1	1	1	0	0	0	26	credit_card
t2	0	0	1	1	0	35	credit_card
t3	0	0	1	1	0	48	cash
t4	0	0	0	0	1	39	credit_card
t5	0	0	1	0	1	46	credit_card
t6	0	1	0	1	0	25	cash
t7	1	1	0	1	0	29	credit_card
t8	1	1	0	1	1	24	credit_card
t9	0	1	0	0	0	28	credit_card
t10	1	0	0	0	0	41	credit_card
t11	1	0	0	1	1	36	cash

attributes as are the possible items that a customer can purchase. We obtain the relation in Table 2.

This data trasformation puts in evidence the main weakness of `MSQL`. `MSQL` is designed to discover the propositional rules satisfied by the values of the attributes inside a tuple of a table. If the number of possible items on which a propositional rule must be generated is very large (as, for instance the number of different products in markets stores) the obtained input table is very large, not easily maintainable and user-readable because it contains for each transaction all the possible items even if they have not been bought. Boolean table is an important fact to take into consideration because its presence is necessary for `MSQL` language, otherwise it cannot work (and so this language is not very much flexible in its input); furthermore, boolean table requires a data transformation which is expensive (especially considering that the volume of tables is huge) and must be performed each time a new problem/source table is submitted.

Pre-processing Step 1: Selection of the Subset of Data to be Mined. We are interested only in clients paying with a credit card. `MSQL` requires that we make a selection of the subset of data to be mined, before the extraction task. The relation on which we will work is supposed to have been correctly selected from the pre-existing set of data in Table 2, by means of a view, named *View_on_Sales*.

Pre-processing Step 2: Encoding Age. `MSQL` provides methods to declare encodings on some attributes. It is important to note that `MSQL` is able to do discretization "on the fly", so that the intermediate encoded value will not appear in the final results. The following query will encode the `age` attribute:

```
CREATE ENCODING e_age ON View_on_Sales.customer_age AS
BEGIN
    (MIN, 9, 0), (10, 19, 1), (20, 29, 2), (30, 39, 3),     (1)
    (40, 49, 4), (50, 59, 5), (60, 69, 6), (70, MAX,7), 0
END;
```

The relation obtained after the two pre-processing steps is shown in Table 3.

Table 3. `View_on_Sales` transactional table after the pre-processing phase

t_id	ski_pants	hiking_boots	col_shirts	brown_boots	jackets	e_age	payment
t1	1	1	0	0	0	2	credit_card
t2	0	0	1	1	0	3	credit_card
t4	0	0	0	0	1	3	credit_card
t5	0	0	1	0	1	4	credit_card
t7	1	1	0	1	0	2	credit_card
t8	1	1	0	1	1	2	credit_card
t9	0	1	0	0	0	2	credit_card
t10	1	0	0	0	0	4	credit_card

Rules Extraction over a Set of Items and Customers' Age. We want to extract rules associating a set of items to the customer's age and having a support over 2 and a confidence over (or equal to) 50%.

```
GETRULES(View_on_Sales) INTO SalesRB
WHERE BODY has {(ski_pants=1) OR (hiking_boots=1) OR        (2)
  (col_shirts=1) OR (brown_boots=1) OR (jackets=1)} AND
  Consequent is {(Age = *)} AND support>2 AND confidence>=0.5
USING e_age FOR customer_age
```

This example puts in evidence a limit of `MSQL`: if the number of items is high, the number of predicates in the `WHERE` clause increases correspondingly! The resulting rules are shown in Table 4.

Table 4. Table `SalesRB` produced by `MSQL` in the first rule extraction phase

Body	Consequent	Support	Confidence
(ski_pants=1)	(customer_age=[20,29])	3	75%
(hiking_boots=1)	(customer_age=[20,29])	4	100%
(brown_boots=1)	(customer_age=[20,29])	3	66%
(ski_pants=1) ∧ (hinking_boots=1)	(customer_age=[20,29])	3	100%

Crossing-over: Looking for Exceptions in the Original Data. We select tuples from *View_on_Sales* that violate all the extracted rules with `ski_pants` in the antecedent (the first and last rule in Table 4).

```
INSERT INTO Sales2 AS
SELECT * FROM View_on_Sales
WHERE VIOLATES ALL (                                          (3)
      SELECTRULES(SalesRB) WHERE BODY HAS {(ski_pants=1)})
```

We obtain results given in Table 5.

Rules Extraction over Two Sets of Items. `MSQL` does not support a conjunction of an arbitrary number of descriptors in the consequent. Therefore, in this step we can extract only association rules between one set of items in the antecedent and a single item in the consequent. The resulting rule set is only $(brown\_boots = 1) \Rightarrow (color\_shirts = 1)$ with support=1 and confidence=100%.

Table 5. Tuples (in Sales2) violating all rules (in SalesRB) with ski_pants in the antecedent

t_id	ski_pants	hiking_boots	col_shirts	brown_boots	jackets	e_age
t2	0	0	1	1	0	3
t4	0	0	0	0	1	3
t5	0	0	1	0	1	4
t9	0	1	0	0	0	2
t10	1	0	0	0	0	4

```
GETRULES(Sales2) INTO SalesRB2
WHERE (Body has {(hiking_boots=1) OR (col_shirts=1)
 OR (brown_boots=1)}
              AND Consequent is {(jackets=1)}
 OR Body has {(col_shirts=1) OR (brown_boots=1) OR (jackets=1)}
              AND Consequent is {(hiking_boots=1)}
 OR Body has {(brown_boots=1) OR (jackets=1)                    (4)
   OR (hiking_boots=1)}
              AND Consequent is {(col_shirts=1)}
 OR Body has {(jackets=1) OR (hiking_boots=1)
  OR (col_shirts=1)}
              AND Consequent is {(brown_boots=1)})
 AND support>=0.0 AND confidence>=0.9
USING e_age FOR customer_age
```

Notice that in this statement the `WHERE` clause allows several different conditions on the `Body` and on the `Consequent`, because we wanted to allow in the `Body` a proposition on every possible attribute except one that is allowed to appear in the `Consequent`. Writing this statement was possible because the total number of items is small in this toy example but would be impossible for a real example in which the number of propositions in the `WHERE` clause explodes.

Post-processing Step 1: Manipulation of Rules. Select the rules with 2 items in the body.

As `MSQL` extracts rules with one item in the consequent and it provides the primitive `length` applied to the itemsets originating rules, we specify that the total length of the rules is 3.

```
SelectRules(SalesRB) where length=3                         (5)
```

The only rule satisfying this condition is:
$(ski\_pants = 1) \wedge (hiking\_boots = 1) \Rightarrow (customer\_age = [20; 29])$

Post-processing Step 2: Extraction of Rules with a Maximal Body. It is equivalent to require that there is no pair of rules with the same consequent, such that the body of the first rule is included in the body of the second one.

```
SELECTRULES(SalesRB) AS R1
WHERE NOT EXISTS (SELECTRULES(SalesRB) AS R2
                  WHERE R2.body has R1.body                    (6)
                    AND NOT (R2.body is R1.body)
                    AND R2.consequent is R1.consequent )
```

There are two rules satisfying this condition:
$(ski\_pants = 1) \wedge (hiking\_boots = 1) \Rightarrow (customer\_age = [20;29])$
$(brown\_boots = 1) \Rightarrow (customer\_age = [30,39])$

Pros and Cons of MSQL. Clearly, the main advantage of MSQL is that it is possible to query rules as well as data, by using **SelectRules** on rulebases and **GetRules** on data. Another good point is that MSQL has been designed to be an extension of classical SQL, making the language quite easy to understand. For example, it is quite simple to test rules against a dataset and to make crossing-over between the original data and query results, by using SATISFIES and VIOLATES. To be considered as a good candidate language for inductive databases, it is clear that MSQL, which is essentially built around the extraction phase, should be extended, particularly with a better handling of pre- and post-processing steps. For instance, even if it provides some pre-processing operators like ENCODE for discretization of quantitative attributes, it does not provide any support for complex pre-processing operations, like sampling. Moreover, tuples on which the extraction task must be performed are supposed to have been selected in advance. Concerning the extraction phase, the user can specify some constraints on rules to be extracted (e.g., inclusion of an item in the body or in the head, rule's length, mutually exclusive items, etc) and the support and confidence thresholds. It would be useful however to have the possibility to specify more complex constraints and interest measures, for instance user defined ones.

4.2 MINE RULE

MINE RULE does not require a specific format for the input table. Therefore we can suppose to receive data either in the set of normalized relations *Sales*, *Transactions* and *Customers* of Figure 1 or in a view obtained joining them. This view is named *SalesView* and is shown in Table 6 and we assume it is the input of the mining task. Using a view is not necessary but it allows to make SQL querying easier by gathering all the necessary information in one table eventhough all these data are initially scattered in different tables. Thus, the user can focus the query writing on the constraints useful for its mining task.

Pre-processing Step 1: Selection of the Subset of Data to be Mined. In contrast to MSQL, MINE RULE does not require to apply some pre-defined view on the original data. As it is designed as an extension to SQL, it perfectly nests SQL, and thus, it is possible to select the relevant subset of data to be mined by specifying it in the FROM.. WHERE.. clauses of the query.

Table 6. SalesView view obtained joining the input relations

transaction_id	item	customer_age	payment
1	ski_pants	26	credit_card
1	hiking_boots	26	credit_card
2	col_shirts	35	credit_card
2	brown_boots	35	credit_card
3	col_shirts	48	cash
3	brown_boots	48	cash
4	jackets	39	credit_card
5	col_shirts	46	credit_card
5	jackets	46	credit_card
6	hiking_boots	25	cash
6	brown_boots	25	cash
7	ski_pants	29	credit_card
7	hiking_boots	29	credit_card
7	brown_boots	29	credit_card
8	ski_pants	24	credit_card
8	hiking_boots	24	credit_card
8	brown_boots	24	credit_card
8	jackets	24	credit_card
9	hiking_boots	28	credit_card
10	ski_pants	48	credit_card
11	ski_pants	35	cash
11	brown_boots	35	cash
11	jackets	35	cash

Pre-processing Step 2: Encoding Age. Since MINE RULE does not have an encoding operator for performing pre-processing tasks, we must discretize the interval values.

Rules Extraction over a Set of Items and Customers' Age. In MINE RULE, we specify that we are looking for rules associating one or more items (rule's body) and customer's age (rule's head):

```
MINE RULE SalesRB AS
SELECT DISTINCT 1..n item AS BODY, 1..1 customer_age AS HEAD,
       SUPPORT, CONFIDENCE
FROM SalesView WHERE payment='credit_card'                    (7)
GROUP BY t_id
EXTRACTING RULES WITH SUPPORT: 0.25, CONFIDENCE: 0.5
```

If we want to store results in a database supporting the relational model, extracted rules are stored into the table $SalesRB(r\_id, b\_id, h\_id, sup, conf)$ where r_id, b_id, h_id are respectively the identifiers assigned to rules, body itemsets and head itemsets. The body and head itemsets are stored respectively in tables $SalesRB\_B(b\_id, item)$ and $SalesRB\_H(h\_id, customer\_age)$. Tables $SalesRB$, $SalesRB\_B$ and $SalesRB\_H$ are shown in Figure 2.

Rule_id	Body_id	Head_id	Support	Confidence
1	1	5	37.5%	75%
2	2	5	50%	100%
3	3	5	37.5%	66%
4	4	5	37.5%	100%

Body_id	item
1	ski_pants
2	hiking_boots
3	brown_boots
4	ski_pants
4	hinking_boots

Head_id	customer_age
5	[20,29]

Fig. 2. Normalized tables containing rules produced by MINE RULE in the first rule extraction phase

Crossing-over: Looking for Exceptions in the Original Data. We want to find transactions of the original relation whose tuples violate all rules with ski_pants in the body. As rule components (bodies and heads) are stored in relational tables, we use an SQL query to manipulate itemsets. The corresponding query is the following:

```
SELECT * FROM SalesView AS S1 WHERE NOT EXISTS
  (SELECT * FROM SalesRB AS R1,
                SalesRB_B AS R1_B, SalesRB_H AS R1_H
   WHERE R1.b_id=R1_B.b_id AND R1.h_id=R1_H.h_id AND
   S1.customer_age=R1_H.customer_age AND S1.item=R1_B.item (8)
   AND EXISTS(SELECT * FROM SalesRB_B AS R2_B
       WHERE R2_B.b_id=R1_B.b_id AND R2_B.item='ski_pants')
   AND NOT EXISTS
         (SELECT * FROM SalesRB_B AS R3_B
          WHERE R1_B.b_id=R3_B.b_id AND NOT EXISTS
            (SELECT * FROM SalesView AS S2
             WHERE S2.t_id=S1.t_id AND S2.item=R3_B.item)))
```

This query is hard to write and to understand. It aims at selecting tuples of the original *SalesView* relation, renamed S1, such that there are no rules with ski_pants in the antecedent, that hold on them. These properties are verified by the first two nested SELECT clauses. Furthermore, we want to be sure that the above rules are satisfied by tuples belonging to the same transaction of the original tuple in S1. In other words, that there are no elements of the body of the rule that are not satisfied by tuples of the same original transaction. Therefore, we verify that each body element in the rule is satisfied by a tuple of the *SalesView* relation (renamed S2) in the same transaction of the tuple in S1 we are considering for the output.

Rules Extraction over Two Sets of Items. This is the classical example of extraction of association rules, formed by two sets of items. Using MINE RULE it is specified as follows:

```
MINE RULE SalesRB2 AS
SELECT DISTINCT 1..n item AS BODY, 1..n item AS HEAD,
       SUPPORT, CONFIDENCE                                  (9)
FROM Sales2
GROUP BY t_id
EXTRACTING RULES WITH SUPPORT: 0.0, CONFIDENCE: 0.9
```

In this simple toy database the result coincides with the one generated by MSQL.

Post-processing Step 1: Manipulation of Rules. Once again, as itemsets corresponding to rule's components are stored in tables (*SalesRB_B*, *SalesRB_H*), we can select rules having two items in the body with a simple SQL query.

```
SELECT * FROM SalesRB AS R1 WHERE 2=                        (10)
  (SELECT COUNT(*) FROM SalesRB_B R2 WHERE R1.b_id=R2.b_id)
```

Post-processing Step 2: Selection of Rules with a Maximal Body. We select rules with a maximal body for a given consequent. As rules' components are stored in relational tables, we use again a SQL query to perform such a task.

```
SELECT * FROM SalesRB AS R1      # We select the rules in R1
WHERE NOT EXISTS                 # such that there are no
 (SELECT * FROM SalesRB AS R2    # other rules (in R2) with
  WHERE R2.h_id=R1.h_id          # the same head, a different
    AND NOT R2.b_id=R1.b_id      # body such that it has no
    AND NOT EXISTS (SELECT *     # items that do not occur in
      FROM SalesRB_B AS B1       # the body of the R1 rule
      WHERE R1.b_id=B1.b_id AND NOT EXISTS (SELECT *
           FROM SalesRB_B AS B2                             (11)
           WHERE B2.b_id=R2.b_id AND B2.item=B1.item)))
```

This rather complex query aims at selecting rules such that there are no rules with the same consequent and a body that strictly includes the body of the former rule. The two inner sub-queries are used to check that rule body in R1 is a superset of the rule body in R2. These post-processing queries probably could be simpler if SQL-3 standard for the ouput of the rules were adopted.

Pros and Cons of MINE RULE. The first advantage of MINE RULE is that it has been designed as an extension to SQL. Moreover, as it perfectly nests SQL, it is possible to use classical statements to pre-process the data, and, for instance, select the subset of data to be mined. Like MSQL, data pre-processing is limited to operations that can be expressed in SQL: it is not possible to sample data before extraction, and the discretization must be done by the user. Notice however, that, by using the CLUSTER BY keyword, we can specify on which subgroups of a group association rules must be found. Like MSQL, MINE RULE allows the user to specify some constraints on rules to be extracted (on items belonging to head or body, their cardinality as well as more complex constraints based on the use of a taxonomy). The interested reader is invited to read [12,13] to have an illustration of these latter capabilities. Like `MSQL`, `MINE RULE` is essentially designed around the extraction step, and it does not provide much support for the other KDD steps (e.g., post-processing tasks must be done with SQL queries). Finally, according to our knowledge, `MINE RULE` is one of the few languages that have a well defined semantics [13] for all its operations. Indeed, it is clear that a clean theoretical background is a key issue to allow the generation of efficient optimizers.

4.3 DMQL

`DMQL` can work with traditional databases, so it can receive as input either the source relations *Sales*, *Transactions* and *Customers* shown in Figure 1 or the view obtained by joining them and shown in Table 6. As already done with the examples on `MINE RULE`, let us consider that the view *SalesView* is given as input, so that the reader's attention is more focused on the constraints that are strictly necessary for the mining task.

Pre-processing Step 1: Selection of the Subset of Data to be Mined. Like `MINE RULE`, `DMQL` nests SQL for relational manipulations. So the selection of the relevant subset of data (i.e. clients buying products with their credit card) will be done via the use of the `WHERE` clause of the extraction query.

Pre-processing Step 2: Encoding Age. `DMQL` does not provide primitives to encode data like `MSQL`. However, it allows us to define a hierarchy to specify ranges of values for customer's age, as follows:

```
define hierarchy age_hierarchy for customer_age on SalesView as
level1:{min...9}$<$level0:all
level1:{10...19}$<$level0:all                                  (12)
      ...
level1:{60...69}$<$level0:all
level1:{70...max}$<$level0:all
```

Rules Extraction over a Set of Items and Customers' Age. `DMQL` allows the user to specify templates of rules to be discovered, called *metapatterns*, by using the **matching** keyword. These metapatterns can be used to impose strong syntactic constraints on rules to be discovered. So we can specify that we are looking for rule bodies relative to bought items and rule heads relative to customer's age. Moreover, we can specify that we desire to use the predefined hierarchy for the age attribute.

use database Sales_db
use hierarchy age_hierarchy **for** customer_age
mine association as SalesRB
matching with $item^{+}(X, \{I\}) \Rightarrow customer\_age(X, A)$ (13)
from SalesView
where payment='credit_card'
group by t_id
with support threshold=25%
with confidence threshold=50%

where the above metarule with the notation $\{I\}$ matches with rules with repeated *item* predicate like $item(X, I_1) \wedge item(X, I_2) \cdots item(X, I_j)$ where $\{I_1, I_2, \cdots I_j\} = I$ are different elements of the I set obtained as input by the WHERE predicate clause. The result is shown in Table 7.

Table 7. Results produced by DMQL in the first rule extraction phase (SalesRB)

item$^+$ (X,{I})	customer_age(X,A)	Support	Confidence
item(X,ski_pants)	customer_age(X,20...29)	37.5%	75%
item(X,hiking_boots)	customer_age(X,20...29)	50%	100%
item(X,brown_boots)	customer_age(X,20...29)	37.5%	66%
item(X,ski_pants)∧item(X,hiking_boots)	customer_age(X,20...29)	37.5%	100%

Crossing-over: Looking for Exceptions in the Original Data. Like MINE RULE, DMQL does not provide support for crossing-over patterns and data: it requires SQL queries as already shown with MINE RULE (query (8)).

Rules Extraction over Two Sets of Items. This phase is performed by the following DMQL statement:

use database Sales_db
mine association as SalesRB2
matching with $item^+(X, \{I\}) \Rightarrow item^+(X, \{J\})$ (14)
from Sales2
group by t_id
with confidence threshold=90%

Post-processing Step 1: Selection of the Rules with Two Items in the Body. Like MINE RULE, DMQL does not provide support for operations of rules manipulation. As we do not have direct access the rules and thus do not the exact storage format of rules, we make the assumption the rules are stored in the same way than in MINE RULE, and that allows us to compare the languages in the same conditions of storage format. So, for this step, an SQL query similar to query (10) shown in the examples of MINE RULE is therefore needed.

Post-processing Step 2: Selection of the Rules with a Maximal Body. Like MINE RULE, DMQL does not provide support for operations of rules manipulation such as the selection of the most general rules. For the same reason as the previous post-processing step, an SQL query analogous to query (11) is therefore required.

Pros and Cons of DMQL. Like MINE RULE, one of the main advantages of DQML is that it completely nests classical SQL, and so it is quite easy for a new user to learn and use the language. Moreover, DMQL is designed to work with traditional databases and datacubes. Concerning the extraction step, DMQL allows to impose strong syntactic contraints on patterns to be extracted, by means of metapatterns allowing the user to specify the form of extracted rules. Another advantage of DMQL is that we can include some background knowledge in the process, by defining hierarchies on items occurring in the database and mining rules across different levels of hierarchies. Once rules are extracted, we can perform roll-up and drill-down manipulations on extracted rules. Clearly, analogously to the other languages studied so far, the main drawback of DMQL is that the language capabilities are essentially centered around the extraction phase, and the language relies on SQL or additional tools to perform pre- and post-processing operations. Finally, we can notice that, beyond association rules, DMQL can perform other mining tasks, such as classification.

4.4 OLE DB DM

OLE DB DM is designed for a simple use of relational data already available via OLE DB. Thus, it can work with relational data. Creating a view is not necessary because putting the data in the right format is exactly one of the purposes of the definition and population of the mining model.

Pre-processing Step 1: Selection of the Subset of Data to be Mined. In the OLE DB DM framework, selection of data to be mined is done in the definition of the data mining model and in the following insertion of data in it. Conceptually, it is very similar to the creation of a view. Here the mining model is named [SalesRB] in analogy to the previous examples for the other languages.

Creation of the mining model:

```
CREATE MINING MODEL [SalesRB](
[transaction_id] LONG KEY,
[customer_age]   LONG DISCRETIZED PREDICT,
[items]          TABLE (
   [item] TEXT KEY
   )
)
USING [My_assoc_Algo] (min_support=2, min_confidence=0.5)
```

Notice that we used a nested table [items] to specify bought items by a customer in a transaction and make reference to a mining algorithm, My_assoc_Algo, for the extraction of association rules.

Insertion of data in the data mining model:

```
INSERT INTO [SalesRB]
([transaction_id],[customer_age],[items])
 SHAPE
 {SELECT [transaction_id],[customer_age]
    FROM Transactions,Customers
    WHERE Transactions.customer=Customers.customer_id
           AND Transactions.payment="credit_card"
  APPEND(
  {SELECT [item] FROM Sales
   ORDER BY [tr_id]}
   RELATE [transaction_id] TO [tr_id])
   AS [items]}
```

Notice that selection of the interesting source data (purchases made by credit card) is done in this step. Notice also that APPEND keyword builds the nested table [items] containing items in source relation Sales purchased in a transaction. The relationship between the transaction identifier in Sales and the analogous identifier in the model is done by means of the RELATE keyword.

Pre-processing Step 2: Encoding Age. The definition of the data mining model allows specification of discretized attributes and of discretization method used. However, discretization itself must be provided by the data mining algorithm provider.

Rules Extraction over a Set of Items and Customer's Age. In SQL Server 2000, no algorithm for association rule mining is currently available, but the specification of OLE DB DM claims that association rule mining algorithm can be supported. Here, we supposed that the user has implemented an association rule mining algorithm, named My_assoc_Algo, which takes as input parameters of minimal support and confidence and refers to the content of the [items] nested tables to elaborate association rules.

The results of the association rule mining process could be stored by the algorithm in a relational table and described by the following PMML representation.

```
<Item id="1" value="ski_pants" />
<Item id="2" value="hiking_boots" />
<Item id="3" value="brown_boots" />

<Itemset id="1" support="0.5" numberOfItems="1">
  <ItemRef itemRef="1">
</Itemset>
<Itemset id="2" support="0.5" numberOfItems="1">
  <ItemRef itemRef="2">
</Itemset>
<Itemset id="3" support="0.375" numberOfItems="1">
  <ItemRef itemRef="3">
</Itemset>
<Itemset id="4" support="0.375" numberOfItems="2">
     <ItemRef itemRef="1" />
     <ItemRef itemRef="2" />
</Itemset>

<Item id="4" value="[20,29]" />

<Itemset id="5" support="0.5" numberOfItems="1">
     <ItemRef itemRef="4" />
</Itemset>
```

```
<AssociationRule support="0.375" confidence="0.75"
 antecedent="1" consequent="5" />
<AssociationRule support="0.50" confidence="1.0"
 antecedent="2" consequent="5" />
<AssociationRule support="0.375" confidence="0.66"
 antecedent="3" consequent="5" />
<AssociationRule support="0.375" confidence="1.0"
 antecedent="4" consequent="5" />
```

Notice that such a PMML description is very similar to the rules storage structure of MINE RULE.

Crossing-over: Looking for Exceptions in the Original Data. For this task, we must write a query in classical SQL. Since the association rules produced by the algorithm could be stored in the PMML format, which is quite close of the storage format of MINE RULE, we can say that the query will be very similar to the one used with MINE RULE.

Concerning post-processing tasks, or the usage of the rules after their proper extraction, notice that OLE DB DM only provides some facilities for prediction, with PREDICTION JOIN. However, this is not useful here.

Rules Extraction over Two Sets of Items. We want to perform a new mining task here, so we must define a new mining model. This one is analogous to the model used in previous step with the exception of customers' age that is not needed in this case. Indeed, the difference of this mining task with respect to previous one lies in the proper execution of the mining algorithm that associates an itemset to another itemset and not to the customers' age. For sake of space we do not report this new model here.

Post-processing Step 1: Manipulation of Rules. Here again, we need to access rules' components. Since the OLE DB DM suggests that bodies and heads of rules are stored following the PMML format, the query will be very similar to the one used with MINE RULE.

Post-processing Step 2: Selection of Rules with a Maximal Body. Again, since the rules could be stored following the PMML format, we can use the same kind of queries used for MINE RULE.

Pros and Cons of OLE DB for DM. The first advantage of OLE DB DM is that it is a first temptative of industrial standard and that it begins to be implemented in some commercial application (like SQL Server 2000). It is designed as an extension to SQL and so a DBA can write queries that are similar to classical SQL queries and that define and populate data mining models. But, the main problem is that the language of OLE DB DM is not really a language for Data Mining like the other three. It is particularly targeted at making the communication between relational databases and data mining algorithms easier. So it can work with a lot of different algorithms, provided that the algorithms are compliant to OLE DB DM mining model. However, it provides no facilities to handle typical constraints of the association rule mining problem, such as constraints on items, frequency and confidence. More generally, all these types of constraints must be given as parameters to the mining algorithm. Moreover, accessing the mining results and browsing of extracted patterns must be managed by the algorithm provider, which makes a general method for post-processing difficult to define. Finally, there is no formal semantics like in `MINE RULE`.

5 Conclusions

We have considered three languages, `MSQL`, `MINE RULE` and `DMQL` and an `API` for data mining, `OLE DB DM`, with an SQL-like language for the deployment of a data mining model. All of them request the extraction from a relational database of data mining patterns, and in particular of association rules. They satisfy the "closure property", a crucial property for inductive databases. We have compared the various features of these languages with the desired properties of an ideal query language for inductive databases dedicated to association rules. We have prepared a benchmark and tested the languages against it. The benchmark is constituted by an hypothetical KDD scenario, taken from the data mining practice, in which we have formulated a collection of queries. We have tested the possibility and the ease for the user to express the chosen queries in the above mentioned languages. The outcome is that no language presents all the desired properties. `MSQL` seems the one that offers the larger number of primitives tailored for post-processing and an on-the-fly encoding, specifically designed for efficiency. `DMQL` allows the extraction of different data patterns, the definition and use of hierarchies, and some visualization primitives. `MINE RULE` is the only one that allows to dynamically partition the source relation into a first and a second level grouping (the clusters) from which more sophisticated rule constraints can be applied. Furthermore, to the best of our knowledge, it looks as the only language having an algebraic semantics, an important factor for an in-depth study of optimization issues. `OLE DB DM` is an API, that allows any application to access by means of SQL-like queries to a relational data source, and to be coupled with specialized mining algorithms. The main motivation of the design of `OLE DB DM` is to ease the communication between a data mining application, the DBMS providing data and a set of available, advanced data mining algorithms. However, at the moment, it does not provide any specific feature tailored to any particular data mining task that is not predictive.

However, it is clear that one of the main limits of all the proposed languages is the weak support of rule post-processing. In particular, in all the languages post-processing capabilities are limited to a few predefined built-in primitives. Instead, it would be desirable that the grammar of the languages would accept a certain degree of extensibility. Indeed, for instance, it is not possible to introduce user-defined functions in the statements. These ones would allow the user to provide the implementation of user-defined sophisticated constraints, based, for instance, on new pattern evaluation measures.

Furthermore, the research on condensed representations for frequent itemsets [2,3] has been proved useful not only for mining frequent itemsets and frequent association rules from dense databases but also for sophisticated post-processing [1,15]. Indeed, one of the problems in association rule mining from real-life data is the huge number of extracted rules. However, many of the rules are redundant and might be useless. Thus, a condensed representation would help visualizing the result and focusing the user attention on the relevant rules. For example, Bastide et al., [1], presents an algorithm to extract a minimal cover of the set of frequent association rules.

Another crucial issue relative to query language for data mining is the optimization for sequences of queries (e.g., deciding of query containment). To the best of our knowledge, the materialization of condensed representations of the frequent itemsets seems to be quite useful [9,4] but still needs further work.

Last but not least, an important issue is the simplicity of the language and its ease of use. Indeed, we think that a good candidate language for data mining should be flexible enough to specify a variety of different mining tasks in a declarative fashion. To the best of our knowledge, the implementation of these languages tackles the mentioned problems (including the lack of instruments dedicated to post-processing) by being embedded in a data mining system, which provides a graphical front end to the language.

References

1. Bastide, Y., Pasquier, N., Taouil, R., Stumme, G., Lakhal, L.: Mining minimal non-redundant association rules using frequent closed itemsets. Proc. CL'00 (2000), London (UK). Springer-Verlag LNCS 1861. pp. 972–986.
2. Boulicaut, J-F., Bykowski, A.: Frequent closures as a concise representation for binary data mining. Proc. PAKDD'00 (2000), Kyoto (JP). Springer-Verlag LNAI 1805. pp. 62–73.
3. Boulicaut J-F., Bykowski, A., Rigotti, C.: Free-sets: a condensed representation of boolean data for the approximation of frequency queries. Data Mining and Knowledge Discovery (2003). *7(1)*5–22.
4. Giacometti, A., Laurent, D., Diop, C.T.: Condensed representations for sets of mining queries. Proc. KDID'02 (2002), Helsinki (FIN). An extended version appears in this volume.
5. Imielinski, T., Mannila, H.: A Database Perspective on Knowledge Discovery. Communications of the ACM (1996). *3(4)*58–64.
6. Imielinski, T., Virmani, A., Abdulghani, A.: DataMine: Application Programming Interface and Query Language for Database Mining. Proc. KDD'96 (1996), Portland (USA). AAAI Press. pp. 256–261.
7. Imielinski, T., Virmani, A.: MSQL: A Query Language for Database Mining. Data Mining and Knowledge Discovery (1999). *3(4)*373–408.
8. Jeudy, B., Boulicaut, J-F.: Optimization of association rule mining queries. Intelligent Data Analysis (2002). *6(4)*341–357.
9. Jeudy, B., Boulicaut, J-F.: Using condensed representations for interactive association rule mining. Proc. PKDD'02 (2002), Helsinki (FIN). Springer-Verlag LNAI 2431. pp. 225–236.

10. Han, J., Fu, Y., Wang, W., Koperski, K., Zaiane, O.: DMQL: A Data Mining Query Language for Relational Databases. Proc. of SIGMOD Workshop DMKD'96 (1996), Montreal (Canada). pp. 27–34.
11. Han, J., Kamber, M.: Data Mining – Concepts and Techniques. Morgan Kaufmann Publishers (2001).
12. Meo, R., Psaila, G., Ceri, S.: A New SQL-like Operator for Mining Association Rules. Proc. VLDB'96 (1996), Bombay (India). Morgan Kaufmann. pp. 122–133.
13. Meo, R., Psaila, G., Ceri, S.: An Extension to SQL for Mining Association Rules. Data Mining and Knowledge Discovery (1998). *2(2)*195–224.
14. Virmani, A.: Second Generation Data Mining. PhD Thesis, Rutgers University, 1998.
15. Zaki, M.J.: Generating non-redundant association rules. Proc. SIGKDD'00 (2000), Boston (USA). ACM Press. pp. 34–43.
16. Netz, A., Chaudhuri, S., Fayyad, U., Bernhardt, J.:Integrating Data Mining with SQL Databases: OLE DB for Data Mining. Proc ICDE'01 (2001), Heidelberg (Germany). IEEE Computer Society. pp. 379–387
17. OLEDB for Data Mining specifications, available at http://www.microsoft.com/data/oledb/dm/
18. Predictive Model Mark-up Language, available at http://www.dmg.org/pmmlv2-0.htm

Declarative Data Mining Using SQL3

Hasan M. Jamil

Department of Computer Science and Engineering
Mississippi State University, USA
jamil@cse.msstate.edu

Abstract. Researchers convincingly argue that the ability to declaratively mine and analyze relational databases using SQL for decision support is a critical requirement for the success of the acclaimed data mining technology. Although there have been several encouraging attempts at developing methods for data mining using SQL, simplicity and efficiency still remain significant impediments for further development. In this article, we propose a significantly new approach and show that any object relational database can be mined for association rules without any restructuring or preprocessing using only basic SQL3 constructs and functions, and hence no additional machineries are necessary. In particular, we show that the cost of computing association rules for a given database does not depend on support and confidence thresholds. More precisely, the set of large items can be computed using one simple join query and an aggregation once the set of all possible meets (least fixpoint) of item set patterns in the input table is known. We believe that this is an encouraging discovery especially compared to the well known SQL based methods in the literature. Finally, we capture the functionality of our proposed mining method in a mine by SQL3 operator for general use in any relational database.

1 Introduction

In recent years, mining association rules has been a popular way of discovering hidden knowledge from large databases. Most efforts have focused on developing novel algorithms and data structures to aid efficient computation of such rules. Despite major efforts, the complexity of the best known methods remain high. While several efficient algorithms have been reported [1,4,10,22,19,12,21,18,23], overall efficiency continues to be a major issue. In particular, in paradigms other than association rules such as ratio rules [11], chi square method [3], and so on, efficiency remains one of the biggest challenges.

The motivation, importance, and the need for integrating data mining with relational databases has been addressed in several articles such as [16,17]. They convincingly argue that without such integration, data mining technology may not find itself in a viable position in the years to come. To be a successful and feasible tool for the analysis of business data in relational databases, such technology must be made available as part of database engines as well as part of its declarative query language.

R. Meo et al.(Eds.): Database Support for Data Mining Applications,LNAI 2682, pp. 52–75, 2004.

While research into procedural computation of association rules has been extensive, fewer attempts have been made to use relational machinery or SQL for *declarative* rule discovery barring a few exceptions such as [13,25,21,8,20,14]. Most of these works follow an apriori like approach by mimicking its functionality and rely on generating candidate sets and consequently suffer from high computational overhead. While it is certainly possible to adapt any of the various procedural algorithms for rule mining as a special mining operator, the opportunity for using existing technology and constructs is preferable if it proves to be more beneficial. Some of the benefits of using existing relational machinery may include opportunity for query optimization, declarative language support, selective mining, mining from non-transactional databases, and so on. From these standpoints, it appears that research into data mining using SQL or SQL-like languages bear merit and warrant attention. But before we proceed any further, we would like to briefly summarize the concept of association rules as follows for the readers unfamiliar with the subject.

Let $\mathcal{I} = \{i_1, i_2, \ldots, i_m\}$ be a set of item identifiers. Let $\mathcal{T}$ be a transaction table such that every tuple in $\mathcal{T}$ is a pair, called the *transaction*, of the form $\langle t_{id}, X\rangle$ such that t_{id} is a unique transaction ID and $X \subseteq \mathcal{I}$ is a set of item identifiers (or items). A transaction is usually identified by its transaction ID t_{id}, and said to contain the item set X. An *association rule* is an implication of the form $X \rightarrow Y$, where $X, Y \subseteq \mathcal{I}$, and $X \cap Y = \emptyset$. Association rules are assigned a *support* (written as δ) and *confidence* (written as η) measure, and denoted $X \rightarrow Y\langle\delta, \eta\rangle$. The rule $X \rightarrow Y$ has a support δ, denoted $sup(X \rightarrow Y)$, in the transaction table $\mathcal{T}$ if δ% of the transactions in $\mathcal{T}$ contain $X \cup Y$. In other words, $sup(X \rightarrow Y) \equiv sup(X \cup Y) = \delta = \frac{|\{t|t\in\mathcal{T} \wedge X\cup Y\subseteq t[I]\}|}{|\mathcal{T}|}$, where $I \subseteq \mathcal{I}$ is a set of items. On the other hand, the rule $X \rightarrow Y$ is said to have a confidence η, denoted $con(X \rightarrow Y)$, in the transaction table $\mathcal{T}$ if η% of the transactions in $\mathcal{T}$ that contain X also contains Y. So, the confidence of a rule is given by $con(X \rightarrow Y) = \eta = \frac{sup(X\cup Y)}{sup(X)}$.

Given a transaction table $\mathcal{T}$, the problem of mining association rules is to generate a set of quadruples $\mathcal{R}$ (a table) of the form $\langle X, Y, \delta, \eta\rangle$ such that $X, Y \subseteq \mathcal{I}$, $X \cap Y = \emptyset$, $\delta \geq \delta_m$, and $\eta \geq \eta_m$, where δ_m and η_m are user supplied minimum support and confidence *thresholds*, respectively. The clarity of the definitions and the simplicity of the problem is actually deceptive. As mentioned before, to be able to compute the rules $\mathcal{R}$, we must first compute the frequent item sets. A set of items X is called a frequent item set if its support δ is greater than the minimum support δ_m.

1.1 Related Research

Declarative computation of association rules were investigated in works such as [13,25,9,21,8,20,14,7,5].Meo et al. [14] proposes an SQL like declarative query language for association rule mining. The language proposed appears to be too oriented towards transaction databases, and may not be suitable for general association rule mining. It is worth mentioning that association rules may be computed for virtually any type of database, transaction or not. In their extended language, they blend a rule mine operator with SQL and other additional features. The series of research reported in [25,21,20] led by IBM researchers, mostly addressed the mining issue itself. They attempted to compute the large item sets by generating candidate sets testing for their admissibility based on their MC model, combination, and GatherJoin operators. Essentially, these works proposed a method for implementing apriori using SQL. In our opinion, by trying to faithfully copy a procedural concept into a declarative representation they retain the drawbacks and inefficiencies of apriori in the model.

The mine rule operator proposed in [13] is perhaps the closest idea to ours. The operator has significant strengths in terms of expressive power. But it also requires a whole suit of new algebraic operators. These operators basically simulate the counting process using a set of predefined functions such as *CountAllGroups*, *MakeClusterPairs*, *ExtractBodies*, and *ExtractRules*. These functions use a fairly good number of new operators proposed by the authors, some of which use looping constructs. Unfortunately, no optimization techniques for these operators are available, resulting in doubts, in our opinion, about the computational viability of this approach.

In this article, we will demonstrate that there is a simpler SQL3 expression for association rule mining that does not require candidate generation such as in [25,21,20] or any implementation of new specialized operators such as in [13,14]. We also show that we can simply add an operator similar to cube by operator proposed for data warehousing applications with an optional having clause to facilitate filtering of unwanted derivations. The striking feature of our proposal is that we can exploit the vast array of optimization techniques that already exists and possibly develop newer ones for better performance. These are some of the advantages of our proposal over previous research in addition to its simplicity and intuitive appeal.

1.2 Contributions of this Article and Plan for the Presentation

We summarize the contributions of this article as follows to give the reader an idea in advance. We present a different view of transaction databases and identify several properties that they satisfy in general in section 2. We exploit these properties to develop a purely SQL3 based solution for association rule mining that uses the idea of least fix point computation. We rely on SQL3 standard as it supports complex structures such as sets and complex operations such as nesting and unnesting. We also anticipate the availability of several set processing functions such as **intersect** ($\cap$) and **setminus** ($\setminus$), set relational operators such as **subset** ($\subset$) and **superset** ($\supset$), nested relational operations such as nest by, etc. Finally, we also exploit SQL3's create view recursive and with constructs to implement our least fixpoint operator for association rule mining.

We follow the tradition of separating the large item set counting from actual mining and propose two operators – one to compute the large item sets from a source table, another one to compute the rules from the large item sets. We provide optional mechanisms to specify support and confidence thresholds and a few additional constraints that the user may wish the mining process to satisfy. We also define a single operator version of mine by operator to demonstrate that it is possible to do so even within our current framework, even though we prefer the two stage approach.

The other implicit contribution of our proposal is that it opens up the opportunity for query optimization, something that was not practically possible until now in mining applications. Finally, it is now possible to use any relational database for mining in which one need not satisfy input restrictions similar to the ones that various mining algorithms require. Consequently, the developments in this article eliminates the need for any traditional preprocessing of input data.

In section 3, we present a discussion on a set theoretic perspective of data mining problem. This discussion builds upon the general properties of transaction tables presented in section 2. In this section, we demonstrate through illustrative examples that we can solve the mining problem just using set, lattice and aggregate operations if we adopt the idea of the so called *non-redundant* large item sets. Once the problem is understood on intuitive grounds, the rest of the development follows in fairly straightforward ways. In section 4 we present a series of SQL3 expressions that capture the spirit of the procedure presented in section 3. One can also verify that these expressions really produce the solution we develop in the illustrative example in this section. We then discuss the key idea we have exploited in developing the solution in section 5. The mining operator is presented in section 6 that is an abstraction of the series of SQL3 expressions in section 4. Several optimization opportunities and related details are discussed in section 7. Before we conclude in section 9, we present a comparative analysis of our method with other representative proposals in section 8.

2 Properties of Transaction Tables

In this section, we identify some of the basic properties shared by all transaction tables. We explain these properties using a synthetic transaction table in relational data model [24]. In the next section, we will introduce the relational solution to the association rule mining problem.

Let $\mathcal{I}$ be a set of items, $\mathcal{P}(\mathcal{I})$ be all possible item sets, T be a set of identifiers, and δ_m be a threshold. Then an *item set* table $\mathcal{S}$ with scheme {Tid, Items} is given by

$$\mathcal{S} \subseteq T \times \mathcal{P}(\mathcal{I})$$

such that $m = |\mathcal{S}|$. An item set table $\mathcal{S}$ is *admissible* if for every tuple $t \in \mathcal{S}$, and for every subset $s \in \mathcal{P}(t[Items])$, there exists a tuple $t' \in \mathcal{S}$ such that $s = t'[Items]$. In other words, every possible subset of items in a tuple is also a member of $\mathcal{S}$. The *frequency* table of an admissible item set table $\mathcal{S}$ can be obtained as

$$F =_{Items} \mathcal{G}_{\mathbf{count}(*)}({}_{Tid}\mathcal{G}(\mathcal{S}))$$

which has the scheme {Items, Count}. A *frequent item set* table F_f is a set of tuples that satisfies the count threshold property as follows.

$$F_f = \sigma_{Count/m \geq \delta_m}(F)$$

F_f satisfies some additional interesting properties. Suppose $I = t[Items]$ is an item set for any tuple $t \in F_f$. Then, for any $X, Y \subset I$, there exists t_1 and t_2 in F_f such that $t_1[Items] = X$, $t_2[Items] = Y$, $t_1[Count] \geq t[Count]$, $t_2[Count] \geq t[Count]$, and $t[Count] \leq min(t_1[Count], t_2[Count])$. The converse, however, is not true. That is, for any two tuples t_1 and t_2 in F_f, it is not necessarily true that there exists a tuple $t \in F_f$ such that $t[Items] = t_1[Items] \cup t_2[Items]$. But if such a t exists then the relation $t[Count] \leq min(t_1[Count], t_2[Count])$ is always true. Such a relationship is called *anti-transitive.*

The goal of the first stage of apriori like algorithms has been to generate the frequent item set table described above from a transaction table $\mathcal{T}$. Note that a transaction table, as defined above, is, in reality, not admissible. But the first stage of apriori mimics admissibility by constructing the k item sets at every kth iteration step.

Once the frequent item set table is available, the association rule table $\mathcal{R}$ can be computed as[1]

$$\mathcal{R} = \Pi_{F_{f_1}.Items, F_{f_2}.Items \setminus F_{f_1}.Items, \frac{F_{f_1}.Count}{m}, \frac{F_{f_2}.Count}{F_{f_1}.Count}} \times (\sigma_{F_{f_1}.Items \subset F_{f_2}.Items}(F_{f_1} \times F_{f_2}))$$

This expression, however, produces all possible rules, some of which are even redundant. For example, let $a \rightarrow b\langle \frac{s_{ab}}{m}, \frac{s_{ab}}{s_a} \rangle$ and $ab \rightarrow c\langle \frac{s_{abc}}{m}, \frac{s_{abc}}{s_{ab}} \rangle$ be two rules discovered from F_f, where s_X and m represent respectively the frequency of an item set X in the item set table (i.e., $t \in F_f$, and $t[Count] = \frac{s_X}{m}$), and number of transactions in the item set table $\mathcal{S}$. Then it is also the case that $\mathcal{R}$ contains another rule (transitive implication) $a \rightarrow bc\langle \frac{s_{abc}}{m}, \frac{s_{abc}}{s_a} \rangle$. Notice that this last rule is a logical consequence of the first two rules that can be derived using the following inference rule, where $X, Y, Z \subset \mathcal{I}$ are item sets.

$$\frac{X \rightarrow Y\langle \frac{s_{X \cup Y}}{m}, \frac{s_{X \cup Y}}{s_X} \rangle \qquad X \cup Y \rightarrow Z\langle \frac{s_{X \cup Y \cup Z}}{m}, \frac{s_{X \cup Y \cup Z}}{s_{X \cup Y}} \rangle}{X \rightarrow Y \cup Z\langle \frac{s_{X \cup Y \cup Z}}{m}, \frac{s_{X \cup Y \cup Z}}{s_X} \rangle}$$

Written differently, using only symbols for support (δ) and confidence (η), the inference rule reads as follows.

$$\frac{X \rightarrow Y\langle \delta_1, \eta_1 \rangle \qquad X \cup Y \rightarrow Z\langle \delta_2, \eta_2 \rangle}{X \rightarrow Y \cup Z\langle \delta_2, \eta_1 * \eta_2 \rangle}$$

Formally, if $X, Y, Z \subseteq \mathcal{I}$ be sets of items, and $X \rightarrow Y\langle \delta_1, \eta_1 \rangle$, $X \cup Y \rightarrow Z\langle \delta_2, \eta_2 \rangle$ and $X \rightarrow Y \cup Z\langle \delta_3, \eta_3 \rangle$ hold, then we say that $X \rightarrow Y \cup Z\langle \delta_3, \eta_3 \rangle$ is an *anti-transitive* rule. Also, if $r = X \rightarrow Y\langle \delta, \eta \rangle$ be a rule, and δ_m and η_m respectively be the minimum support and confidence requirements, then r is *redundant* if it is derivable from other rules, or if $\delta < \delta_m$ or $\eta < \eta_m$.

[1] Assuming that two copies of F_f are available as F_{f_1} and F_{f_2}.

It is possible to show that for any minimum support and confidence thresholds δ_m and η_m respectively, if the rules $X \rightarrow Y\langle\delta_1, \eta_1\rangle$, and $X \cup Y \rightarrow Z\langle\delta_2, \eta_2\rangle$ hold, then the rule $X \rightarrow Y \cup Z\langle\delta_3, \eta_3\rangle$ also holds such that $\delta_3 = \delta_2 \geq \delta_m$, and $\eta_3 = \eta_1 * \eta_2 \leq min(\eta_2, \eta_1)$. Notice that $\eta_3 = \eta_1 * \eta_2$ could be less then the confidence threshold η_m, even though $\eta_1 \geq \eta_m$ and $\eta_2 \geq \eta_m$. In other words, $\eta_1 \geq \eta_m \wedge \eta_2 \geq \eta_m \not\Rightarrow \eta_1 * \eta_2 \geq \eta_m$. Furthermore, we can show that any rule $r = X \rightarrow Y\langle\delta, \eta\rangle$ is redundant if it is anti-transitive. It is interesting to observe that the redundancy of rules is a side effect of redundancy of large item sets. Intuitively, for any given pair of large item sets l_1 and l_2, l_1 is *redundant* if $l_1 \subseteq l_2$, and the support s_{l_1} of l_1 is equal to the support s_{l_2} of l_2. Otherwise, it is *non-redundant*. Intuitively, l_1 is redundant because its support s_{l_1} can be computed from s_{l_2} just by copying. A more formal treatment of the concept of large itemsets and redundant large itemsets may be found in section 5.1.

Since in the frequent item set table, every item set is a member of a chain that differs by only one element, the following modification for $\mathcal{R}$ will compute the rules that satisfies given support and confidence thresholds and avoids generating all such redundant rules[2].

$$\begin{aligned}\mathcal{R} = \sigma_{Conf \geq \eta_m}(&\rho_{r(Ant,Cons,Sup,Conf)} \\ &\times(\Pi_{F_{f_1}.Items, F_{f_2}.Items \setminus F_{f_1}.Items, \frac{F_{f_1}.Count}{m}, \frac{F_{f_2}.Count}{F_{f_1}.Count}} \\ &\times(\sigma_{F_{f_1}.Items \subset F_{f_2}.Items \wedge (|F_{f_2}.Items| - |F_{f_1}.Items| = 1)}(F_{f_1} \times F_{f_2}))))\end{aligned}$$

2.1 The Challenge

The preceding discussion was aimed to demonstrate that a relational computation of association rules is possible. However, we used an explicit generation of the power set of the items in $\mathcal{I}$ to be able to compute the frequent item set table F_f from the item set table $\mathcal{S}$. This is a huge space overhead, and consequently, imposes a substantial computational burden on the method. Furthermore, we required that the item set table $\mathcal{S}$ be admissible, another significant restriction on the input transaction table. These are just some of the difficulties faced when a set theoretic or relational characterization of data mining is considered. The procedurality involved acts as a major bottleneck. So, the challenge we undertake is to admit any arbitrary transaction table, yet be able to compute the association rules "without" explicit generation of candidate item sets from a relational database, and compute the relation $\mathcal{R}$ as introduced before using existing SQL3 constructs and machineries.

3 A Set Theoretic Perspective of Data Mining

In this section, we present our idea of a SQL mine operator on intuitive grounds using a detailed example. The expectation here is that once we intuitively understand the issues related to the operator, it should be relatively easier to follow

[2] ρ is a relation renaming operator defined in [24].

the technical developments in the later sections. Also, this simple explanation will serve as the basis for a more general relational mining operator we plan to present at the end of this article.

Consider a database, called the transaction table, **T** as shown in figure 1. Following the traditional understanding of association rule mining, and also from the discussion in section 2, from the source table **T** we expect to obtain the large item set table (l_table) and the rules table (r_table) shown in figure 1 below once we set the support threshold at 25%. The reasoning process of reaching to the large item set and rules tables can be explained as follows.

t_table

Tranid	Items
t_1	a
t_1	b
t_1	c
t_2	b
t_2	c
t_2	f
t_3	b
t_3	f
t_4	a
t_4	b
t_4	c
t_5	b
t_5	e
t_6	d
t_6	f
t_7	d

transaction table

l_table

Items	Support
{a, b, c}	.29
{b, f}	.29
{b, c}	.38
{f}	.43
{d}	.29
{b}	.71

large item set table

r_table

Ant	Cons	Support	Conf
{b}	{c}	0.38	0.60
{f}	{b}	0.29	0.66
{b}	{f}	0.29	0.40
{b,c}	{a}	0.29	0.66

association rules table

Fig. 1. Source transaction database **T** is shown as t_table, large item set table as l_table, and finally the association rules as r_table

We can think of **T** as the set of complex tuples shown in nested table (n_table) in figure 2 once we nest the items on transaction numbers. If we use a group by on the Items column and count the transactions, we will compute the frequency table (f_table) in figure 2 that will show how many times a single item set pattern appears in the transaction table (t_table) in figure 1. Then, let us assume that we took a cross product of the frequency table with itself, and selected the rows for which

- the Items column in the first table is a proper subset of the Items column in the second table, and finally projected out the Items column of the first table and Support column of the second table[3], or
- the Items columns are not subset of one another, and we took the intersection of the Items of both the tables, created a new relation (int_table, called the

[3] This will give us $< \{b, f\}, 1 >$ and $< \{d\}, 1 >$.

n_table

Tranid	Items
t_1	{a,b,c}
t_2	{b,c,f}
t_3	{b,f}
t_4	{a,b,c}
t_5	{b,e}
t_6	{d,f}
t_7	{d}

nested table

i_table

Items	Support
{b,c}	3
{b,f}	1
{b}	5
{f}	3
{d}	1

inheritance table

f_table

Items	Support
{a, b, c}	2
{b,c,f}	1
{b, e}	1
{b, f}	1
{d}	1
{d,f}	1

frequency table

c_table

Items	Support
{a, b, c}	2
{b,c,f}	1
{b,c}	3
{b, f}	2
{b, e}	1
{d, f}	1
{b}	5
{f}	3
{d}	2

count table

Fig. 2. n_table: t_table after nesting on Tranid, f_table: n_table after grouping on Items and counting, i_table: generated from f_table, and c_table: grouping on Items and sum on Support on the union of i_table and f_table

intersection table) with distinct tuples of such Items with Support 0, and then finally computed the support counts as explained in step 1 now with the frequency table and intersection table[4].

This will give us the inheritance table (i_table) as shown in figure 2. Finally, if we took a union of the frequency table and the inheritance table, and then do a group by on the Items column and sum the Support column, we would obtain the count table (c_table) of figure 2.

The entire process of large item set and association rule generation can be conveniently explained using the so called item set lattice found in the literature once we enhance it with some additional information. Intuitively, consider placing the transactions with item set u appearing in the frequency table with their support count t as a node in the lattice $\mathcal{L}$ as shown in figure 3. Notice that in the lattice, each node is represented as u_c^t, where it denotes the fact that u appears in exactly t transactions in the source table, and that u also appears as a subset of other transactions n number of times such that $c = n + t$. t is called

[4] The result of this will be tuple $< \{b, c\}, 3 >$, $< \{b\}, 5 >$, and $< \{f\}, 3 >$ in this example. Note that the intersection table will contain the tuples $< \{b, c\}, 0 >$, $< \{b\}, 0 >$, and $< \{f\}, 0 >$, and that these patterns are not part of the frequency table in figure 2. The union of step 1, and step 2 processed with the intersection table will now produce the inheritance table in figure 2.

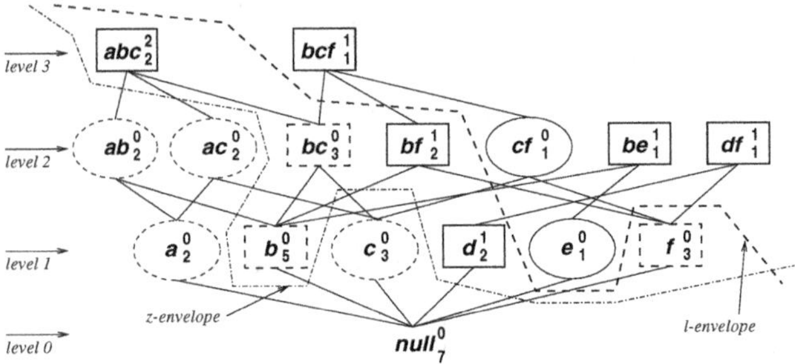

Fig. 3. Lattice representation of item set database **T**

the *transaction count*, or *frequency count*, and c is called the *total count* of item set u. The elements or nodes in $\mathcal{L}$ also satisfy additional interesting properties. A node v at level l differs from its child u at level $l-1$ by exactly 1 element, and that $u \subset v$. For any two children u and w of a node v, $v = u \cup w$. For any two nodes[5] $u_{c_u}^{t_u}$ and $v_{c_c}^{t_v}$ at any level l, their *join* is defined as $(u \cap v)_{c_j}^{t_j}$, and the *meet* as $(u \cup v)_{c_m}^{t_m}$, such that $c_j \leq min(c_u, c_v)$ and $c_m \geq max(c_u, c_v)$.

Note that in figure 3, the nodes marked with a solid rectangle are the nodes (or the item sets) in **T**, nodes identified with dotted rectangles are the *intersection*[6] nodes or the *virtual*[7] nodes, and the nodes marked with ellipses (dotted or solid) are *redundant*. The nodes below the dotted line, called the large item set envelope, or *l-envelope*, are the large item sets. Notice that the node *bc* is a large item set but is not a member of **T**, while *bcf*, *df* and *be* are in **T**, yet they are not included in the set of large item sets of **T**. We are assuming here a support threshold of 25%. So, basically, we would like to compute only the nodes *abc*, *bc*, *bf*, *b*, *d* and *f* from **T**. This set is identified by the *sandwich* formed by the l-envelope and the zero-envelope, or the *z-envelope*, that marks the lowest level nodes in the lattice. If we remove the non-essential, or redundant, nodes from the lattice in figure 3, we are left with the lattice shown in figure 4. It is possible to show that the lattice shown in figure 4 is the set of non-redundant large item sets of **T** at a support threshold 25%. The issue now is how to read this lattice. In other words, can we infer all the large item sets that an apriori like algorithm will yield on **T**? The answer is yes, but in a somewhat different manner. This is demonstrated in the following way.

Notice that there are five large 1-items – namely a, b, c, d and f. But only three, b, d and f, are listed in the lattice. The reason for not listing the other large 1-items is that they are implied by one of the nodes in the lattice. For

---

[5] For any node u, the notations t_u and c_u mean respectively the transaction count and total count of u.

[6] Nodes that share items in multiple upper level nodes and have a total count higher than any of the upper level nodes.

[7] Nodes with itemsets that do not appear in **T**, and also have total count equal to all the nodes above them.

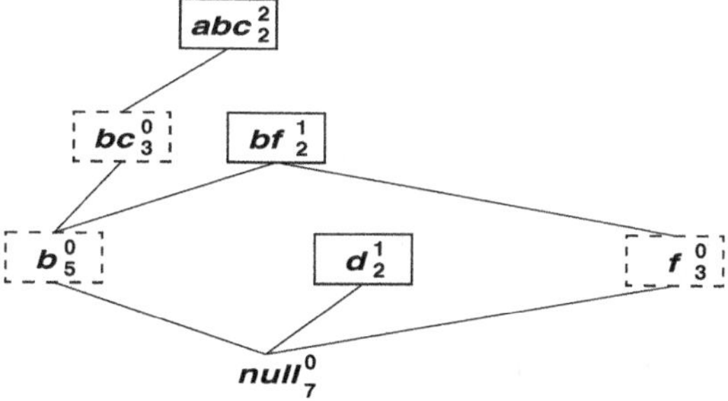

Fig. 4. Non-redundant large item sets of **T** when $\delta_m = 0.25$

example, c is implied by bc for which the count is 3. The nodes b and bc should be read as follows – b alone appears in **T** 5 times, whereas bc appears 3 times. Since c appears a maximum of 3 times with b (2 times in abc and 1 time in bcf actually), its total count can be derived from bc's count. Similarly, a's count can be derived from abc – 2. Hence, the lattice in figure 4 does not include them and considers them as redundant information. This view point has another important implication. We are now able to remove (or prune) redundant association rules too. We will list $b \rightarrow c\langle .38, .60\rangle$ and $bc \rightarrow a\langle .29, .66\rangle$, among several others, as two association rules that satisfy the 25% support and 40% confidence thresholds. Notice that we do not derive the rule $b \rightarrow ac\langle .29, .40\rangle$ in particular. The reason is simple – it is redundant for it can be derived from the rules $b \rightarrow c\langle .38, .60\rangle$ and $bc \rightarrow a\langle .29, .66\rangle$ using the following inference rule. Notice that if we accept the concept of redundancy we propose for rules, computing $b \rightarrow ac\langle .29, .40\rangle$ does not strengthen the information content of the discovery in any way.

$$\frac{X \rightarrow Y\langle \delta_1, \eta_1\rangle \qquad X \cup Y \rightarrow Z\langle \delta_2, \eta_2\rangle}{X \rightarrow Y \cup Z\langle \delta_2, \eta_1 * \eta_2\rangle} = \frac{b \rightarrow c\langle .38, .60\rangle \qquad bc \rightarrow a\langle .29, .66\rangle}{b \rightarrow ac\langle .29, .40\rangle}$$

Finally, we would like to point to an important observation. Take the case of the rules $b \rightarrow f\langle .29, .40\rangle$ and $f \rightarrow b\langle .29, .66\rangle$. These two rules serve as an important reminder that $X \rightarrow Y\langle s_1, c_1\rangle$, and $Y \rightarrow X\langle s_2, c_2\rangle \not\Rightarrow c_1 = c_2$, and that $X \rightarrow Y\langle s_1, c_1\rangle$, and $Y \rightarrow X\langle s_2, c_2\rangle \Rightarrow s_1 = s_2$. But in systems such as apriori, where all the large item sets are generated without considering redundancy, it would be difficult to prune rules based on this observation as we do not know which one to prune. For example, for the set of large item sets $\{bc_3, b_3, c_3\}$, we must derive rules $b \rightarrow c\langle \frac{3}{m}, 1\rangle$ and $c \rightarrow b\langle \frac{3}{m}, 1\rangle$[8] and cannot prune them without any additional processing. Instead, we just do not generate them at all.

[8] Assuming there are m number of transactions.

4 Computing Item Set Lattice Using SQL3

Now that we have explained what non-redundant large item sets and association rules mean in our framework, we are ready to discuss computing them using SQL. The reader may recall from our discussion in the previous section that we have already given this problem a relational face by presenting them in terms of (nested) tables. We will now present a set of SQL3 sentences to compute the tables we have discussed earlier. We must mention here that it is possible to evaluate the final table in figure 1 by mimicking the process using a lesser number of expressions than what we present below. But we prefer to include them all separately for the sake of clarity. In a later section, we will discuss how these series of SQL sentences can be replaced by an operator, the actual subject of this article.

For the purpose of this discussion, we will assume that several functions that we are going to use in our expressions are available in some SQL3 implementation, such as Oracle, DB2 or Informix. Recall that SQL3 standard requires or implies that, in some form or other, these functions are supported[9]. In particular, we have used a nest by clause that functions like a group by on the listed attributes, but returns a nested relation as opposed to a first normal form relation returned by group by. We have also assumed that SQL3 can perform group by on nested columns (columns with set values). Finally, we have also used set comparators in where clause, and set functions such as **intersect** and **setminus** in the select clause, which we think are natural additions to SQL3 once nested tuples are supported. As we have mentioned before, we have, for now, used user defined functions (UDFs) by treating set of items as a string of labels to implement these features in Oracle.

The following two view definitions prepare any first normal form transaction table for the mining process. Note that these view definitions act as idempotent functions on their source. So, redoing them does not harm the process if the source table is already in one of these forms. These two views compute the n_table and the f_table of figure 2.

```
create view n_table as
   (select Tranid, Items
   from t_table
   nest by Tranid)

create view f_table as
   (select Items, count(*) as Support
   from n_table
   group by Items)
```

Before we can compute the i_table, we need to know what nodes in the imaginary lattice will inherit transaction counts from some of the transaction

[9] Although some of these functions are not supported right now, once they are, we will be in a better shape. Until then, we can use PL/SQL codes to realize these functions.

nodes in the lattice – Support value of Items in the f_table. Recall that nodes that are subset of another node in the lattice, inherit the transaction count of the superset node towards its total count. We also know that only those (non-redundant) nodes which appear in the f_table, or are in the least fixpoint of the nodes in f_table will inherit them. So, we compute first the set of intersection nodes implied by f_table using the newly proposed SQL3 create view recursive statement as follows.

```
create view recursive int_table as
  ((select distinct intersect(t.Items, p.Items), 0
  from f_table as t, f_table as p
  where t.Items ⊄ p.Items and p.Items ⊄ t.Items
    and not exists
      (select *
      from f_table as f
      where f.Items = intersect(t.Items, p.Items)))
  union
  (select distinct intersect(t.Items, p.Items), 0
  from int_table as t, int_table as p
  where t.Items ⊄ p.Items and p.Items ⊄ t.Items
    and not exists
      (select *
      from f_table as f
      where f.Items = intersect(t.Items, p.Items))))
```

We would like to mention here again that we have implemented this feature using PL/SQL in Oracle. Notice that we did not list the int_table we create below in figure 1 or 2 because it is regarded as a transient table needed for the computation of i_table.

It is really important that we create only distinct set of intersection items and only those ones that do not appear in the f_table for the purpose of accuracy in support counting. Take for example three transactions in a new frequency table, f′_table, represented as $\{abc_0^1, bcd_0^1, bcf_0^1, bc_0^1\}$. Assume that we compute the set of intersections of the entries in this table. If we do not guard against the cautions we have mentioned, we will produce the set $\{bc_0^0, bc_0^0, bc_0^0\}$ using the view expression for int_table – which is not desirable. Because, these three will inherit Support from $\{abc_0^1, bcd_0^1, bcf_0^1\}$ giving a total count of 10, i.e., bc_{10}^1. The correct total count should have been bc_4^1. If we just ensure the uniqueness of a newly generated item set (but not its absence in the f_table) through meet computation, we still derive $\{bc_0^0\}$ instead of an empty set, which is also incorrect. This means that not including the following condition in the above SQL expression will be a serious mistake.

```
not exists (select *
  from f_table as f
  where f.Items = intersect(t.Items, p.Items)
```

Once we have computed the int_table, the rest of the task is pretty simple. The i_table view is computed by copying the Support of a tuple in f_table for any tuple in the collection of f_table and int_table which is a subset of the tuple in the f_table. Intuitively, these are the nodes that need to inherit the transaction counts of their ancestors (in f_table).

```
create view i_table as
   (select t.Items, p.Support
   from f_table as p,
       ((select *
       from f_table)
       union
       (select *
       from int_table)) as t,
   where t.Items ⊂ p.Items)
```

From the i_table, a simple grouping and sum operation as shown below will give us the count table, or the c_table, of figure 2.

```
create view c_table as
   (select t.Items, sum(t.Support) as Support
   from ((select *
       from f_table)
       union
       (select *
       from i_table)) as t
   group by t.Items)
```

The large item sets of l_table in figure 1 can now be generated by just selecting on the c_table tuples as shown next. Notice that we could have combined this step with the c_table expression above with the help of a having clause.

```
create view l_table as
   (select Items, Support
   from c_table
   where Support ≥ δm)
```

Finally, the (non-redundant) association rules of figure 1 are computed using the r_table view below. The functionality of this view can be explained as follows. Two item sets $u[Items]$ and $v[Items]$ in a pair of tuple u and v in the l_table implies an association rule of the form $u[Items] \rightarrow v[Items] \setminus u[Items]\langle v[Support], \frac{v[Support]}{u[Support]}\rangle$ only if $u[Items] \subset v[Items]$ and there does not exist any intervening item set x in the l_table such that x is a superset of $u[Items]$ and is a subset of $v[Items]$ as well. In other words, in the lattice, $v[Items]$ is one of the immediate ancestors of $u[Items]$. In addition, the ratio of the Supports, for example, $\frac{v[Support]}{u[Support]}$ must be at least equal to the minimum confidence threshold η_m.

```
create view r_table as
  (select a.Items, c.Items\a.Items, c.Support, c.Support/a.Support
  from l_table as a, l_table as c
  where a.Items ⊂ c.Items and c.Items/a.Items ≥ η_m and not exists
    (select Items
    from l_table as i
    where a.Items ⊂ i.Items and i.Items ⊂ c.Items))
```

The readers may verify that these are the only "generic" SQL3 expressions (or their equivalent) that are needed to mine any relational database (assuming proper name adaptations for tables and columns). The essence of this relational interpretation of the problem of mining, as demonstrated by the SQL3 expressions above, is that we do not need to think in terms of iterations, candidate generation, space time overhead, and so on. Instead, we can now express our mining problems on any relational database in declarative ways, and leave the optimization issues with the system and let the system process the query using the best available method to it, recognizing the fact that depending on the instance of the database, the choice of best methods may now vary widely.

5 An Enabling Observation

Level wise algorithms such as apriori essentially have three distinct steps at each pass k: (i) scan the database and count length k candidate item sets against the database, (ii) test and discard the ones that are not large item sets, and (iii) generate candidate item sets of length $k+1$ from the length k large items sets just generated and continue to next iteration level. The purpose of the second step is to prune potential candidates that are not going to generate any large item sets. This heuristic is called the *anti-monotonicity* property of large item sets. While this heuristic saves an enormous amount of space and time in large item set computation and virtually makes association rule mining feasible, further improvements are possible. We make an observation that apriori fails to remove *redundant* large item sets that really do not contribute anything new, and not generating the redundant large item sets do not cause any adverse effect on the discovery of the set of association rules implied by the database. In other words, apriori fails to potentially recognize another very important optimization opportunity. Perhaps the most significant and striking contribution of this new optimization opportunity is its side effect on the declarative computation of association rules using languages such as SQL which is the subject of this article. This observation of optimization opportunity helps us avoid thinking level wise and allows us to break free from the expensive idea of candidate generation and testing even in SQL like set up such as in [25,21,20]. As mentioned earlier, methods such as [4,27] have already achieved significant performance improvements over apriori by not requiring to generate candidate item sets.

To explain the idea we have on intuitive grounds, let us consider the simple transaction table t_table in figure 1. Apriori will produce the database in figure 5(a) in three iterations when the given support threshold is $\approx 25\%$, or 2 out of 7 transactions.

Items	Support
a	2
b	5
c	3
d	2
f	3
a, b	2
a, c	2
b, c	3
b, f	2
a, b, c	2

(a)

Candidates
a
b
c
d
e
f

(b)

Candidates
a, b
a, c
a, d
a, f
b, c
b, d
b, f
c, d
c, f
d, f

(c)

Candidates
a, b, c
a, b, f
b, c, f

(d)

Items	Support
a	2
c	3
a, b	2
a, c	2

(e)

Fig. 5. (a) Frequent item set table generated by apriori and other major algorithms such as FP-tree. Candidate sets generated by a very *smart* apriori at iterations (b) $k = 1$, (c) $k = 2$, and at (d) $k = 3$

Although apriori generates the table in figure 5(a), it needs to generate the candidate sets in figure 5(b) through 5(d) to test. Notice that although the candidates $\{a, d\}$, $\{a, f\}$, $\{b, d\}$, $\{c, d\}$, $\{c, f\}$, $\{d, f\}$, $\{a, b, f\}$ and $\{b, c, f\}$ were generated (figures 5(c) and 5(d)), they did not meet the minimum support threshold and were never included in the frequent item set table in figure 5(a). Also notice that these are the candidate item sets generated by a very smart apriori algorithm that scans the large items sets generated at pass k in order to guess the possible set of large item sets it could find in pass $k + 1$, and selects the guessed sets as the candidate item sets. A naive algorithm on the other hand would generate all the candidate sets from the large item sets generated at pass k exhaustively by assuming that they are all possible. Depending on the instances, they both have advantages and disadvantages. But no matter which technique is used, apriori must generate a set of candidates, store them in some structures to be able to access them conveniently and check them against the transaction table to see if they become large item sets at the next iteration step. Even by conservatively generating a set of candidate item sets as shown in figures 5(b) through 5(d) for the database in figure 1(t_table), it wastes (for the candidate sets that never made it to the large item sets) time and space for some of the candidate sets. Depending on the transaction databases, the wastage could be significant. The question now is, could we generate candidate sets that will have a better chance to become a large item set? In other words, could we generate the absolute minimum set of candidates that are surely a large item set? In some way, we think the answer is in the positive as we explain in the next section.

5.1 Implications

We take the position and claim that the table l_table shown in figure 1 is an information equivalent table of figure 5(a) which essentially means that these tables faithfully imply one another (assuming identical support thresholds, $\approx 25\%$). Let us now examine what this means in the context of association rule mining on intuitive grounds.

First of all, notice that the tuples (i.e., large item sets) missing in table of figure 1 are listed in table of figure 5(e), i.e., the union of these two tables gives us the table in figure 5(a). Now the question is why do we separate them and why do we deem the tuples in figure 5(e) redundant? Before we present the reasoning, we would like to define the notion of redundancy of large item sets in order to keep our discussion in perspective.

Definition 5.1. Let $\mathcal{T}$ be a transaction table over item sets $\mathcal{I}$, $I \subseteq \mathcal{I}$ be an item set, and n be a positive integer. Also let n represent the frequency of the item set I with which it appears in $\mathcal{T}$. Then the pair $\langle I, n\rangle$ is called a *frequent* item set, and the pair is called a *large* item set if $n \geq \delta_m$, where δ_m is the minimum support threshold.

We define redundancy of large item sets as follows. If for any large item set I, its frequency n can be determined from other large item sets, then I is redundant. Formally,

Definition 5.2 (Redundancy of Large Item Sets). Let L be a set of large item sets of tuples of the form $\langle I_u, n_u\rangle$ such that $\forall x, y(x = \langle I_x, n_x\rangle, y = \langle I_y, n_y\rangle \in L \wedge I_x = I_y \Rightarrow n_x = n_y)$, and let $u = \langle I_u, n_u\rangle$ be such a tuple. Then u is *redundant* in L if $\exists v(v \in L, v = \langle I_v, n_v\rangle, I_u \subseteq I_v \Rightarrow n_u = n_v)$.

The importance of the definition 5.1 may be highlighted as follows. For any given set of large item sets L, and an element $l = \langle I_l, n_l\rangle \in L$, I_l is unique in L. The implication of anti-monotonicity is that for any other $v = \langle I_v, n_v\rangle \in L$ such that $I_l \subset I_v$ holds, $n_v \leq n_l$ because an item set cannot appear in a transaction database less number of times than any of its supersets. But the important case is when $n_v = n_l$ yet $I_l \subset I_v$. This implies that I_l never appears in a transaction alone, i.e., it always appeared with other items. It also implies for all large item sets $s = \langle I_s, n_s\rangle \in L$ of I_v such that $I_s \supset I_v$, if it exists, $n_v = n_s$ too. As if not, n_l should be different than n_v, which it is not, according to our assumption. It also implies that I_l is not involved in any other sub-superset relationship chains other that I_v. There are several other formal and interesting properties that the large item sets satisfy some of which we will present in a later section.

The importance of the equality of frequency counts of large item sets that are related via sub-superset relationships is significant. This observation offers us another opportunity to optimize the computation process of large item sets. It tells us that there is no need to compute the large item sets for which there exists another large item set which is a superset and has identical frequency count. For example, for an item set $S = \{a, b, c, d, e, f, g, h\}$, apriori will iterate eight times if S is a large item set and generate $|\mathcal{P}(S)|$ subsets of S with identical frequency counts when, say, S is the only distinct item set in a transaction table. A hypothetical smart algorithm armed with definition 5.1 will only iterate once and stop. Now, if needed, the other large item sets computed by apriori can be computed from S just by generating all possible subsets of S and copying the frequency count of S. If S is not a large item set, so cannot be any subset of S. Apriori will discover it during the first iteration and stop, only if S is not a large item set, and so will the smart algorithm.

Going back to our example table in figure 5(a), and its equivalent table l_table in figure 1, using definition 5.1 we can easily conclude that the set of large item sets in table 5(e) are redundant. For example, the item set $\{a\}$ is a subset of $\{a, b, c\}$ and both have frequency or support count 2. This implies that there exists no other transactions that contribute to the count of a. And hence, it is redundant. On the other hand, $\{b\}$ is not redundant because conditions of definition 5.1 does not apply. And indeed we can see that $\{b\}$ is a required and non-redundant large item set because if we delete it, we cannot hope to infer its frequency count from any other large item sets in table l_table of figure 1. Similar arguments hold for other tuples in l_table of figure 1.

The non-redundant large item set table in figure 1 unearth two striking and significant facts. All the item sets that are found to be large either appear as a transaction in the t_table in figure 1, e.g., $\{b, f\}$ and $\{a, b, c\}$, or are intersections of two or more item sets of the source table not related via sub-superset relationships, e.g., $\{b\}$, which is an intersection of $\{b, e\}$ and $\{b, c, f\}$, and $\{b, e\} \not\subset \{b, c, f\}$ and $\{b, e\} \not\supset \{b, c, f\}$.

We would like to point out here that depending on the database instances it is possible that apriori will generate an optimal set of candidate sets and no amount of optimization is possible. Because in that situation, all the candidate sets that were generated would contribute towards other large item sets and hence, were required. This implies that the candidate sets were themselves large item sets by way of anti-monotonicity property of large item sets. This will happen if there are no redundant large item sets. But the issue here is that when there are redundant large item sets, apriori will fail to recognize that. In fact, FP-tree [4] and CHARM [27] gains performance advantage over apriori when there are long chains and low support threshold due to this fact. Apriori must generate the redundant set of large item sets to actually compute the non-redundant ones while the others don't.

This observation is important because it sends the following messages:

- Only the item sets that appear in a source table can be a large item set, or their meets with any other item set in the source table can be a large item set, if ever.
- There is no need to consider any other item set that is not in the source table or can be generated from the source table by computing the least fixpoint of the meets of the source item sets, as the others are invariably redundant, even if they are large item sets.
- The support count for any large item set can be obtained by adding the frequency counts of its ancestors (superset item sets) in the source table with its own frequency count.
- No item set in the item set lattice will ever contribute to the support count of any item set other than the source item sets (transaction nodes/records).

These observations readily suggest the approach we adopted here in developing a relational solution to the mining problem. All we needed was to apply a least fixpoint computation of the source items sets to find their meets. Then we applied the idea of inheritance of frequency counts of ancestors (source item sets) to other source item sets as well as to the newly generated meet item sets. It is evident that the least fixpoint of the meets we need to compute is only for the set of items that are not related by a subset superset relationship in the item set lattice.

6 An SQL3 Mining Operator

We are now ready to discuss our proposal for a mining operator for SQL3. We already know that the (non-redundant) large item sets and the (non-redundant) rules can be computed using SQL3 for which we have discussed a series of examples and expressions. We also know from our previous discussion that the method we have adopted is sound. We further know that the set of rules computed by our method is identical to the set computed by non-redundant apriori, or are equivalent to rules computed by naive apriori. So, it is perfectly all right to abstract the idea into an operator for generic use.

The mine by operator shown below will generate the l_table in figure 1. Basically, its semantics translates to the set of view definitions (or their equivalents) for n_table, f_table, int_table, i_table, c_table and l_table. However, only l_table view is returned to the user as a response of the mining query, and all the other tables remain hidden (used by the system and discarded). Notice that we have supplied two column names to the mine by operator – *Tranid* and *Items*. The *Tranid* column name instructs the system that the nesting should be done on this column and thus the support count comes from the count of *Tranids* for any given set of *Items*. The *Items* column name suggests that the equivalent of the l_table shown in figure 1 should be constructed for the *Items* column. Essentially, this mine by expression will produce the l_table of figure 1 once we set $\delta_m = 0.25$.

select *Items*, **sup**(*Tranid*) as *Support*
from *t_table*
mine by *Tranid* for *Items*
having **sup**(*Tranid*) $\geq \delta_m$

We have also used a having clause for the mine by operator in a way similar to the having clause in SQL group by operator. It uses a function called **sup**. This function, for every tuple in the c_table, generates the ratio of the *Support* to the total number of distinct transactions in the t_table. Consequently, the having option with the condition as shown filters unwanted tuples (large item sets). The select clause allows only a subset of the column names listed in the mine by clause along with any aggregate/mine operations on them. In this case, we are computing support for every item set using the **sup** function just discussed.

For the purpose of generating the association rules, we propose the so called *extract rules using* operator. This operator requires a list of column names, for example *Items*, using which it derives the rules. Basically the expression below produces the r_table of figure 1 for $\eta_m = 0.40$. Notice that we have used a having clause and a mine function called **conf** that computes the confidence of the rule. Recall that the confidence of a rule can be computed from the support values in the l_table – it is the (appropriately taken) ratio of the two supports.

select **ant**(*t.Items*) as *Ant*, **cons**(*t.Items*) as *Conseq*, *t.Support*, **conf**(*t.Support*)
from (select *Items*, **sup**(*Tranid*) as *Support*
    from *t_table*
    mine by *Tranid* for *Items*
    having **sup**(*Tranid*) $\geq \delta_m$) as *t*
extract rules using *t.Items* on *t.Support*
having **conf**(*t.Support*) $\geq \eta_m$

Notice that this query is equivalent to the view definition for r_table in section 4. Consequently here is what this syntax entails. The extract rules using clause forces a Cartesian product of the relation (or the list of relations) named in the from clause. Naturally, the two attribute names mentioned in the extract rules using clause will have two copies in two columns. As explained as part of the r_table view discussion in section 4, from these four attributes all the necessary attributes of r_table can be computed even though we mention only two of the four attributes without any confusion (see r_table view definition). All this clause needs to know is which two attributes it must use from the relation in the from clause, and among them which one has the support values. The rest is trivial.

The mine functions **ant** and **cons** generates the antecedent and consequent of a rule from the source column included as the argument. Recall that rule extraction is done on a source relation by pairing its tuples (Cartesian product) and checking for conditions of a valid rule. It must be mentioned here that the **ant** and **cons** functions can also be used in the having clause. For example if we were interested in finding all the rules for which *ab* is in the consequent, we would then rewrite the above rule as follows:

select **ant**(*t.Items*) as *Ant*, **cons**(*t.Items*) as *Conseq*, *t.Support*, **conf**(*t.Support*)
from (select *Items*, **sup**(*Tranid*) as *Support*
    from *t_table*
    mine by *Tranid* for *Items*
    having **sup**(*Tranid*) $\geq \delta_m$) as *t*
extract rules using *t.Items* on *t.Support*
having **conf**(*t.Support*) $\geq \eta_m$ and **cons**(*t.Items*) $= \{a, b\}$

It is however possible to adopt a single semantics for mine by operator. To this end we propose a variant of the mine by operator to make a syntactic distinction between the the two, called the mine with operator, as follows. In this approach, the mine with operator computes the rules directly and does not produce the intermediate large item set table l_table. In this case, however, we need to change the syntax a bit as shown below. Notice the change is essentially in the argument of **conf** function. Previously we have used the *Support* column of the l_table, but now we use the *Tranid* column instead. The reason for this choice makes sense since support is computed from this column and that support column is still hidden inside the process and is not known yet, which was not the case for the extract rules using operator. In that case we knew which column to use as an argument for **conf** function. But more so, the *Tranid* column was not even available in the l_table as it was not needed.

select **ant**(*Items*) as *Ant*, **cons**(*Items*) as *Conseq*,
  sup(*Tranid*) as *Support*, **conf**(*Support*) as *Confidence*
from *t_table*
mine with *Tranid* for *Items*
having **sup**(*Tranid*) $\geq \delta_m$ and **conf**(*Tranid*) $\geq \eta_m$

Here too, one can think of appropriately using any of the mine functions in the having clause to filter unwanted derivations. We believe, this modular syntax and customizable semantics brings in strength and agility in our system.

We would like to point out here that while both the approaches are appealing, depending on the situations, we prefer the first approach – breaking the process in two steps. The first approach may make it possible to use large item sets for other kind of computations that were not identified yet. Conversely speaking, the single semantics approach makes it difficult to construct the large item sets for any sort of analysis which we believe has applications in other system of rule mining.

7 Optimization Issues

While it was intellectually challenging and satisfying to develop a declarative expression for association rule mining from relational databases using only existing (or standard) object relational machinery, we did not address the issue related to query optimization in this article. We address this issue in a separate article [6] for the want of space and also because it falls outside the scope of the current article. We would like to point out here that several non-trivial optimization opportunities exist for our mining operator and set value based queries we have exploited. Fortunately though, there has been a vast body of research in optimizing relational databases, and hence, the new questions and research challenges that this proposal raises for declarative mining may exploit some of these advances.

There are several open issues with some hopes for resolution. In the worst case, the least fixpoint needs to generate n^2 tuples in the first pass alone when the database size is n - which is quite high. Theoretically, this can happen only when each transaction in the database produces an intersection node, and when they are not related by subset-superset relationship. In the second pass, we need to do n^4 computations, and so on. The question now is, can we avoid generating, and perhaps scanning, some of these combinations as they will not lead to useful intersections? For example, the node c_3^0 in figure 3 is redundant. In other words, can we only generate the nodes within the sandwich and never generate any node that we would not need? A significant difference with apriori like systems is that our system generates all the item sets top down (in the lattice) without taking their candidacy as a large item set into consideration. Apriori, on the other hand, does not generate any node if their subsets are not large item sets themselves, and thereby prunes a large set of nodes. Optimization techniques that exploit this so called "anti-monotonicity" property of item set lattices similar to apriori could make all the difference in our setup. The key issue would be how we push the selection threshold (minimum support) inside the top down computation of the nodes in the lattice in our method. Technically, we should be able to combine view int_table with i_table, c_table and l_table and somehow not generate a virtual node that is outside the sandwich (below the z-envelope in figure 3). This will require pushing selection condition inside aggregate operations where the condition involves the aggregate operation itself.

For the present, and for the sake of this discussion, let us consider a higher support threshold of 45% (3 out of 7 transactions) for the database **T** of figure 1. Now the l-envelope will be moving even closer to the z-envelop and the nodes bf_1^2 and d_2^1 will be outside this sandwich. This raises the question, is it possible to utilize the support and confidence thresholds provided in the query and prune candidates for intersection any further? Ideas similar to magic sets transformation [2,26] and relational magic sets [15] may be borrowed to address this issue. The only problem is that pruning of any node depends on its support count which may come at a later stage. By then all nodes may already have been computed. Specifically, pushing selection conditions inside aggregate operator may become challenging. Special data structures and indexes may perhaps aid in developing faster methods to compute efficient *intersection joins* that we have utilized in this article. We leave these questions as open issues that should be taken up in the future.

Needless to emphasize, a declarative method, preferably a formal one, is desirable because once we understand the functioning of the system, we will then be able to select appropriate procedures depending on the database instances to compute the relational queries involving mining operators which we know is intended once we establish the equivalence of declarative and procedural semantics of the system. Fortunately, we have numerous procedural methods for computing association rules which complement each other in terms of speed and database instances. In fact, that is what declarative systems (or declarativity) buy us – a choice for the most efficient and accurate processing possible.

8 Comparison with Related Research

We would like to end our discussion by highlighting some of the contrasts between our proposal and the proposals in [25,21,13,14]. It is possible to compute the rules using the mine rule operator of [13,14] as follows.

mine rule *simpleassociation* as
select distinct 1..n *Items* as body, 1..n *Items* as head, support, confidence
from *t_table*
group by *Tranid*
extracting rules with support: δ_m, confidence: η_m

As we mentioned before, this expression is fine, but may be considered rigid and does not offer the flexibilities offered by our syntax. But the main difference is in its implementation which we have already highlighted in section 1.1.

It is relatively difficult to compare our work with the set of works in [25,21,20] as the focus there is somewhat different, we believe. Our goal has been to define a relational operator for rule mining and develop the formal basis of the operator. Theirs, we believe, was to develop implementation strategies of apriori in SQL and present some performance metrics. As long as the process is carried out by a system, it is not too difficult to develop a front end operator that can be implemented using their technique. Even then, the only comparison point with our method would be the execution efficiency. In fact, it might become possible to implement our operator using their technique.

9 Conclusions

It was our goal to demonstrate that association rules can be computed using existing SQL3 machineries, which we believe we have done successfully. We have, of course, used some built-in functions for set operations that current SQL systems do not possibly support, but we believe that future enhancements of SQL will. These functions can be easily implemented using SQL's create function statements as we have done. We have utilized SQL's create view recursive clause to generate the intersection nodes which was implemented in PL/SQL.

If one compares the SQL3 expressions presented in this article with the series of SQL expressions presented in any of the works in [25,21,13,14] that involve multiple new operators and update expressions, the simplicity and strength of our least fixpoint based computation will be apparent. Hence, we believe that the idea proposed in this article is novel because to our knowledge, association rule mining using standard SQL/SQL3 is unprecedented. By that, we mean SQL without any extended set of operators such as combination and GatherJoin [25,21], or the *CountAllGroups, MakeClusterPairs, ExtractBodies,* and *ExtractRules* operators in [13,14].

Our mine by operator should not be confused with the set of operators in [25,21] and [13,14]. These operators are essential for their framework to function whereas the mine by operator in our framework is not necessary for our mining queries to be functional. It is merely an abstraction (and a convenience) of the series of views we need to compute for association rule mining. The method proposed is soundly grounded on formal treatment of the concepts, and its correctness may be established easily. We did not attempt to prove the correctness for the sake of conciseness and for want of space, but we hope that readers may have already observed that these are just a matter of details, and are somewhat intuitive too.

The mine by operator proposed here is simple and modular. The flexibilities offered by it can potentially be exploited in real applications in many ways. The operators proposed can be immediately implemented using the existing object relational technology and exploit existing optimization techniques by simply mapping the queries containing the operators to equivalent view definitions as discussed in section 4. These are significant in terms of viability of our proposed framework.

As a future extension of the current research, we are developing an efficient algorithm for top-down procedural computation of the non-redundant large item sets and an improved SQL3 expression for computing such a set. We believe that a new technique for computing item set join (join based on subset condition as shown in the view definition for int_table) based on set indexing would be useful and efficient. In this connection, we are also looking into query optimization issues in our framework.

References

1. Rakesh Agrawal and Ramakrishnan Srikant. Fast algorithms for mining association rules in large databases. In *VLDB*, pages 487–499, 1994.
2. C. Beeri and R. Ramakrishnan. On the power of magic. In *ACM PODS*, pages 269–283, 1987.
3. Sergey Brin, Rajeev Motwani, and Craig Silverstein. Beyond market baskets: Generalizing association rules to correlations. In *Proc. ACM SIGMOD*, pages 265–276, 1997.
4. Jiawei Han, Jian Pei, and Yiwen Yin. Mining frequent patterns without candidate generation. In *Proc. ACM SIGMOD*, pages 1–12, 2000.
5. H. M. Jamil. Mining first-order knowledge bases for association rules. In *Proceedings of the 13th IEEE International Conference on Tools with Artificial Intelligence (ICTAI)*, pages 218–227, Dallas, Texas, 2001. IEEE Press.
6. H. M. Jamil. A new indexing scheme for set-valued keys. Technical report, Department of Computer Science, MSU, USA, June 2001.
7. Hasan M. Jamil. Ad hoc association rule mining as SQL3 queries. In *Proceedings of the IEEE International Conference on Data Mining*, pages 609–612, San Jose, California, 2001. IEEE Press.
8. Hasan M. Jamil. On the equivalence of top-down and bottom-up data mining in relational databases. In *Proc. of the 3rd International Conference on Data Warehousing and Knowledge Discovery (DaWaK 01)*, pages 41–50, Munich, Germany, 2001.
9. Hasan M. Jamil. Bottom-up association rule mining in relational databases. *Journal of Intelligent Information Systems*, 19(2):191–206, 2002.
10. Mika Klemettinen, Heikki Mannila, Pirjo Ronkainen, Hannu Toivonen, and Inkeri Verkamo. Finding interesting rules from large sets of discovered association rules. In *CIKM*, pages 401–407, 1994.
11. Flip Korn, Alexandros Labrinidis, Yannis Kotidis, and Christos Faloutsos. Ratio rules: A new paradigm for fast, quantifiable data mining. In *Proc of 24th VLDB*, pages 582–593, 1998.
12. Brian Lent, Arun N. Swami, and Jennifer Widom. Clustering association rules. In *Proc of the 3th ICDE*, pages 220–231, 1997.
13. Rosa Meo, Giuseppe Psaila, and Stefano Ceri. A new SQL-like operator for mining association rules. In *Proc of 22nd VLDB*, pages 122–133, 1996.

14. Rosa Meo, Giuseppe Psaila, and Stefano Ceri. An extension to SQL for mining association rules. *DMKD*, 2(2):195–224, 1998.
15. Inderpal Singh Mumick and Hamid Pirahesh. Implementation of magic-sets in a relational database system. In *ACM SIGMOD*, pages 103–114, 1994.
16. Amir Netz, Surajit Chaudhuri, Jeff Bernhardt, and Usama M. Fayyad. Integration of data mining with database technology. In *Proceedings of 26th VLDB*, pages 719–722, 2000.
17. Amir Netz, Surajit Chaudhuri, Usama M. Fayyad, and Jeff Bernhardt. Integrating data mining with SQL databases. In *IEEE ICDE*, 2001.
18. Raymond T. Ng, Laks V. S. Lakshmanan, Jiawei Han, and Alex Pang. Exploratory mining and pruning optimizations of constrained association rules. In *Proc. ACM SIGMOD*, pages 13–24, 1998.
19. Jong Soo Park, Ming-Syan Chen, and Philip S. Yu. An effective hash based algorithm for mining association rules. In *Proc. ACM SIGMOD*, pages 175–186, 1995.
20. Karthick Rajamani, Alan Cox, Bala Iyer, and Atul Chadha. Efficient mining for association rules with relational database systems. In *IDEAS*, pages 148–155, 1999.
21. Sunita Sarawagi, Shiby Thomas, and Rakesh Agrawal. Integrating mining with relational database systems: Alternatives and implications. In *Proc. ACM SIGMOD*, pages 343–354, 1998.
22. Ashoka Savasere, Edward Omiecinski, and Shamkant B. Navathe. An efficient algorithm for mining association rules in large databases. In *Proc of 21th VLDB*, pages 432–444, 1995.
23. Pradeep Shenoy, Jayant R. Haritsa, S. Sudarshan, Gaurav Bhalotia, Mayank Bawa, and Devavrat Shah. Turbo-charging vertical mining of large databases. In *ACM SIGMOD*, pages 22–33, 2000.
24. Abraham Silberschatz, Henry F. Korth, and S. Sudarshan. *Database System Concepts*. McGraw-Hill, third edition, 1996.
25. Shiby Thomas and Sunita Sarawagi. Mining generalized association rules and sequential patterns using SQL queries. In *KDD*, pages 344–348, 1998.
26. J. D. Ullman. *Principles of Database and Knowledge-base Systems, Part I & II*. Computer Science Press, 1988.
27. Mohammed J. Zaki. Generating non-redundant association rules. In *Proc. of the 6th ACM SIGKDD Intl. Conf.*, Boston, MA, August 2000.

Towards a Logic Query Language for Data Mining

Fosca Giannotti[1], Giuseppe Manco[2], and Franco Turini[3]

[1] CNUCE-CNR - Via Alfieri 1 - I56010 Ghezzano (PI), Italy
Fosca.Giannotti@cnuce.cnr.it
[2] ICAR-CNR - Via Bucci 41c - I87036 Rende (CS), Italy
manco@icar.cnr.it
[3] Department of Computer Science, Univ. Pisa - C.so Italia 40 - I56125 Pisa, Italy
turini@di.unipi.it

Abstract. We present a logic database language with elementary data mining mechanisms to model the relevant aspects of knowledge discovery, and to provide a support for both the iterative and interactive features of the knowledge discovery process. We adopt the notion of user-defined aggregate to model typical data mining tasks as operations unveiling unseen knowledge. We illustrate the use of aggregates to model specific data mining tasks, such as frequent pattern discovery, classification, data discretization and clustering, and show how the resulting data mining query language allows the modeling of typical steps of the knowledge discovery process, that range from data preparation to knowledge extraction and evaluation.

1 Introduction and Motivations

Research in data mining and knowledge discovery in databases has mostly concentrated on algorithmic issues, assuming a naive model of interaction in which data is first extracted from a database and transformed in a suitable format, and next processed by a specialized inductive engine. Such an approach has the main drawback of proposing a fixed paradigm of interaction. Although it may at first sound appealing to have an autonomous data mining system, it is practically unfeasible to let the data mining algorithm "run loose" into the data in the hope to find some valuable knowledge. Blind search into a database can easily bring to the discovery of an overwhelming large set of patterns, many of which could be irrelevant, difficult to understand, or simply not valid: in one word, uninteresting.

On the other side, current applications of data mining techniques highlight the need for flexible knowledge discovery systems, capable of supporting the user in specifying and refining mining objectives, combining multiple strategies, and defining the quality of the extracted knowledge. A key issue is the definition of *Knowledge Discovery Support Environment* [16], i.e., a query system capable of obtaining, maintaining, representing and using high level knowledge in a unified framework. This comprises representation of domain knowledge, extraction of

R. Meo et al.(Eds.): Database Support for Data Mining Applications,LNAI 2682, pp. 76–94, 2004.

new knowledge and its organization in ontologies. In this respect, a knowledge discovery support environment should be an integrated mining and querying system capable of

- rigorous definition of user interaction during the search process,
- separation of concerns between the specification and the mapping to the underlying databases and data mining tools, and
- understandable representations for the knowledge.

Such an environment is expected to increase the programmer productivity of KDD applications. However, such capabilities require higher-order expressive features capable of providing a tight-coupling between knowledge mining and the exploitation of domain knowledge to share the mining itself.

A suitable approach can be the definition of a set of *data mining primitives*, i.e., a small number of constructs capable of supporting a vast majority of KDD applications. The main idea (outlined in [13]) is to combine relational query languages with data mining primitives in an overall framework capable of specifying data mining problems as complex queries involving KDD objects (rules, clustering, classifiers, or simply tuples). In this way, the mined KDD objects become available for further querying. The principle that query answers can be queried further is typically referred to as *closure*, and is an essential feature of SQL. KDD queries can thus generate new knowledge or retrieve previously generated knowledge. This allows for interactive data mining sessions, where users cross boundaries between mining and querying. Query optimization and execution techniques in such a query system will typically rely on advanced data mining algorithms.

Today, it is still an open question how to realize such features. Recently, the problem of defining a suitable knowledge discovery query formalism has interestend many researchers, both from the database perspective [1,2,13,19,14,11] and the logic programming perspective [21,22,20]. The approaches devised, however, do not explicitly model in a uniform way features such as closure, knowledge extration and representation, background knowledge and interestingness measures. Rather, they are often presented as "ad-hoc" proposals, particularly suitable only for subsets of the described peculiarities.

In such a context, the idea of integrating data mining algorithms in a deductive environment is very powerful, since it allows the direct exploitation of domain knowledge within the specification of the queries, the specification of ad-hoc interest measures that can help in evaluating the extracted knowledge, the modelization of the interactive and iterative features of knowledge discovery in a uniform way. In [5,7] we propose two specific models based on the notion of user-defined aggregate, namely for association rules and bayesian classification. In these approaches we adopt aggregates as an interface to mining tasks in a deductive database, obtaining a powerful amalgamation between inferred and induced knowledge. Moreover, in [4,8] we show that efficiency issues can be takled in an efficient way, by providing a suitable specification of aggregates.

In this paper we generalize the above approaches. We present a logic database language with elementary data mining mechanisms to model the relevant aspects of knowledge discovery, and to provide a support for both the iterative and interactive features of the knowledge discovery process. We shall refer to the notion of *inductive database* introduced in [17,18,1], and provide a logic database system capable of representing inductive theories. The resulting data mining query language shall incorporate all the relevant features that allow the modeling of typical steps of the knowledge discovery process.

The paper is organized as follows. Section 2 provides an introduction to user-defined aggregates. Section 3 introduces the formal model that allows to incorporate both induced and deduced knowledge in a uniform framework. In section 4 we instantiate the model to some relevant mining tasks, namely frequent patterns discovery, classification, clustering and discretization, and show how the resulting framework allows the modeling of the interactive and iterative features of the knowledge discovery process. Finally, section 5 discusses how efficiency issues can be profitably undertaken, in the style of [8].

2 User-Defined Aggregates

In this paper we refer to datalog++ and its current implementation $\mathcal{LDL}$++, a highly expressive language which includes among its features recursion and a powerful form of stratified negation [23], and is viable for efficient query evaluation [6].

A remarkable capability of such a language is that of expressing distributive aggregates, such as `sum` or `count`. For example, the following clause illustrates the use of the `count` aggregate:

$$\mathtt{p(X, count\langle Y\rangle) \leftarrow r(X, Y).}$$

It is worth noting the semantic equivalence of the above clause to the following SQL statement:

```
SELECT X, COUNT(Y)
FROM r
GROUP BY X
```

Specifying User-defined aggregates. Clauses with aggregation are possible mainly because datalog++ supports nondeterminism [9] and XY-stratification [6,23]. This allows the definition of distributive aggregate functions, i.e., aggregate functions defined in an inductive way (S denotes a set and h is a composition operator):

$$\begin{array}{ll} \textit{Base:} & f(\{x\}) := g(x) \\ \textit{Induction:} & f(S \cup \{x\}) := h(f(S), x) \end{array}$$

Users can define aggregates in datalog++ by means of the predicates `single` and `multi`. For example, the `count` aggregate is defined by the unit clauses:

$$\mathtt{single(count, X, 1).}$$
$$\mathtt{multi(count, X, C, C + 1).}$$

The first clause specifies that the count of a set with a single element x is 1. The second clause specifies that the count of $S \cup \{x\}$ is $c+1$ for any set S such that the count of S is c.

Intuitively, a `single` predicate computes the *Base* step, while the `multi` predicate computes the *Inductive* step. The evaluation of clauses containing aggregates consists mainly in *(i)* evaluating the body of the clause, and nondeterministically sorting the tuples resulting from the evaluation, and *(ii)* evaluating the single predicate on the first tuple, and the multi predicate on the other tuples.

The `freturn` and `ereturn` predicates [25] allow building complex aggregate functions from simpler ones. For example, the average of a set S is obtained by dividing the sum of its elements by the number of elements. If S contains c elements whose sum is s, then $S \cup \{x\}$ contains $c+1$ elements whose sum is $s+x$. This leads to the following definition of the `avg` function:

$$\begin{aligned}&\mathtt{single(avg, X, (X, 1)).}\\&\mathtt{multi(avg, X, (S, C), (S + X, C + 1)).}\\&\mathtt{freturn(avg, (S, C), S/C).}\end{aligned}$$

Iterative User-Defined Aggregates. Distributive aggregates are easy to define by means of the user-defined predicates `single` and `multi` in that they simply require a single scan of the available data. In many cases, however, even simple aggregates require multiple scans of the data. As an example, the absolute deviation $S_n = \sum_x |\overline{x} - x|$ of a set of n elements is defined as the sum of the absolute difference of each element with the average value $\overline{x} = 1/n \sum_x x$ of the set. In order to compute such an aggregate, we need to scan the available data twice: first, to compute the average, and second, to compute the sum of the absolute difference.

However, the datalog++ language is powerful enough to cope with multiple scans in the evaluation of user-defined aggregates. Indeed, in [4] we extend the semantics of datalog++ by introducing the `iterate` predicate, which can be exploited to impose some user-defined conditions for iterating the scans over the data. More specifically, the evaluation of a clause

$$\mathtt{p(X, aggr\langle Y\rangle) \leftarrow r(X, Y).} \tag{1}$$

(where `aggr` denotes the name of a user-defined aggregate) can be done according to the following schema:

- Evaluate `r(X, Y)`, and group the resulting values associated with `Y` into subsets $S_1, \ldots, S_n$ according to the different values associated with `X`.
- For each subset $S_i = \{x_1, \ldots, x_n\}$, compute the aggregation value r as follows:
  1. evaluate $\mathtt{single(aggr}, x_1, \mathtt{C)}$. Let c be the resulting value associated with `C`.
  2. Evaluate $\mathtt{multi(aggr}, x_i, c, \mathtt{C)}$, for each i (by updating c to the resulting value associated with `C`).

3. Evaluate $\mathtt{iterate}(\mathtt{aggr}, c, \mathtt{C})$. If the evaluation is successful, then update c to the resulting value associated with C, and return to step 2. Otherwise, evaluate $\mathtt{freturn}(\mathtt{aggr}, c, \mathtt{R})$ and return the resulting value r associated with R.

The following example shows how iterative aggregates can be defined in the datalog++ framework. Further details on the approach can be found in [16,5].

Example 1. By exploiting the **iterate** predicate, the **abserr** aggregate for computing the absolute deviation S_n can be defined as follows:

$$\mathtt{single}(\mathtt{abserr}, \mathtt{X}, (\mathtt{avg}, \mathtt{X}, 1)).$$

$$\mathtt{multi}(\mathtt{abserr}, (\mathtt{nil}, \mathtt{S}, \mathtt{C}), \mathtt{X}, (\mathtt{avg}, \mathtt{S} + \mathtt{X}, \mathtt{C} + 1)).$$

$$\mathtt{multi}(\mathtt{abserr}, (\mathtt{M}, \mathtt{D}), \mathtt{X}, (\mathtt{M}, \mathtt{D} + (\mathtt{M} - \mathtt{X}))) \leftarrow \mathtt{M} > \mathtt{X}.$$
$$\mathtt{multi}(\mathtt{abserr}, (\mathtt{M}, \mathtt{D}), \mathtt{X}, (\mathtt{M}, \mathtt{D} + (\mathtt{X} - \mathtt{M}))) \leftarrow \mathtt{M} \leq \mathtt{X}.$$

$$\mathtt{iterate}(\mathtt{abserr}, (\mathtt{avg}, \mathtt{S}, \mathtt{C}), (\mathtt{S}/\mathtt{C}, 0)).$$

$$\mathtt{freturn}(\mathtt{abserr}, (\mathtt{M}, \mathtt{D}), \mathtt{D}).$$

The first two clauses compute the average of the tuples under examination, in a way similar to the computation of the **avg** aggregate. The remaining clauses assume that the average value has already been computed, and are mainly used in the incremental computation of the sum of the absolute difference with the average. Notice how the combined use of **multi** and **iterate** allows the definition of two scans over the data. □

3 Logic-Based Inductive Databases

A suitable conceptual model that summarizes the relevant aspects discussed in section 1 is the notion of *inductive database* [1,17], that is is a first attempt to formalize the notion of interactive mining process. In the following definition, proposed by Mannila [18,17], the term inductive database refers to a relational database plus the set of all sentences from a specified class of sentences that are true of the data.

Definition 1. *Given an instance* $\mathbf{r}$ *of a relation* $\mathbf{R}$, *a class* $\mathcal{L}$ *of sentences (patterns), and a selection predicate* q, *a pattern discovery task is to find a theory*

$$Th(\mathcal{L}, \mathbf{r}, q) = \{s \in \mathcal{L} | q(\mathbf{r}, s) \text{ is true}\}$$

□

The main idea here is to provide a unified and transparent view of both deductive knowledge, and all the derived patterns, (the induced knowledge) over the data. The user does not care about whether he/she is dealing with inferred or induced knowledge, and whether the requested knowledge is materialized or not. The only detail he/she is interested in is the high-level specification of the query involving both deductive and inductive knowledge, according to some interestingness quality measure (which in turn can be either objective or subjective).

The notion of Inductive Database fits naturally in rule-based languages, such as Deductive Databases [7,5]. A deductive database can easily represent both extensional and intensional data, thus allowing a higher degree of expressiveness than traditional relational algebra. Such capability makes it viable for suitable representation of domain knowledge and support of the various steps of the KDD process.

The main problem in a deductive approach is how to choose a suitable representation formalism for the inductive part, enabling a tight integration with the deductive part. More specifically, the problem is how to formalize the specification of the set $\mathcal{L}$ of patterns in a way such that each pattern $s \in \mathcal{T}h(\mathcal{L}, \mathbf{r}, q)$ is represented as an independent (logical) entity (i.e., a predicate) and each manipulation of $\mathbf{r}$ results in a corresponding change in s. To cope with such a problem, we introduce the notion of *inductive clauses*, i.e., clauses that formalize the dependency between the inductive and the deductive part of an inductive database.

Definition 2. *Given an inductive database theory $\mathcal{T}h(\mathcal{L}, \mathbf{r}, q)$, an inductive clause for the theory is a clause (denoted by s)*

$$H \leftarrow B_1, \ldots B_n$$

such that

- *The evaluation of $B_1, \ldots B_n$ in the computed stable model[1] $M_{s \cup \mathbf{r}}$ correspond to the extension $\mathbf{r}$;*
- *there exist an injective function ϕ mapping each ground instance p of H in $\mathcal{L}$;*
- *$\mathcal{T}h(\mathcal{L}, \mathbf{r}, q)$ corresponds to the model $M_{s \cup \mathbf{r}}$ i.e.,*

$$p \in M_{s \cup \mathbf{r}} \Longleftrightarrow \phi(p) \in \mathcal{T}h(\mathcal{L}, \mathbf{r}, q)$$

□

As a consequence of the above definition, we can formalize the notion of logic-based knowledge discovery support environment, as a deductive database programming language capable of expressing both inductive clauses and deductive clauses.

Definition 3. *A logic-based knowledge discovery support environment is a deductive database language capable of specifying:*

[1] See [24,6] for further details.

- *relational extensions;*
- *intensional predicates, by means of deductive clauses;*
- *inductive predicates, by means of inductive clauses.*

□

The main idea of the previous definition is that of providing a simple way for modeling the key aspects of a data mining query language:

- the source data is represented by the relational extensions;
- intensional predicates provide a way of dealing with background knowledge;
- inductive predicates provide a representation of both the extracted knowledge and the interestingness measures.

In order to formalize the notion of inductive clauses, the first fact that is worth observing is that it is particularly easy to deal with data mining tasks in a deductive framework, if we use aggregates as a basic tool.

Example 2. A frequent itemset $\{I_1, I_2\}$ is a database pattern with a validity specified by the estimation of the posterior probability $\Pr(I_1, I_2|\mathbf{r})$ (i.e., the probability that items I_1 and I_2 appear together according to $\mathbf{r}$). Such a probability can be estimated by means of *iceberg queries*: an iceberg query is a query containing an aggregate, in which a constraint (typically a threshold constraint) over the aggregate is specified. For example, the following query

```
SELECT R1.Item, R2.Item, COUNT(Tid)
FROM r R1, r R2
WHERE R1.Tid = R2.Tid
      AND R1.Item <> R2.Item
GROUP BY R1.Item, R2.Item
HAVING COUNT(Tid) > thresh
```

computes all the pairs of items appearing in a database of transactions with a given frequency. The above query has a straightforward counterpart in datalog++. The following clauses define typical (two-dimensional) association rules by using the `count` aggregate.

$$\begin{array}{ll}\texttt{pair(I1,I2,count}\langle\texttt{T}\rangle\texttt{)} & \leftarrow \texttt{basket(T,I1),basket(T,I2),I1} < \texttt{I2}.\\ \texttt{rules(I1,I2)} & \leftarrow \texttt{pair(I1,I2,C),C} \geq 2.\end{array}$$

The first clause generates and counts all the possible pairs, and the second one selects the pairs with sufficient support (i.e., at least 2). As a result, the predicate `rules` specifies associations, i.e. rules stating that certain combinations of values occur with other combinations of values with a certain frequency. Given the following definitions of the `basket` relation,

```
basket(1,fish).   basket(2,bread).   basket(3,bread).
basket(1,bread).  basket(2,milk).    basket(3,orange).
                  basket(2,onions).  basket(3,milk).
                  basket(2,fish).
```

by querying `rules(X,Y)` we obtain predicates that model the corresponding inductive instances: `rules(bread,milk)` and `rules(bread,fish)`. ◁

Example 3. A classifier is a model that describes a discrete attribute, called the *class*, in terms of other attributes. A classifier is built from a set of objects (the training set) whose class values are known. Practically, starting from a table **r**, we aim at computing the probability $\Pr(C = c|A = a, \mathbf{r})$ for each pair a, c of the attributes A and C [10]. This probability can be roughly estimated by computing some statistics over the data, such as, e.g.:

```
SELECT A, C, COUNT(*)
FROM r
GROUP BY A, C
```

Again, the datalog++ language easily allows the specification of such statistics. Let us consider for example the `playTennis(Out, Temp, Hum, Wind, Play)` table. We would like to predict the probability of playing tennis, given the values of the other attributes. The following clauses specify the computation of the necessary statistics for each attribute of the `playTennis` relation:

$$
\begin{array}{ll}
\mathtt{statistics_{Out}(O, P, count\langle * \rangle)} & \leftarrow \mathtt{playTennis(O, T, H, W, P).} \\
\mathtt{statistics_{Temp}(T, P, count\langle * \rangle)} & \leftarrow \mathtt{playTennis(O, T, H, W, P).} \\
\mathtt{statistics_{Hum}(H, P, count\langle * \rangle)} & \leftarrow \mathtt{playTennis(O, T, H, W, P).} \\
\mathtt{statistics_{Wind}(W, P, count\langle * \rangle)} & \leftarrow \mathtt{playTennis(O, T, H, W, P).} \\
\mathtt{statistics_{Play}(P, count\langle * \rangle)} & \leftarrow \mathtt{playTennis(O, T, H, W, P).}
\end{array}
$$

The results of the evaluation of such clauses can be easily combined in order to obtain the desired classifier. ◁

The above examples show how the simple clauses specifying aggregates can be devised as inductive clauses. Clauses containing aggregates, in languages such as datalog++, satisfy the most desirable property of inductive clauses, i.e., the capability to specify patterns of $\mathcal{L}$ that hold in Th in a "parameterized" way, i.e., according to the tuples of an extension **r**. In this paper, we use aggregates as the means to introduce mining primitives into the query language.

As a matter of fact, an aggregate is a "natural" definition of an inductive database schema, in which patterns correspond to the true facts in the computed stable model, as the following statement shows.

Lemma 1. *An aggregate defines an inductive database.*

Proof. By construction. Let us consider the following clause (denoted by r_p):

$$p(X_1, \ldots, X_n, aggr\langle Y_1, \ldots, Y_m \rangle) \leftarrow r(X_1, \ldots, X_n, Y_1, \ldots, Y_m).$$

We then define

$$\mathcal{L} = \{\langle t_1, \ldots, t_n, s \rangle | p(t_1, \ldots, t_n, s) \text{ is ground}\}$$

and

$$q(\mathbf{r}, \langle t_1, \ldots, t_n, s \rangle) = true \text{ if and only if } p(t_1, \ldots, t_n, s) \in M_{r_p \cup \mathbf{r}}$$

that imposes that the only valid patterns are those belonging to the iterated stable model procedure. □

In such a context, an important issue to investigate is the correspondence between inductive schemas and aggregates, i.e., whether a generic inductive schema can be specified by means of an aggregate. Formally, for given an inductive database $\mathcal{T}h(\mathcal{L}, \mathbf{r}, q)$, we are interested in providing a specification of an aggregate `aggr`, and in defining a clause

$$\mathtt{q}(Z_1, \ldots, Z_k, \mathtt{aggr}\langle X_1, \ldots, X_n\rangle) \leftarrow \mathbf{r}(Y_1, \ldots, Y_m).$$

which should turn out to be a viable inductive clause for $\mathcal{T}h$. The clause should define the format of any valid pattern $s \in \mathcal{L}$. In particular, the correspondence of s with the ground instances of the `q` predicate should be defined by the specification of the aggregate `aggr` (as described in section 2). Moreover, any predicate $\mathtt{q}(\mathtt{t}_1, \ldots, \mathtt{t}_k, \mathtt{s})$ resulting from the evaluation of such a clause and corresponding to s, should model the fact that $s \in \mathcal{T}h(\mathcal{L}, \mathbf{r}, q)$. When such a definition is possible, the "inductive" predicate `q` itself can be used in the definition of more complex queries.

Relating the specification of aggregates with inductive clauses is particularly attractive for two main reasons. First of all, it provides an amalgamation between mining and querying, and hence makes it easy to provide a unique interface capable of specifying source data, knowledge extraction, background knowledge and interestingness specification. Moreover, it allows a good flexibility in the exploitation of mining algorithms for specific tasks. Indeed, we can implement aggregates as simple language interfaces for the algorithms (implemented as separate modules, like in [7]); conversely, we can exploit the notion of iterative aggregate, and explicitly specify the algorithms in datalog++ (like in [8]). The latter approach, in particular, gives two further advantages:

- from a conceptual point of view, it allows the use of background knowledge directly in the exploration of the search space.
- from an efficiency point of view, it provides the opportunity of integrating specific optimizations inside the algorithm.

4 Mining Aggregates

As stated in the previous section, our main aim is the definition of inductive clauses, formally modeled as clauses containing specific user-defined aggregates, and representing specific data mining tasks. The discussion on how to provide efficient implementations of algorithms for such tasks by means of iterative aggregates is given in [8]. The rest of the paper is devoted at showing how the proposed model is suitable to some important knowledge discovery tasks. We shall formalize some inductive clauses by means of aggregates), to formulate data mining tasks in the datalog++ framework. In the resulting framework, the integration of inductive clauses with deductive clauses allows a simple and intuitive formalization of the various steps of the data mining process, in which deductive rules can specify both the preprocessing and the result evaluation phase, while inductive rules can specify the mining phase.

4.1 Frequent Patterns Discovery

Associations are rules that state that certain combinations of values occur with other combinations of values with a certain frequency and certainty. A general definition is the following.

Let $\mathcal{I} = \{a_1, \ldots, a_n\}$ be a set of literals, called *items*. An *itemset* T is a set of items such that $T \subseteq \mathcal{I}$. Given a relation $\mathbf{R} = A_1 \ldots A_n$, a *transaction* in an instance $\mathbf{r}$ of $\mathbf{R}$, associated to attribute A_i (where $dom(A_i) = \mathcal{I}$) according to a *transaction identifier* A_j, is a set of items of the tuples of $\mathbf{r}$ having the same value of A_j.

An association rule is a statement of the form $X \Rightarrow Y$, where $X \subseteq \mathcal{I}$ and $Y \subseteq \mathcal{I}$ are two sets of items. To an association rule we can associate some statistical parameters. The *support* of a rule is the percentage of transactions that contain the set $X \cup Y$, and the *confidence* is the percentage of transactions that contain Y, provided that they contain X.

The problem of association rules mining can be finally stated as follows: given an instance $\mathbf{r}$ of $\mathbf{R}$, find all the association rules from the set of transactions associated to A_i (grouped according to A_j), such that for each rule $A \Rightarrow B[S, C]$, $S \geq \sigma$ and $C \geq \gamma$, where σ is the *support threshold* and γ is the *confidence threshold*.

The following definition provides a formulation of the association rules mining task in terms of inductive databases.

Definition 4. *Let* $\mathbf{r}$ *be an instance of the table* $\mathbf{R} = A_1 \ldots A_n$*, and* $\sigma, \gamma \in [0, 1]$*. For given* $i, j \leq n$*, let*

- $\mathcal{L} = \{A \Rightarrow B | A, B \subseteq dom(\mathbf{R}[A_i])\}$*, and*
- $q(\mathbf{r}, A \Rightarrow B) = true$ *if and only if* $freq(A \cup B, \mathbf{r}) \geq \sigma$ *and* $freq(A \cup B, \mathbf{r})/freq(A, \mathbf{r}) \geq \gamma$*.*

Where $freq(s, \mathbf{r})$ *is the (relative) frequency of* s *in the set of the transactions in* $\mathbf{r}$ *grouped by* A_j*. The theory* $Th(\mathcal{L}, \mathbf{r}, q)$ *defines the* frequent patterns discovery *task.* □

The above definition provides an inductive schema for the frequent pattern discovery task. We now specify a corresponding inductive clause.

Definition 5. *Given a relation* `r`*, the* `patterns` *aggregate is defined by the rule*

$$\mathtt{p}(\mathtt{X}_1, \ldots, \mathtt{X}_n, \mathtt{patterns}\langle(\mathtt{m\_s}, \mathtt{m\_c}, \mathtt{Y})\rangle) \leftarrow \mathtt{r}(\mathtt{Z}_1, \ldots, \mathtt{Z}_m) \tag{2}$$

where the variables $\mathtt{X}_1, \ldots, \mathtt{X}_n, \mathtt{Y}$ are a rearranged subset of the variables $\mathtt{Z}_1, \ldots, \mathtt{Z}_k$ of `r`, and the `Y` variable denotes a set of elements. The aggregate `patterns` computes the set of predicates $\mathtt{p}(\mathtt{t}_1, \ldots, \mathtt{t}_n, \mathtt{l}, \mathtt{r}, \mathtt{f}, \mathtt{c})$ where:

1. $\mathtt{t}_1, \ldots, \mathtt{t}_n$ are distinct instances of the variables $\mathtt{X}_1, \ldots, \mathtt{X}_n$, as resulting from the evaluation of `r`;
2. $\mathtt{l} = \{\mathtt{l}_1, \ldots, \mathtt{l}_k\}$ and $\mathtt{r} = \{\mathtt{r}_1, \ldots, \mathtt{r}_h\}$ are subsets of the value of `Y` in a tuple resulting from the evaluation of `r`;

3. f and c are respectively the support and the confidence of the rule $\mathtt{l} \Rightarrow \mathtt{r}$, such that $f \geq m\_s$ and $c \geq m\_c$.

□

Example 4. Let us consider a sample `transaction` table. The following rule specifies the computation of association rules (with 30% support threshold and 100% confidence threshold) starting from the extension of the table:

$$\begin{array}{ll} \mathtt{transaction(D, C, \langle I \rangle)} & \leftarrow \mathtt{transaction(D, C, I, Q, P)}. \\ \mathtt{rules(patterns\langle(0.3, 1.0, S)\rangle)} & \leftarrow \mathtt{transaction(D, C, S)}. \end{array} \quad (3)$$

The first clause collects all the transactions associated to attribute `I` and grouped by attributes `D` and `C`. The second clause extracts the relevant patterns from the collection of available transactions. The result of the evaluation of the predicate `rules(L, R, S, C)` against such a program yields, e.g., the answer predicate `rules(diapers, beer, 0.5, 1.0)`. ◁

It is easy to see how inductive clauses exploiting the `patterns` aggregate allow the specification of frequent pattern discovery tasks. The evidence of such a correspondence can be obtained by suitably specifying the `patterns` aggregate [8,16].

4.2 (Bayesian) Classification

It is particularly simple to specify Naive Bayes classification by means of an inductive database schema. Let us consider a relation $\mathbf{R}$ with attributes $A_1, \ldots, A_n$ and C. For simplicity, we shall assume that all the attributes represent discrete values. This is not a major problem, since

- it is well-known that classification algorithms perform better with discrete-valued attributes;
- supervised discretization [15] combined with discrete classification allows a more effective approach. As we shall see, the framework based on inductive clauses easily allows the specification of discretization tasks as well.

The bayesian classification task can be summarized as follows. Given an instance $\mathbf{r}$ of $\mathbf{R}$, we aim at computing the function

$$\max_c \ \Pr(C = c | A_1 = a_1, \ldots, A_n = a_n, \mathbf{r}) \quad (4)$$

where $c \in dom(C)$ and $a_i \in dom(A_i)$. By repeated application of Bayes' rule and the assumption that $A_1, \ldots, A_n$ are independent, we obtain

$$\Pr(C = c | A_1 = a_1, \ldots, A_n = a_n, \mathbf{r}) = \Pr(C = c | \mathbf{r}) \times \prod_i \Pr(A_i = a_i | C = c, \mathbf{r})$$

Now, each factor in the above product can be estimated from $\mathbf{r}$ by means of the following equation

$$\tilde{\Pr}(A_j = a_j | C = c, \mathbf{r}) = \frac{freq(c, \sigma_{A_j = a_j}(\mathbf{r}))}{freq(c, \mathbf{r})}$$

Hence, the definition of a classification task can be accomplished by computing some suitable statistics. That is, a suitable inductive theory associates to each possible pair the corresponding statistic.

Definition 6. *Let* $\mathbf{R} = A_1 \ldots A_n C$ *be a relation schema. Given an instance* $\mathbf{r}$ *of* $\mathbf{R}$*, we define*

- $\mathcal{L} = \{\langle A_i = a_i \wedge C = c, n_A, n_C\rangle | a_i \in dom(A_i), c \in dom(C)$ *and* $n_A, n_C \in \mathrm{I\!R}\}$.
- $q(\mathbf{r}, \langle A_i = a_i \wedge C = c, n_A, n_C\rangle) = true$ *if and only if* $n_A = \Pr(A_i = a_i | C = c, \mathbf{r})$ *and* $n_C = \Pr(C = c|\mathbf{r})$.

The resulting theory $\mathcal{T}h(\mathcal{L}, \mathbf{r}, q)$ *formalizes a naive bayesian classification task.*

□

Notice that the datalog++ language easily allows the computation of all the needed statistics, by enumerating all the pairs for which we need to count the occurrences (see, e.g., example 3). However, we can associate the inductive theory with a more powerful user-defined aggregate, in which all the needed statistics can be efficiently computed without resorting to multiple clauses.

Definition 7. *Given a relation* `r`*, the* `nbayes` *aggregate is defined by a rule schema*

$$\mathtt{s}(\mathtt{X_1}, \ldots, \mathtt{X_m}, \mathtt{nbayes}\langle(\{(1, \mathtt{A_1}), \ldots, (\mathtt{n}, \mathtt{A_n})\}, \mathtt{C})\rangle) \leftarrow \mathtt{r}(\mathtt{Z_1}, \ldots, \mathtt{Z_k}).$$

where

- *The variables* $\mathtt{X_1}, \ldots, \mathtt{X_m}, \mathtt{A_1}, \ldots, \mathtt{A_n}, \mathtt{C}$ *are a (possibly rearranged) subset of the values of* $\mathtt{Z_1}, \ldots, \mathtt{Z_k}$ *resulting from the evaluation of* `r`*;*
- *The result of such an evaluation is a predicate* $\mathtt{s}(\mathtt{t_1}, \ldots, \mathtt{t_n}, \mathtt{c}, (\mathtt{i}, \mathtt{a_i}), \mathtt{v_i}, \mathtt{v_c})$*, representing the set of counts of all the possible values* $\mathtt{a_i}$ *of the i-th attribute* $\mathtt{A_i}$*, given any possible value* `c` *of* `C`*. In particular,* $\mathtt{v_i}$ *represents the frequency of the pair* $\mathtt{a_i}, \mathtt{c}$*, and* $\mathtt{v_c}$ *represents the frequency of* `c` *in the extension* `r`*.*

□

Example 5. Let us consider the toy `playTennis` table defined in example 3. The frequencies of the attributes can be obtained by means of the clause

$$\mathtt{classifier}(\mathtt{nbayes}\langle(\{(1, \mathtt{O}), (2, \mathtt{T}), (3, \mathtt{H}), (4, \mathtt{W})\}, \mathtt{P})\rangle) \leftarrow \mathtt{playTennis}(\mathtt{O}, \mathtt{T}, \mathtt{H}, \mathtt{W}, \mathtt{P}).$$

The evaluation of the query $\mathtt{classifier}(\mathtt{C}, \mathtt{F}, \mathtt{C_F}, \mathtt{C_C})$ returns, e.g., the answer predicate `classifier(yes, (1, sunny), 0.6, 0.4)`. ◁

Again, by suitably specifying the aggregate, we obtain a correspondence between the inductive database schema and its deductive counterpart [16,5].

4.3 Clustering

Clustering is perhaps the most straightforward example of data mining task providing a suitable representation of its results in terms of relational tables. In a relation **R** with instance **r**, the main objective of clustering is that of labeling each tuple $\mu \in \mathbf{r}$. In relational terms, this correspond in adding a set of attributes $A_1, \ldots, A_n$ to $\mathbf{R}$, so that a tuple $\langle a_1, \ldots a_n \rangle$ associated to a tuple $\mu \in \mathbf{r}$ represents a cluster assigment for μ. For example, we can enhance $\mathbf{R}$ with two attributes C and M, where C denotes the cluster identifier and M denotes a probability measure. A tuple $\mu \in \mathbf{r}$ is represented in the enhancement of $\mathbf{R}$ by a new tuple μ', where $\mu'[A] = \mu[A]$ for each $A \in \mathbf{R}$, and $\mu'[C]$ represents the cluster to which μ belongs, with probability $\mu'[M]$. In the following defintion, we provide a sample inductive schema formalizing the clustering task.

Definition 8. *Given a relation* $\mathbf{R} = A_1 \ldots A_n$ *with extension* $\mathbf{r}$, *such that tuples in* $\mathbf{r}$ *can be organized in* k *clusters, an inductive database modeling clustering is defined by* $\mathcal{Th}(\mathcal{L}, \mathbf{r}, q)$, *where*

- $\mathcal{L} = \{\langle \mu, i \rangle | \mu \in dom(A_1) \times \ldots \times dom(A_n), i \in \mathbb{N}\}$, *and*
- $q(\mathbf{r}, \langle \mu, i \rangle)$ *is true if and only if* $\mu \in \mathbf{r}$ *is assigned to the i-th cluster.*

□

It is particularly intuitive to specify the clustering data mining task as an aggregate.

Definition 9. *Given a relation* r, *the* cluster *aggregate is defined by the rule schema*

$$\mathtt{p}(\mathtt{X}_1, \ldots, \mathtt{X}_n, \mathtt{clusters}\langle(\mathtt{Y}_1, \ldots, \mathtt{Y}_k)\rangle) \leftarrow \mathtt{r}(\mathtt{Z}_1, \ldots, \mathtt{Z}_m).$$

where the variables $\mathtt{X}_1, \ldots, \mathtt{X}_n, \mathtt{Y}_1, \ldots, \mathtt{Y}_k$ *are a rearranged subset of the variables* $\mathtt{Z}_1, \ldots, \mathtt{Z}_m$ *of* r. clusters *computes the set of predicates* $\mathtt{p}(\mathtt{t}_1, \ldots, \mathtt{t}_n, \mathtt{s}_1, \ldots, \mathtt{s}_k, \mathtt{c})$, *where:*

1. $\mathtt{t}_1, \ldots, \mathtt{t}_n, \mathtt{s}_1, \ldots, \mathtt{s}_k$ *are distinct instances of the variables* $\mathtt{X}_1, \ldots, \mathtt{X}_n, \mathtt{Y}_1, \ldots \mathtt{Y}_k$, *as resulting from the evaluation of* r;
2. c *is a label representing the cluster to which the tuple* $\mathtt{s}_1, \ldots, \mathtt{s}_k$ *is assigned, according to some clustering algorithm.*

□

Example 6. Consider a relation customer(name, address, age, income), storing information about customers, as shown in fig. 1 a). We can define a "clustered" view of such a database by means of the following rule:

$$\mathtt{custCView}(\mathtt{clusters}\langle(\mathtt{N}, \mathtt{AD}, \mathtt{AG}, \mathtt{I})\rangle) \leftarrow \mathtt{customer}(\mathtt{N}, \mathtt{AD}, \mathtt{AG}, \mathtt{I}).$$

The evaluation of such rule, shown in fig. 1 b), produces two clusters. ◁

It is particularly significant to see how the proposed approach allows to directly model closure (i.e., to manipulate the mining results).

Example 7. We can easily exploit the patterns aggregate to find an explanation of such clusters:

$$\mathtt{frqPat}(\mathtt{C}, \mathtt{patterns}\langle(0.6, 1.0, \{\mathtt{f\_a}(\mathtt{AD}), \mathtt{s\_a}(\mathtt{I})\})\rangle) \leftarrow \mathtt{custCView}(\mathtt{N}, \mathtt{AD}, \mathtt{AG}, \mathtt{I}, \mathtt{C}).$$

name	address	age	income
cust1	pisa	50	50K
cust2	rome	30	30K
cust3	pisa	45	48K
cust4	florence	24	30K
cust5	pisa	60	50K
cust6	rome	26	30K

(a)

name	address	age	income	cluster
cust1	pisa	50	50K	1
cust2	rome	30	30K	2
cust3	pisa	45	48K	1
cust4	florence	24	30K	2
cust5	pisa	60	50K	1
cust6	rome	26	30K	2

(b)

cluster	left	right	support	conf
1	{s_a(50K)	f_a(pisa)}	2	1.0
2	{f_(rome)	s_a(30K)}	2	1.0

(c)

Fig. 1. a) Sample `customer` table. Each row represents some relevant features of a customer. b) Cluster assignments: a cluster label is associated to each row in the originary `customer` table. c) Cluster explanations with frequent patterns. The first cluster contains customers with high income and living in `pisa`, while the second cluster contains customers with low income and living in `rome`

Notice how the cluster label `C` is used for separating transactions belonging to different clusters, and mining associations in such clusters separately. Table 1 c) shows the patterns resulting from the evaluation of such a rule. As we can see, cluster 1 is mainly composed by high-income people living in Pisa, while cluster 2 is mainly composed by low-income people living in Rome. ◁

4.4 Data Discretization

Data discretization can be used to reduce the number of values for a given continuous attribute, by dividing the range of the attribute into intervals. Interval labels can be used to replace actual data values.

Given an instance $\mathbf{r}$ of a relation $\mathbf{R}$ with a numeric attribute A, the main idea is to provide a mapping among the values of $dom(A)$ and some given labels. More precisely, we define $\mathcal{L}$ as the pairs $\langle a, i \rangle$, where $a \in dom(A)$ and $i \in \mathcal{V}$ represents an interval label (i.e., $\mathcal{V}$ is a representation of all the intervals $[l, s]$ such that $l, s \in dom(A)$). A discretization task of the tuples of $\mathbf{r}$, formalized as a theory $\mathcal{T}h(\mathcal{L}, \mathbf{r}, q)$, can be defined according to some discretization objective, i.e., a way of relating a value $a \in dom(A)$ to an interval label i.

In such a context, discretization techniques can be distinguished into [12,3] *supervised* or *unsupervised.* The objective of supervised methods is to discretize continuous values in homogeneous intervals, i.e., intervals that preserve a predefined property (which, in practice, is represented by a label associated to each continuous value). The formalization of supervised discretization as an inductive theory can be tuned as follows.

Definition 10. *Let* $\mathbf{r}$ *be an instance of a relation* $\mathbf{R}$. *Given an attribute* $A \in \mathbf{R}$, *and a discrete-valued attribute* $C \in \mathbf{R}$, *an inductive database theory* $\mathcal{T}h(\mathcal{L}, \mathbf{r}, q)$ *defines a supervised discretization task if*

- *either* $dom(A) \in \mathbb{N}$ *or* $dom(A) \in \mathbb{R}$;
- $\mathcal{L} = \{\langle a, C, i\rangle | a \in dom(A), i \in \mathbb{N}\}$;
- $q(\mathbf{r}, \langle a, C, i\rangle) = true$ *if and only if there is a discretization of* $\pi_A(\mathbf{r})$ *in homogeneous intervals with respect to the attribute* C*, such that* a *belongs to the* i*-th interval.*

□

The following definition provides a corresponding inductive clause.

Definition 11. *Given a relation* `r`*, the aggregate* `discr` *defines the rule schema*

$$\mathtt{s}(\mathtt{Y}_1, \ldots, \mathtt{Y}_\mathtt{k}, \mathtt{discr}\langle(\mathtt{Y}, \mathtt{C})\rangle) \leftarrow \mathtt{r}(\mathtt{X}_1, \mathtt{X}_2, \ldots, \mathtt{X}_\mathtt{n}).$$

where

- `Y` *is a continuous-valued variable, and* `C` *is a discrete-valued variable;*
- $\mathtt{Y}_1, \ldots, \mathtt{Y}_\mathtt{k}, \mathtt{Y}, \mathtt{C}$ *are a rearranged subset of the variables* $\mathtt{X}_1, \mathtt{X}_2, \ldots, \mathtt{X}_\mathtt{n}$ *in* `r`.

The result of the evaluation of such rule is given by predicates $\mathtt{p}(\mathtt{t}_1, \ldots, \mathtt{t}_\mathtt{k}, \mathtt{v}, \mathtt{i})$*, where* $\mathtt{t}_1, \ldots, \mathtt{t}_\mathtt{k}, \mathtt{v}$ *are distinct instances of the variables* $\mathtt{Y}_1, \ldots, \mathtt{Y}_\mathtt{k}, \mathtt{Y}$*, and* `i` *is an integer value representing the i-th interval in a supervised discretization of the values of* `Y` *labelled with the values of* `C`*.* □

In the ChiMerge approach to supervised discretization [15], a bottom-up interval generation procedure is adopted: initially, each single value is considered an interval. At each iteraction, two adjacent intervals are chosen and joined, provided that they are sufficiently homogeneous. The degree of homogeneity is measured w.r.t. a class label, and is computed by means a χ^2 statistics. Homogeneity is made parametric to a user-defined significance level α, identifying the probability that two adjacent intervals have independent label distributions. In terms of inductive clauses, this can be formalized as follows:

$$\mathtt{s}(\mathtt{Y}_1, \ldots, \mathtt{Y}_\mathtt{k}, \mathtt{discr}\langle(\mathtt{Y}, \mathtt{C}, \alpha)\rangle) \leftarrow \mathtt{r}(\mathtt{X}_1, \mathtt{X}_2, \ldots, \mathtt{X}_\mathtt{n}).$$

Here, `Y` represents the attribute to discretize, `C` represents the class label and α is the significance level.

Example 8. The following rule

$$\mathtt{intervals}(\mathtt{discr}\langle(\mathtt{Price}, \mathtt{Beer}, 0.9)\rangle) \leftarrow \mathtt{serves}(\_, \mathtt{Beer}, \mathtt{Price}).$$

defines a 0.9 significance level supervised discretization of the `price` attribute, according to the values of `beer`, of the relation `serves` shown in fig. 2. More precisely, we aim at obtaining a discretization of `price` into intervals preserving the values of the `beer` attribute (as they appear associate to `price` in the tuples of the relation `serves`. For example, the values 100 and 117 can be merged into the interval [100, 117], since the tuples of `serves` in which such values occur contain the same value of the `beer` attribute (the `Bud` value). The results of the evaluation of the inductive clause are shown in fig. 2 c). ◁

bar	beer	price
A	Bud	100
A	Becks	120
C	Bud	117
D	Bud	130
D	Bud	150
E	Becks	140
E	Becks	122
F	Bud	121
G	Bud	133
H	Becks	125
H	Bud	160
I	Bud	135

(a)

value	interval
100	1
120	1
117	1
130	3
150	4
140	4
122	2
121	2
133	3
125	2
160	4
135	3

(b)

Fig. 2. a) The `serves` relation. Each row in the relation represents a brand of beer served by a given bar, with the associated price. b) ChiMerge Discretization: each value of the `price` attribute in `serves` is associated with an interval. Intervals contain values which occur in tuples of `serves` presenting similar values of the `beer` attribute

It is particularly interesting to see how the discretization and classification tasks can be combined, by exploiting the respective logical formalizations.

Example 9. A typical dataset used as a benchmark for classification tasks is the *Iris* classification dataset [15]. The dataset can be represented by means of a relation containing 5 attributes:

$$\texttt{iris(Sepal\_length, Sepal\_width, Petal\_length, Petal\_width, Specie)}$$

Each tuple in the dataset describes the relevant features of an iris flower. The first four attributes are continuous-valued attributes, and the `Specie` attribute is a nominal attribute corresponding to the class to which the flower belongs (either *iris-setosa*, *iris-versicolor*, or *iris-virginica*). We would like to characterize species in the `iris` relation according to their features. To this purpose, we may need a preprocessing phase in which continuous attributes are discretized:

$$\begin{array}{l}\texttt{intervals}_{\texttt{SL}}(\texttt{discr}\langle(\texttt{SL}, \texttt{C}, 0.9)\rangle) \leftarrow \texttt{iris}(\texttt{SL}, \texttt{SW}, \texttt{PL}, \texttt{PW}, \texttt{C}).\\ \texttt{intervals}_{\texttt{SW}}(\texttt{discr}\langle(\texttt{SW}, \texttt{C}, 0.9)\rangle) \leftarrow \texttt{iris}(\texttt{SL}, \texttt{SW}, \texttt{PL}, \texttt{PW}, \texttt{C}).\\ \texttt{intervals}_{\texttt{PL}}(\texttt{discr}\langle(\texttt{PL}, \texttt{C}, 0.9)\rangle) \leftarrow \texttt{iris}(\texttt{SL}, \texttt{SW}, \texttt{PL}, \texttt{PW}, \texttt{C}).\\ \texttt{intervals}_{\texttt{PW}}(\texttt{discr}\langle(\texttt{PW}, \texttt{C}, 0.9)\rangle) \leftarrow \texttt{iris}(\texttt{SL}, \texttt{SW}, \texttt{PL}, \texttt{PW}, \texttt{C}).\end{array}$$

The predicates defined by the above clauses provide a mapping of the continuous values to the intervals shown in fig. 3. A classification task can be defined by exploiting such predicates:

$$\begin{array}{rl}\texttt{irisCl}(\texttt{nbayes}\langle(\{\texttt{SL}, \texttt{SW}, \texttt{PL}, \texttt{PW}\}, \texttt{C})\rangle) \leftarrow & \texttt{iris}(\texttt{SL1}, \texttt{SW1}, \texttt{PL1}, \texttt{PW1}, \texttt{C}),\\ & \texttt{intervals}_{\texttt{SW}}(\texttt{SW1}, \texttt{SW}),\\ & \texttt{intervals}_{\texttt{SL}}(\texttt{SL1}, \texttt{SL}),\\ & \texttt{intervals}_{\texttt{PW}}(\texttt{PW1}, \texttt{PW}),\\ & \texttt{intervals}_{\texttt{PL}}(\texttt{PL1}, \texttt{PL}).\end{array}$$

interval	setosa	virginica	versicolor
[4.3,4.9)	16	0	0
[4.9,5)	4	1	1
[5,5.5)	25	5	0
[5.5,5.8)	4	15	2
[5.8,6.3)	1	15	10
[6.3,7.1)	0	14	25
[7.1,7.9]	0	0	12

sepal_length

interval	setosa	virginica	versicolor
[2,2.5)	1	9	1
[2.5 ,2.9)	0	18	18
[2.9,3)	1	7	2
[3,3.4)	18	15	24
[3.4,4.4]	30	1	5

sepal_width

interval	setosa	virginica	versicolor
[1,3)	50	0	0
[3,4.8)	0	44	1
[4.8,5.2)	0	6	15
[5.2,6.9]	0	0	34

petal_length

interval	setosa	virginica	versicolor
[0.1,1)	50	0	0
[1,1.4)	0	28	0
[1.4,1.8)	0	21	5
[1.8,2.5]	0	1	45

petal_width

Fig. 3. Bayesian statistics using ChiMerge and the `nbayes` aggregate. For each attribute of the `iris` relation, a set of intervals is obtained, and the distribution of the classes within such intervals is computed. For example, interval $[4.3, 4.9)$ of the `sepal_length` attribute contains 16 tuples labelled as `setosa`, while interval $[7.1, 7.9)$ of the same attribute contains 12 tuples labelled as `versicolor`

The statistics resulting from the evaluation of the predicate defined by the above rule are shown in fig. 3. ◁

5 Conclusions

The main purpose of flexible knowledge discovery systems is to obtain, maintain, represent, and utilize high-level knowledge. This includes representation and organization of domain and extracted knowledge, its creation through specialized algorithms, and its utilization for context recognition, disambiguation, and needs identification. Current knowledge discovery systems provide a fixed paradigm that does not sufficiently supports such features in a coherent formalism. On the contrary, logic-based databases languages provide a flexible model of interaction that actually supports most of the above features in a powerful, simple and versatile formalism. This motivated the study of a logic-based framework for intelligent data analysis.

The main contribution of this paper was the development of a logic database language with elementary data mining mechanisms to model extraction, representation and utilization of both induced and deduced knowledge. In particular, we have shown that aggregates provide a standard interface for the specification of data mining tasks in the deductive environment: i.e., they allow to model mining tasks as operations unveiling pre-existing knowledge. We used such main features to model a set of data mining primitives: frequent pattern discovery, Bayesian classification, clustering and discretization.

The main drawback of a deductive approach to data mining query languages concerns efficiency: a data mining algorithm can be worth substantial optimizations that come both from a smart constraining of the search space, and from the exploitation of efficient data structures. In this case, the adoption of the datalog++ logic database language has the advantage of allowing a direct specification of mining algorithms, thus allowing specific optimizations. Practically, we can directly specify data mining algorithms by means of iterative user-defined aggregates, and implement the most computationally intensive operations by means of *hot-spot* refinements [8]. Such a feature allows to modularize data mining algorithms and integrate domain knowledge in the right points, thus allowing crucial domain-oriented optimizations.

References

1. J-F. Boulicaut, M. Klemettinen, and H. Mannila. Querying Inductive Databases: A Case Study on the MINE RULE Operator. In *Procs. 2nd European Conf. on Principles and Practice of Knowledge Discovery in Databases (PKDD98)*, LNCS 1510, pages 194–202, 1998.
2. M.S. Chen, J. Han, and P.S. Yu. Data Mining: An Overview from a Database Perspective. *IEEE Trans. on Knowledge and Data Engineering*, 8(6):866–883, 1996.
3. J. Dougherty, R. Kohavi, and M. Sahami. Supervised and Unsupervised Discretization of Continuous Features. In *Procs. 12th International Conference on Machine Learning*, pages 194–202, 1995.
4. F. Giannotti and G. Manco. Declarative knowledge extraction with iterative user-defined aggregates. *AI*IA Notizie*, 13(4), December 2000.
5. F. Giannotti and G. Manco. Making Knowledge Extraction and Reasoning Closer. In *Procs. 4th Pacific-Asia Conference on Knowledge Discovery and Data Mining (PAKDD 2000)*, LNAI 1805, pages 360–371, 2000.
6. F. Giannotti, G. Manco, M. Nanni, and D. Pedreschi. Nondeterministic, Nonmonotonic Logic Databases. *IEEE Trans. on Knowledge and Data Engineering*, 13(5):813–823, 2001.
7. F. Giannotti, G. Manco, D. Pedreschi, and F. Turini. Experiences with a Logic-Based Knowledge Discovery Support Environment. In *Selected Papers of the Sixth congress of the Italian Congress of Artificial Intellingence*, LNCS 1792, 2000.
8. F. Giannotti, G. Manco, and F. Turini. Specifying Mining Algorithms with Iterative User-Defined Aggregates: A Case Study. In *Procs. 5th European Conference on Principles and Practice of Knowledge Discovery in Databases (PKDD 2001)*, LNAI 2168, pages 128–139, 2001.
9. F. Giannotti, D. Pedreschi, and C. Zaniolo. Semantics and Expressive Power of Non Deterministic Constructs for Deductive Databases. *Journal of Computer and Systems Sciences*, 62(1):15–42, 2001.
10. G. Graefe, U. Fayyad, and S. Chaudhuri. On the Efficient Gathering of Sufficient Statistics for Classification from Large SQL Databases. In *Proc. 4th Int. Conf. on Knowledge Discovery and Data Mining (KDD98)*, pages 204–208, 1998.
11. J. Han, Y. Fu, K. Koperski, W. Wang, and O. Zaiane. DMQL: A Data Mining Query Language for Relational Databases. In *SIGMOD'96 Workshop on Research Issues on Data Mining and Knowledge Discovery (DMKD'96)*, 1996.
12. F. Hussain, H. Liu, C. Tan, and M. Dash. Discretization: An Enabling Technique. *Journal of Knowledge Discovery and Data Mining*, 6(4):393–423, 2002.

13. T. Imielinski and H. Mannila. A Database Perspective on Knowledge Discovery. *Communications of the ACM*, 39(11):58–64, 1996.
14. T. Imielinski and A. Virmani. MSQL: A Query Language for Database Mining. *Journal of Knowledge Discovery and Data Mining*, 3(4):373–408, 1999.
15. R. Kerber. ChiMerge: Discretization of Numeric Attributes. In *Proc. 10th National Conference on Artificial Intelligence (AAAI92)*, pages 123–127. The MIT Press, 1992.
16. G. Manco. *Foundations of a Logic-Based Framework for Intelligent Data Analysis.* PhD thesis, Department of Computer Science, University of Pisa, April 2001.
17. H. Mannila. Inductive databases and condensed representations for data mining. In *International Logic Programming Symposium*, pages 21–30, 1997.
18. H. Mannila and H. Toivonen. Levelwise Search and Border of Theories in Knowledge Discovery. *Journal of Knowledge Discovery and Data Mining*, 3:241–258, 1997.
19. R. Meo, G. Psaila, and S. Ceri. A Tightly-Coupled Architecture for Data Mining. In *International Conference on Data Engineering (ICDE98)*, pages 316–323, 1998.
20. L. De Raedt. Data mining as constraint logic programming. In *Procs. Int. Conf. on Inductive Logic Programming*, 2000.
21. W. Shen, K. Ong, B. Mitbander, and C. Zaniolo. Metaqueries for Data Mining. In *Advances in Knowledge Discovery and Data Mining*, pages 375–398. AAAI Press/The MIT Press, 1996.
22. D. Tsur et al. Query Flocks: A Generalization of Association-Rule Mining. In *Proc. ACM Conf. on Management of Data (Sigmod98)*, pages 1–12, 1998.
23. C. Zaniolo, N. Arni, and K. Ong. Negation and Aggregates in Recursive Rules: The $\mathcal{LDL}$++ Approach. In *Proc. 3rd Int. Conf. on Deductive and Object-Oriented Databases (DOOD93)*, LNCS 760, 1993.
24. C. Zaniolo, S. Ceri, C. Faloutsos, R.T Snodgrass, V.S. Subrahmanian, and R. Zicari. *Advanced Database Systems.* Morgan Kaufman, 1997.
25. C. Zaniolo and H. Wang. Logic-Based User-Defined Aggregates for the Next Generation of Database Systems. In *The Logic Programming Paradigm: Current Trends and Future Directions*, Springer Verlag, 1998.

A Data Mining Query Language for Knowledge Discovery in a Geographical Information System

Donato Malerba, Annalisa Appice, and Michelangelo Ceci

Dipartimento di Informatica, Università degli Studi di Bari
via Orabona 4, 70125 Bari, Italy
{malerba, appice, ceci}@di.uniba.it

Abstract. Spatial data mining is a process used to discover interesting but not explicitly available, highly usable patterns embedded in both spatial and non-spatial data, which are possibly stored in a spatial database. An important application of spatial data mining methods is the extraction of knowledge from a Geographic Information System (GIS). INGENS (INductive GEographic iNformation System) is a prototype GIS which integrates data mining tools to assist users in their task of topographic map interpretation. The spatial data mining process is aimed at a user who controls the parameters of the process by means of a query written in a mining query language. In this paper, we present SDMOQL (Spatial Data Mining Object Query Language), a spatial data mining query language used in INGENS, whose design is based on the standard OQL (Object Query Language). Currently, SDMOQL supports two data mining tasks: inducing classification rules and discovering association rules. For both tasks the language permits the specification of the task-relevant data, the kind of knowledge to be mined, the background knowledge and the hierarchies, the interestingness measures and the visualization for discovered patterns. Some constraints on the query language are identified by the particular mining task. The syntax of the query language is described and the application to a real repository of maps is briefly reported.

1 Introduction

Spatial data are important in many applications, such as computer-aided design, image processing, VLSI, and GIS. This steady growth of spatial data is outpacing the human ability to interpret them. There is a pressing need for new techniques and tools to find implicit regularities hidden in the spatial data.

Advances in spatial data structures [7], spatial reasoning [3], and computational geometry [22] have paved the way for the study of knowledge discovery in spatial data, and, more specifically, in geo-referenced data. *Spatial data mining* methods have been proposed for *the extraction of implicit knowledge, spatial relations, or other patterns not explicitly stored in spatial databases* [15]. Generally speaking, a *spatial pattern* is a pattern showing the interaction between two or more spatial objects or space-dependent attributes, according to a particular spacing or set of arrangements [1].

Knowledge discovered from spatial data may include classification rules, which describe the partition of the database into a given set of classes [14], clusters of spatial objects ([11], [24]), patterns describing spatial trends, that is, regular changes of one or more non-spatial attributes when moving away from a given start object [6], and

R. Meo et al. (Eds.): Database Support for Data Mining Applications, LNAI 2682, pp. 95–116, 2004.

subgroup patterns, which identify subgroups of spatial objects with an unusual, an unexpected, or a deviating distribution of a target variable [13]. The problem of mining spatial association rules has been tackled by [14], who implemented the module Geo-associator of the spatial data mining system GeoMiner [9].

A database perspective on spatial data mining is given in the work by Ester *et al.* [6], who define a small set of database primitives for the manipulation of neighbourhood graphs and paths used in some spatial data mining systems. An Inductive Logic Programming (ILP) perspective on spatial data mining is reported in [21], which proposes a logical framework for spatial association rule mining.

GIS offers an important application area where spatial data mining techniques can be effectively used. In the work by Malerba *et al.* [20], it can be seen how some classification patterns, induced from georeferenced data, can be used in topographic map interpretation tasks. A prototype of GIS, named INGENS [19], has been built around this application. In INGENS the geographical data collection is organized according to an object-oriented data model and is stored in a commercial Object Oriented DBMS (ODBMS).

INGENS data mining facilities support sophisticated end users in their topographic map interpretation tasks. In INGENS, each time a user wants to query its database on some geographical objects not explicitly modelled, he/she can prospectively train the system to recognize such objects and to create a special user view. Training is based on a set of examples and counterexamples of geographical concepts of interest to the user (e.g., ravine or steep slopes). Such concepts are not explicitly modelled in the map legends, so they cannot be retrieved by simple queries. Furthermore, the user has serious difficulty formalizing their operational definitions. Therefore, it is necessary to rely on the support of a knowledge discovery system that generates some plausible "definitions". The sophisticated user is simply asked to provide a set of (counter-) examples (e.g., map cells) and a number of parameters that define the data mining task more precisely.

An INGENS user should not have problems, due to the integration of different technologies, such as data mining, OODBMS, and GIS. In general, to solve any such problems the use of *data mining languages* has been proposed, which interface users with the whole system and hide the different technologies [10]. However, the problem of designing a spatial mining language has received little attention in the literature. To our knowledge, the only spatial mining language is GMQL (Geo Mining Query Language) [16], which is based on DMQL (Data Mining Query Language) [8]. These languages have both been developed for mining knowledge from *relational* databases, so SQL remains the milestone on which their syntax and semantics are built.

This paper presents SDMOQL (Spatial Data Mining Object Query Language) a spatial mining query language for INGENS sophisticated users. Its main characteristics are the following:

- It is based on OQL, the standard defined by ODMG (Object Database Management Group) for designing object oriented models (www.odmg.org).
- It interfaces relational data mining systems [2] that work with first-order representations of input data and output patterns.
- It separates the logical representation of spatial objects from their physical or geometrical representation.

The paper is organized as follows. INGENS architecture and conceptual database schema are described in the next section, while in Section 3 the spatial data mining process in INGENS is introduced. In Section 4 the syntax of SDMOQL is presented, while in Section 5 complete example of SDMOQL's use in INGENS is described. Finally, related works are discussed in Section 6.

2 INGENS Architecture and Conceptual Database Schema

The three-layered architecture of INGENS is illustrated in Fig. 1. The interface layer implements a *Graphical User Interface* (GUI), a java applet which allows the system to be accessed by the following four categories of users:

- Administrators, who are responsible for GIS management.
- Map *maintenance* users, whose main task is updating the Map Repository.
- Sophisticated end users, who can ask the system to learn operational definitions of geographical *objects* not explicitly modelled in the database.
- Casual end users, who occasionally access the database and may need different information *each* time. Casual users cannot train INGENS.

Only sophisticated end-users are allowed to discover new patterns by using SDMOQL.

The application enablers layer makes several facilities available to the four categories of INGENS users. In particular, the *Map Descriptor* is the application enabler responsible for the automated generation of first-order logic descriptions of some geographical objects. Descriptors generated by a Map Descriptor are called *operational*. The *Data Mining Server* provides a suite of data mining systems that can be run concurrently by multiple users to discover previously unknown, useful patterns in geographical data. In particular, the Data Mining Server provides sophisticated

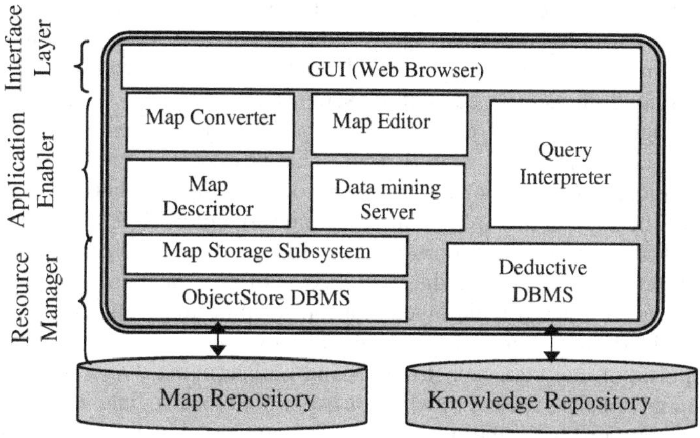

Fig. 1. INGENS three-layered software architecture

users with an inductive learning system, named ATRE [18], which can generate models of geographical objects from a set of training examples and counter-examples.

The *Query Interpreter* allows any user to formulate a query in SDMOQL. The query can refer to a specific map and can contain both predefined predicates and predicates, whose operational definition has already been learned. Therefore, it is the Query Interpreter's responsibility to select the objects involved from the Map Repository, to ask the Map Descriptor to generate their logical descriptions and to invoke the inference engine of the Deductive Database, in order to check conditions expressed by both predefined and learned predicates. The *Map Converter* is a suite of tools which supports the acquisition of maps from external sources. Currently, INGENS can export maps in Drawing Interchange Format (DXF) by Autodesk Inc. (www.autodesk.com) and can automatically acquire information from vectorized maps in the MAP87 format, defined by the Italian Military Geographic Institute (IGMI) (www.nettuno.it/fiera/igmi/igmit.htm). Since IGMI's maps contain static information on orographic, hydrographic and administrative *boundaries* alone, a *Map Editor* is required to integrate and/or modify this information.

The resource layer controls the access to both the *Knowledge Repository* and the *Map Repository*. The former contains the operational definitions of geographical objects induced by the Data Mining Server. In INGENS, different users can have different definitions of the same geographical object. Knowledge is expressed according to a relational representation paradigm and managed by an XSB-based deductive relational DBMS [23]. The *Map Repository* is the database instance that contains the actual collection of maps stored in the GIS. Geographic data are organized according to an object-oriented data model. The object-oriented DBMS used to store data is a commercial one (ObjectStore 5.0 by Object Design, Inc.), so that full use is made of a well-developed, technologically mature non-spatial DBMS. Moreover, an object-oriented technology facilitates the extension of the DBMS to accommodate management of geographical objects. The *Map Storage Subsystem* is involved in storing, updating and retrieving items in and from the map collection. As a *resource manager*, it represents the only access path to the data contained in the Map Repository and which are accessed by multiple, concurrent clients.

Each map is stored according to a hybrid tessellation – topological model. The tessellation model follows the usual topographic practice of superimposing a regular grid on a map, in order to simplify the localization process. Indeed, each map in the repository is divided into square cells of the same size.

In the topological model of each cell it is possible to distinguish two different structural hierarchies: *physical* and *logical*. The physical hierarchy describes the geographical objects by means of the most appropriate physical entity, that is: point, line or region. The logical hierarchy expresses the semantics of geographical objects, independent of their physical representation. In the Map Repository, the logical hierarchy is represented by eight distinct classes of the database schema, each of which correspond to a geographic layer in a topographic map, namely hydrography, orography, land administration, vegetation, administrative (or political) boundary, ground transportation network, construction and built-up area. Objects of a layer are instances of more specific classes to which it is possible to associate a unique physical representation. For instance, the administrative boundary layer describes objects of one of the following subclasses: city, province, county or state.

Finally, each geographical object in the map has both a *physical structure* and a *logical structure*. The former is concerned with the representation of the object on some media by means of a point, a line or a region. Therefore, the physical structure of a cell associates the content of a cell with the physical hierarchy. On the other hand, the logical structure of a cell associates the content with the logical hierarchy, such as river, city, and so on. The logical structure is related to the map legend.

3 Spatial Data Mining Process in INGENS

The spatial data mining process in INGENS (see Fig. 2) is aimed at a user who controls the parameters of the process. Initially, the query written in SDMOQL is syntactically and semantically analyzed. Then the Map Descriptor generates a highly conceptual qualitative representation of the raw data stored in the object-oriented database (see Fig. 3). This representation is a conjunctive formula in a first-order logic language, whose atoms have the following syntax: $f(t_1,\ldots,t_n) = value$, where f is a function symbol called *descriptor*, t_i are terms and the *value* is taken from the range of f. A set of descriptors used in INGENS is reported in Table 1. They can be roughly classified in *spatial* and *non-spatial*.

According to their *nature*, it is possible to spatial descriptors as follows:

Geometrical, if they depend on the computation of some metric/distance. Their domain is typically numeric. Examples are *line_shape* and *extension*.

Topological, if they are relations that are invariant under the topological transformations (translation, rotation, and scaling). The type of their domain is nominal. Examples are *region_to_region* and *point_to_region*.

Directional, if they concern orientation. The type of their domain can be either numerical or nominal. An example is *geographic_direction*.

Locational, if they concern the location of objects. Locations are represented by numeric values that express co-ordinates. There are no examples of locational descriptors in Table 1.

Some spatial descriptors are *hybrid*, in the sense that they merge properties of two or more categories above. For instance, the descriptor *line_to_line* that expresses conditions of parallelism and perpendicularity is both topological (it is invariant with respect to translation, rotation and scaling) and geometrical (it is based on the angle of incidence).

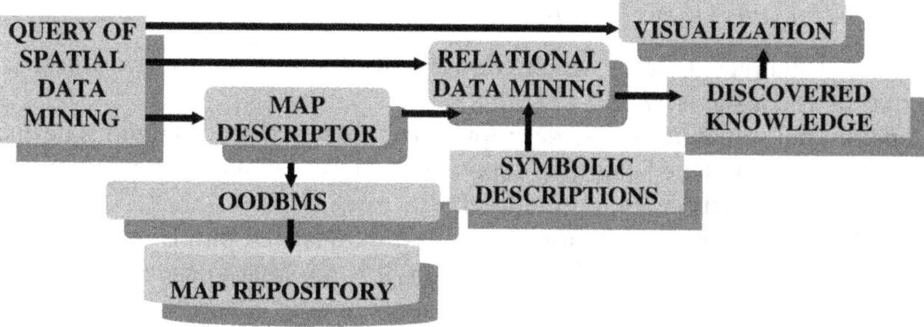

Fig. 2. Spatial data mining process in INGENS

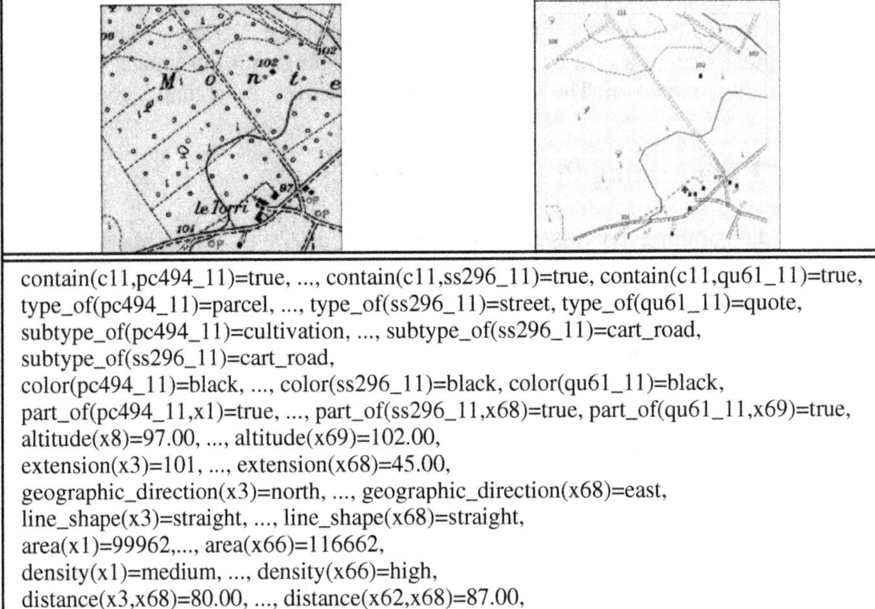

contain(c11,pc494_11)=true, ..., contain(c11,ss296_11)=true, contain(c11,qu61_11)=true,
type_of(pc494_11)=parcel, ..., type_of(ss296_11)=street, type_of(qu61_11)=quote,
subtype_of(pc494_11)=cultivation, ..., subtype_of(ss296_11)=cart_road,
subtype_of(ss296_11)=cart_road,
color(pc494_11)=black, ..., color(ss296_11)=black, color(qu61_11)=black,
part_of(pc494_11,x1)=true, ..., part_of(ss296_11,x68)=true, part_of(qu61_11,x69)=true,
altitude(x8)=97.00, ..., altitude(x69)=102.00,
extension(x3)=101, ..., extension(x68)=45.00,
geographic_direction(x3)=north, ..., geographic_direction(x68)=east,
line_shape(x3)=straight, ..., line_shape(x68)=straight,
area(x1)=99962,..., area(x66)=116662,
density(x1)=medium, ..., density(x66)=high,
distance(x3,x68)=80.00, ..., distance(x62,x68)=87.00,
line_to_line(x3,x68)=almost_parallel, ..., line_to_line(x62,x68)=almost_parallel,
region_to_region(x1,x9)=disjoint, ..., region_to_region(x45,x66)=disjoint,
line_to_region(x16,x1)=adjacent, ..., line_to_region(x67,x66)=intersect,
point_to_region(x2,x1)=outside, ..., point_to_region(x69,x66)=outside

Fig. 3. Raster and vector representation (above) and symbolic description of cell 11 (below). The cell is an example of a territory where there is a system of farms. The cell is extracted from a topographic chart (Canosa di Puglia 176 IV SW - Series M891) produced by the Italian Geographic Military Institute (IGMI) at scale 1:25,000 and stored in INGENS

In INGENS geo-referenced objects can also be described by three non-spatial descriptors are *color*, *type_of* and *subtype_of*. Finally, the descriptor *part_of* associates the physical structure to a logical object. For instance, in the description:

type_of(s1)=street , part_of(s1,x1)=true, part_of(s1,x2)=true

the constant *s1* denotes a street which is physically represented by two lines, which are referred to as constants *x1* and *x2*.

The operational semantics of the descriptors is based on a set of methods defined in the object-oriented model of the map repository. More details on the computational methods for their extraction are reported in [17].

This qualitative data representation can be easily translated into Datalog with built-in predicates [18]. Thanks to this transformation, it is possible to use the output of the Map Descriptor module in many relational data mining algorithms, which return spatial patterns expressed in a first-order language. Finally, the results of the mining process are presented to the user. The graphical feedback is very important in the analysis of the results.

Table 1. Set of descriptors extracted by the map descriptor module in INGENS

Feature	Meaning	Type	Domain	
			Type	**Values**
contain(C,L)	Cell C contains a logical object L	Topological relation	Boolean	true, false
part_of(L,F)	Logical object L is composed by physical object F	Topological relation	Boolean	true, false
type_of(L)	Type of L	Non-spatial attribute	Nominal	33 nominal values
subtype_of(L)	Specialization of the type of L	Non-spatial attribute	Nominal	101 nominal values specializing the type_of domain
color(L)	Color of L	Non-spatial attribute	Nominal	blue, brown, black
area(F)	Area of F	Geometrical attribute	Linear	[0..MAX_AREA]
density(F)	Density of F	Geometrical attribute	Ordinal	Symbolic names chosen by an expert
extension(F)	Extension of F	Geometrical attribute	Linear	[0..MAX_EXT]
geographic_direction(F)	Geographic direction of F	Directional attribute	Nominal	north, east, north_west, north_east
line_shape(F)	Shape of the linear object F	Geometrical attribute	Nominal	straight, cuspidal, curvilinear
altitude(F)	Altitude of F	Geometrical attribute	Linear	[0.. MAX_ALT]
line_to_line(F_1,F_2)	Spatial relation between two lines F_1 and F_2	Hybrid relation	Nominal	almost parallel, almost perpendicular
distance(F_1,F_2)	Distance between two lines F_1 and F_2	Geometrical relation	Linear	[0..MAX_DIST]
region_to_region(F_1,F_2)	Spatial relation between two regions F_1 and F_2	Topological relation	Nominal	disjoint, meet, overlap, covers, contains, equal, covered_by, inside
line_to_region(F_1,F_2)	Spatial relation between a line F_1 and a region F_2	Hybrid relation	Nominal	along_edge, intersect
point_to_region(F_1, F_2)	Spatial relation between a point F_1 and a region F_2	Topological relation	Nominal	inside, outside, on_boundary, on_vertex

4 Design of a Data Mining Language for INGENS

SDMOQL is designed to support the interactive data mining process in INGENS. Designing a comprehensive data mining language is a challenging problem because data mining covers a wide spectrum of tasks, from data classification to mining association rules. The design of an effective data mining query language requires a deep understanding of the power, limitation and underlying mechanisms of the various kinds of data mining tasks. A data mining query language must incorporate a set of data mining primitives designed to facilitate efficient, fruitful knowledge discovery. Seven primitives have been considered as guidelines for the design of SDMOQL. They are:

1. the set of objects relevant to a data mining task,
2. the kind of knowledge to be mined,
3. the set of descriptors to be extracted from a digital map
4. the set of descriptors to be used for pattern description
5. the background knowledge to be used in the discovery process,
6. the concept hierarchies,
7. the interestingness measures and thresholds for pattern evaluation, and
8. the expected representation for visualizing the discovered patterns.

These primitives correspond directly to as many non-terminal symbols of the definition of an SDMOQL statement, according to an extended BNF grammar. Indeed, the SDMOQL top-level syntax is the following:

<SDMOQL> ::= <SDMOQL_Statement>; { <SDMOQL_Statement>;}
<SDMOQL_Statement> ::=<Spatial_Data_Mining_Statement>
| *<Background_Knowledge>*
| *<Hierarchy>*
| *<Result_Displaying>*
<Spatial_Data_Mining_Statement> ::= <Object_Specification_Query>
mine *<Kind_of_Pattern>*
analyze *<Primitive_descriptors>*
with descriptors *<Pattern_descriptors>*
[<Background_Knowledge>]
{<Hierarchy>}
*[**with** <Interestingness_Measures>]*
[<Result_Displaying>]

where "[]" represents 0 or one occurrence and "{}" represents 0 or more occurrences, and words in bold type represent keywords. In sections 4.2 to 4.8 the detailed syntax for each data mining primitive is both formally specified and explained through various examples of possible mining problems.

4.1 Data Specification: General Principles

The first step in defining a data mining task is the specification of the data on which mining is to be performed. Data mining query languages presented in the literature allows the user to specify, through an SQL query, a single data table, where each row represents one *unit of analysis*[1] while each column corresponds to a variable (i.e. an attribute) of the unit of analysis. Generally, no interaction between units of analysis is assumed.

The situation is more complex in spatial data mining. First, units of analysis are spatial objects, which means that they are characterized, among other things, by spatial properties.

Second, attributes of some spatial objects in the neighborhood of, or contained in, a unit of analysis may affect the values taken by attributes of the unit of analysis. Therefore, we need to distinguish units of analysis, which are the *reference* objects of an analysis, from other *task-relevant* spatial objects, and we need to represent interactions between them. In this context, the single table representation supported by traditional data mining query languages is totally inadequate, since different geographical objects may have different properties, which can be properly modelled by as many data tables as the number of object types.

Third, in traditional data mining relatively simple transformations are required to obtain units of analysis from the *units of observation*[2] explicitly stored in the database. The unit of observation is often the same as the unit of analysis, in which case no trasformation at all is required. On the contrary, in GIS research, the wealth of secondary data sources creates opportunities to conduct analyses with data from multiple units of observation. For instance, a major national study uses a form that collects information about each person in a dwelling and information about the housing structure, hence it collects data for two units of observation: persons and housing structures. From these data, different units of analysis may be constructed: household could be examined as a unit of analysis by combining data from people living in the same dwelling or family could be treated as the unit of analysis by combining data from all members in a dwelling sharing a familial relationship. Units of analysis can be constructed from units of observation by making explicit the spatial relations such as topological, distance and direction relations, which are implicitly defined by the location and the extension of spatial objects.

Fourth, working at the level of stored data, that is, geometrical representations (points, lines and regions) of geographical objects, is often undesirable. The GIS user is interested in working at higher conceptual levels, where human-interpretable properties and relations between geographical objects are expressed. A typical example is represented by the possible relations between two roads, which either cross each other, or run parallel, or can be confluent, independently of the fact that they are physically represented as "lines" or "regions" in a map.

To solve these problems, in SDMOQL the specification of the *geographical objects* (both reference and task-relevant) of interest for the data mining task (first primitive) is separated from the description of the units of analysis (third and fourth primitives). Each unit of analysis is described by means of both (non-)spatial properties and spatial relations between selected objects. First-order logic is adopted as representation language, since it overcomes the limitations of the single table representation. Some basic descriptors (see Table 1) are generated by the Map

---

[1] In statistics, the unit of analysis is the basic entity or object about which generalizations are to be made based on an analysis and for which data are collected in the form of variables.

[2] The unit of observation is the entity in primary research that is observed and about which information is systematically collected.

Descriptor, to support complex transformations of the data stored in the database into descriptions of units of analysis. Their specification is given in the third primitive. However, these descriptors refer to the physical representation of geographical objects of interest. To produce high-level conceptual descriptions involving objects of the logical hierarchy, the user can specify a set of new descriptors on the basis of those extracted by the Map descriptors. New descriptors are specified in the background knowledge (fifth primitive) by means of logic programs. Moreover, it is possible to specify that final patterns should be described by means of these new descriptors in the fourth primitive.

4.2 Task-Relevant Object Specification

The selection of geographical objects is performed by means of simplified OQL queries with a SELECT-FROM-WHERE structure, namely:

<Data_Specification_Query> ::= *<Query_Statement>*
*{**UNION** <Query_Statement>}*

<Query_Statement>::= ***SELECT*** *<Object> {, <Object>}*
FROM *<Class> {, <Class>}*
[WHERE <Conditions>]

The SELECT clause should return objects of a class in the database schema corresponding to a cell, a layer or a type of logical object. Therefore, the selection of object properties such as the attribute *river_name* of a river, is not permitted. Moreover, the selected objects must belong to the same symbolic level *(cell, layer or logic object)*. More formally the FROM clause can contain either a group of *Cells* or a set of *Layers*, or a set of *Logic Objects*, but never a mixture of them. Whenever the generation of the descriptions of objects belonging to different symbolic levels is necessary, the user can obtain it by means of the UNION operator. The following are examples of valid data queries:

Example 1:Cell-level query. The user selects *cell* 26 from the topographic map of Canosa (Apulia) and the Map Descriptor generates the description of all the objects in this cell.

SELECT x
FROM x in **Cell**
WHERE x->**num_cell** = 26 AND x->part_map->**map_name** = "Canosa"

Example 2: Layer-level query. The user selects the *Orography layer* from the topographic map of Canosa and the *Construction layer* from any map. The Map Descriptor generates the description of the objects in these layers.

SELECT x, y
FROM x in **Horography**, y in **Construction**
WHERE x->part_map->**map_name** = "Canosa"

Example 3: Object-level query. The user selects the *objects* of the logic class River and the objects of type motorway (instances of the class Road), from cell 26 of the topographic map of Canosa. The Map Descriptor generates the description of these objects.

```
SELECT x, y
FROM x in River, y in Road
WHERE x->part_map->map_name = "Canosa"
AND  y->part_map->map_name = "Canosa"
AND  x->log_incell->num_cell = 26 AND  y->log_incell->num_cell = 26
AND  y->type_road = "motorway"
```

The above queries do not present semantic problems. However, the next example is an OQL query which is syntactically correct but selects data that cannot be a valid input to the Map Descriptor.

Example 4: Semantically ambiguous query.

```
SELECT x, y
FROM x in Cell, y in River
WHERE x->num_cell  = 26 AND  y->log_incell->num_cell = 26
```

This query selects the object cell 26 and all rivers in it. However, it is unclear whether the Map Descriptor should describe the entire cell 26 or only the rivers in it, or both. In the first case, a cell-level query must be formulated (see example 1). In the second case, an object-level query produces the desired results (see example 3). In the (unusual) case that both kinds of descriptions have to be generated, the problem can be solved by the UNION operator, applied to the cell-level query and the object-level query. Therefore, the following constraint is imposed on SDMOQL: the selected data must belong to the same symbolic level (cell, layer or logic object). More formally the FROM clause can contain either a group of Cells or a set of Layers, or a set of Logic Objects, but never a mixture of them.

The next example is useful to present the constraints imposed on the SELECT clause.

Example 5: Attributes in the SELECT clause.

```
SELECT x.name_river
FROM x in River
```

The query selects the names of all the rivers stored in the database. The result set contains attributes and not geographic objects to be described by a set of attributes and relations. In order to select proper input data for the Map Descriptor, the SELECT clause should return objects of a class in the database schema corresponding to a cell, a layer or a type of logical object. It might be observed that the presence of an attribute in the SELECT clause can be justified when its type corresponds to a class. For instance, the following query:

```
SELECT x->River
FROM x in Cell
WHERE x->num_cell = 26
```

concerns all rivers in cell 26. Nevertheless, thanks to inverse relations (inverse members) characterizing an object model, it is possible to reformulate it as follows:

```
SELECT x
FROM x in River
WHERE x->log_incell->num_cell = 26
```

In this way, all the above constraints should be respected.

4.3 The Kind of Knowledge to be Mined

The kind of knowledge to be mined determines the data mining task in hand. For instance, classification rules or decision trees are used in classification tasks, while association rules or complex correlation coefficients are extracted in association tasks. Currently, SDMOQL supports the generation of either classification rules[3] or association rules, which means that only two different mining problems can be solved in INGENS: the former has a predictive nature, while the latter is descriptive. The top-level syntax is defined below:

<Kind_of_Pattern> ::= *<Classification_Rules>* | *<Association_Rules>*

The non-terminal *<Classification_Rules>* specifies that patterns to be mined concern a classification task

<Classification_Rules> ::= ***classification as*** *<Pattern_Name>*
for *<Classification_Concept> {, <Classification_Concept> }*

In a classification task, the user may be interested in inducing a set of classification rules for a *subset* of the classes (or concepts) to which training examples belong. Typically, the user specifies both "positive" and "negative" examples, that is, he/she specifies examples of two different classes, but he/she is interested in classification rules for the "positive" class alone. In this case, the subset of interest for the user is specified in the *<Classification_Concept>* list.

In SDMOQL, spatial association rule mining tasks are specified as follows:

<Association_Rules> ::= ***association as*** *<Pattern_Name>*
key is *<Descriptor>*

As pointed out, spatial association rules define spatial patterns involving both *reference objects* and *task-relevant objects* [21]. For instance, a user may be interested in describing a given area by finding associations between large towns (reference objects) and spatial objects in the road network, hydrography, and administrative boundary layers (task-relevant objects). The atom denoting the reference objects is called *key atom*. The predicate name of the key atom is specified in the **key is** clause.

4.4 Specification of Primitive and Pattern Descriptors

The **analyze** clause specifies what descriptors, among those automatically generated by the Map Descriptor, can be used to describe the geographical objects extracted by means of the first primitive. The syntax of the analyze clause is the following:

analyze *<Primitive_descriptors>*

where:

<Primitive_descriptors> ::= *<Descriptor> {, <Descriptor>}*
parameters *<Parameter_specs>{, <Parameter_specs>}*

<Descriptor> ::= *<Predicate>/<Arity>*

<Parameter_specs> ::= *<Parameter_name>* **threshold** *<Integer>*

---

[3] Here, the term classification rule denotes the result of a *supervised discrimination* process. On the contrary, Han & Kamber [9,10] use the same term to denote the result of an *unsupervised clustering* process.

The specification of a set of parameters is required by the Map Descriptor to automatically generate some primitive descriptors.

The language used to describe generated patters is specified by means of the following clause:

with descriptors *<Pattern_descriptors>*

where:

<Pattern_descriptors> ::= *<Descriptor_specification>*
{; <Descriptor_specification>}
<Descriptor_specification> ::= *<Descriptor>* [***cost*** *<Integer>*] |
<Descriptor> [***with*** *<Terms_Spec>*]
<Terms_Spec>::= <Term_Spec>{,<Term_Spec>}
<Term_Spec> ::= <Constant_Type> |*<Variable_Type>*
<Constant_Type> ::= ***constant*** *[<Value>]*
<Variable_Type> ::= ***variable mode*** *<Variable_Mode>* ***role*** *< Variable_Role>*
<Variable_Mode> ::= ***old*** | ***new*** | ***diff***
<Variable_Role> ::= ***ro*** | ***tro***

The specification of descriptors to be used in the high-level conceptual descriptions can be of two types: either the name of the descriptor and its relative cost, or the name of the descriptor and the full specification of its arguments. The former is appropriate for classification tasks, while the latter is required by association rule mining tasks.

An example of a classification task activated by an SDMOQL statement is reported in Fig. 4. In this case, the Map Descriptor generates a symbolic description of some cells by using the predicates listed in the analyze clause. These are four concepts to be learned, namely *class(_)=system_of_farms*, *class(_)=fluvial_ landscape*, *class(_)=.royal_cattle_track*, and *class(_)=.system_of_cliffs*. Here the function symbol *class* is unary and "_" denotes the anonymous variables *à la* Prolog. The user can provide examples of these four classes, as well as of other classes. Examples of systems of farms are considered to be positive for the first concept in the list and negative for the others. The converse is true for examples of fluvial landscapes, royal cattle track and system of cliffs. Examples of other classes are considered to be counterexamples of all classes for which rules will be generated. The only requirement for the INGENS user is the ability to detect and mark some cells that are instances of a class. Indeed, INGENS GUI allows the user both to formulate and run an SDMOQL query and to associate the description of each cell with a class.

Rules generated for the four concepts are expressed by means of descriptors specified in the **with descriptors** list. They are specified by Prolog programs on the basis of descriptors generated by the Map Descriptor. For instance, the descriptor *font_to_parcel*/2 has two arguments which denote two logical objects, a font and a parcel. The topological relation between the two logical objects is defined by means of the clause:

font_to_parcel(Font,Parcel) = Topographic_Relation :-
type_of(Font) = font, part_of(Font,Point) = true,
type_of(Parcel) = parcel, part_of(Parcel,Region) = true,
point_to_region(Point,Region) = Topographic_Relation.

```
SELECT x FROM x in Cell WHERE x->num_cell = 5
UNION SELECT x  FROM x in Cell WHERE x->num_cell = 8
UNION SELECT x FROM x in Cell WHERE x->num_cell = 11
UNION SELECT x FROM x in Cell WHERE x->num_cell = 15
UNION SELECT x FROM x in Cell WHERE x->num_cell = 16
UNION SELECT x FROM x in Cell WHERE x->num_cell = 17
UNION SELECT x FROM x in Cell WHERE x->num_cell = 27
UNION SELECT x FROM x in Cell WHERE x->num_cell = 28
UNION SELECT x FROM x in Cell WHERE x->num_cell = 34
UNION SELECT x FROM x in Cell WHERE x->num_cell = 83
UNION SELECT x FROM x in Cell WHERE x->num_cell = 84
UNION SELECT x FROM x in Cell WHERE x->num_cell = 89
mine classification as MorphologicalElements
for class(_)=.system_of_farms, class(_)=.fluvial_landscape, class(_)=.royal_cattle_track,
class(_)=.system_of_cliffs
analyze contain/2, part_of/2, type_of/1, subtype_of/1, color/1, … ,
        line_to_line/2, distance/2, line_to_region/2, …, point_to_region/2
parameters
        maxValuePointRegionClose              threshold  300,
        minValueLineLong                      threshold  100,
        maxValueLinesClose                    threshold  300,
        minValueRegionLarge                   threshold  5000,
        maxValueRegionClose                   threshold  500
with descriptors contain/2  cost  1; class/1 cost  0; subtype_of/1 cost  0;
        parcel_to_parcel/2  cost  0; slope_to_slope/2 cost 0; …
        canal_to_parcel/2 cost 0; …; font_to_parcel/2 cost 0; …
define knowledge
        font_to_parcel(Font,Parcel) = Topographic_Relation :-
                type_of(Font) = font, part_of(Font,Point) = true,
                type_of(Parcel) = parcel, part_of(Parcel,Region) = true,
                point_to_region(Point,Region) = Topographic_Relation.
…
criteria
intermediate minimize negative_example_covered           with tolerance 0.6,
intermediate maximize positive_example_covered           with tolerance 0.4,
intermediate maximize selectors_of_clause                with tolerance 0.3,
intermediate minimize cost                               with tolerance 0.4
final maximize positive_example_covered          with tolerance 0.0,
final maximize selectors_of_clause               with tolerance 0.0,
final minimize cost                              with tolerance 0.0
maxstar    threshold  25,
consistent threshold  500,
max_ps     threshold  11
recursion  =  off,
verbosity  =  { off, off, on, on, on }
```

Fig. 4. A complete SDMOQL query for a classification task. In this case, the INGENS user marks 5,11,34,42 as instances of the class "system of farms" and the cells 1, 2, 3, 4, 15, 16, 17 as instances of the class "other"

In association rule mining tasks, the specification of pattern descriptors correspond to the specification of a collection of atoms: *predicateName*(t_1,..., t_n), where the name of the predicate corresponds to a *<Descriptor>*, while *<Term_Spec>* describes each term t_i, which can be either a constant or a variable. When the term is a variable the mode and role clauses indicate respectively the type of variable to add to the atom and its role in a unification process. Three different modes are possible: *old* when the introduced variable can be unified with an existing variable in the pattern, *new* when it is a not just present in the pattern, *diff* when it is a new variable but its values are different from the values of a similar variable in the same pattern. Furthermore, the variable can fill the role of reference object (*ro*) or task-relevant object (*tro*) in a discovered pattern during the unification process. The *is key* clause specifies the atom which has the key role during the discovery process. The first term of the key object must be a variable with mode new and role ro. The following is an example of specification of pattern descriptors defined by an SDMOQL statement for :

with descriptors
contain/2 **with variable mode old role ro, variable mode new role tro;**
type_of/2 **with variable mode diff role tro, constant;**
is key with variable mode new role ro, constant cultivation;

This specification helps to select only association rules where the descriptors *contain/2* and *type_of/2* occur. The first argument of a *type_of* is always a *diff* variable denoting a spatial object, and it can play the roles of both *ro* and *tro*, whereas the second argument, i.e. the type of object, is the constant 'cultivation', if the first argument is a reference object, otherwise it is any other constant. The predicate contain links the *ro* of type cultivation with other spatial objects contained in the cultivation. The following association rule:

type_of(X,cultivation), contain(X,Y), type_of(Y,olive_tree), X Y → contain(X,Z), type_of(Z,almond_tree), X Z, Y X

satisfies the constraints of the specification and express the co-presence of both almond trees and olive-trees in some extensive cultivations.

4.5 Syntax for Background Knowledge and Concept Hierarchy Specification

Many data mining algorithms use background knowledge or concept hierarchies to discover interesting patterns. Background knowledge is provided by a domain expert on the domain to be mined. It can be useful in the discovery process. The SDMOQL syntax for background knowledge specification is the following:

<Background_Knowledge> ::= [<New_Knowledge>] {<Use_Knowledge>}
<New_Knowledge> ::= ***define knowledge*** *<Clause> {, <Clause>}*
<Use_Knowledge> ::= ***use background knowledge of users*** *<User> {, <User>}*
on *<Descriptor> {, <Descriptor>}*

In INGENS, the user can define a new background knowledge expressed as a set of definite clauses; alternatively, he/she can specify a set of rules explicitly stored in a deductive database and possibly mined in a previous step. The following is an example of a background knowledge specification:

Example 7: Definition of close_to *and import of the definition of* ravine.
close_to(X,Y)=true :- region_to_region(X,Y)=meet.
close_to(X,Y)=true :- close_to(Y,X)=true.
use background knowledge of users UserName1 **on** ravine/1

Concept hierarchies allow knowledge mining at multiple abstraction levels. In order to accommodate the different viewpoints of users regarding the data, there may be more than one concept hierarchy per attribute or dimension. For instance, some users may prefer to organize census districts by wards and districts, while others may prefer to organize them according to their main purpose (industrial area, residential area, and so on). There are four major types of concept hierarchies [8]:

- *Schema hierarchies*, which define total or partial orders among attributes in the database schema.
- *Set-grouping hierarchies*, which organize values for given attributes or dimensions into groups of constants or range values.
- *Operation-derived hierarchies*, which are based on operations specified by experts or data mining systems.
- *Rule-based hierarchies*, which occur when either a whole concept or a portion of it is defined by a set of rules.

In SDMOQL a specific syntax is defined for the first two types of hierarchies:

<Hierarchy> ::= [<New_Hierarchy>] [<Use_Hierarchy>]
<New_Hierarchy> ::= ***define hierarchy*** *<Schema_Hierarchy>*
| ***define hierarchy for*** *<Set_Grouping_Hierarchy>*
<Use_Hierarchy> ::= ***use hierarchy*** *<Name_Hierarchy>* ***of user*** *<User>*

The following example shows how to define some hierarchies in SDMOQL.

Example 8: A definition of a schema hierarchy for some activity-related attributes and a set-grouping hierarchy for the descriptor distance**.**

define hierarchy Activity **as**
level1:{business_activity, other_activity} < level0: Activity;
level2:{low_business_activity,high_business_activity}<level1:business_activity;
define hierarchy Distance for distance/2 **as**
level1:{far, near} < level0: Distance;
level2:{0, 1999} < level1: near;
level2:{2000, +inf} < level1: far;

The activity hierarchy can be used to mine multi-level spatial association rules [21].

4.6 Syntax for Interestingness Measure Specification

The user can control the data mining process by specifying interestingness measures for data patterns and their corresponding thresholds. The SDMOQL syntax is the following:

<Interestingness_Measures> ::= [<Criteria>] [<Thresholds>] {<Settings>}

<Criteria> ::= ***criteria (intermediate | final) (minimize | maximize)*** *<Parameter>* ***with tolerance*** *<Value> {,**(intermediate | final) (minimize | maximize)** <Parameter>* ***with tolerance*** *<Value>}*

<Thresholds> ::=<Parameter> ***threshold*** *<Threshold_Value>*
{, <Parameter> ***threshold*** *<Threshold_Value>}*

<Settings> ::= <Parameter> = <String_Value>

Interestingness measures may include: threshold values, weights, search biases in the hypotheses space and algorithm-specific parameters. In particular the user can bias the search in the hypotheses space by a number of *preference criteria*, such as the maximization of the number of covered examples or the minimization of the number of variables in the body of a learned clause. He/she can also set thresholds such as *confidence*, *support* or *number of learned concepts*. Finally, the user can set the value of a generic input parameter of a data mining algorithm.

4.7 Syntax for Visualization

Data mining results should be displayed using rule visualization tools or some different output forms. SDMOQL provides the following primitives for displaying results in different forms:

<Result_Displaying> ::= ***display as*** *<Form>*
*[****at level*** *<Int_Value>* ***for*** *<Hierarchy_Name>],*

where *<Form>* describes the output form, for example, *if-then* rules or *tree*. Moreover, if a hierarchy is available, mined results can be represented at different concept levels. This is particularly true in the case of multiple-level association rules.

5 Mining Classification Rules for Topographic Map Interpretation

In the previous section, the syntax of SDMOQL has been defined. Here we present a data problem concerning the generation of classification rules for topographic map interpretation. Let us suppose that a GIS user needs to locate a "*sistema poderale*" (system of farms) in the large territory of his/her interest. This geographical object is not present in the GIS model, thus, only the specification of its operational definition will allow the GIS to find cells containing a system of farms in a vectorized map. Who can provide it? The user is not able to do so for a number of reasons.

Firstly, providing the GIS with operational definitions of some environmental concepts is not a trivial task. Often only declarative and abstract definitions are available, which are difficult to compile into database queries.

Secondly, the operational definitions of some geographical objects are strongly dependent on the data model that is adopted by the GIS. Finding relationships between density of vegetation and climate is easier with a *raster data model*, while determining the usual orientation of some morphological elements is simpler in a *topological data model*.

Thirdly, different applications of a GIS require the recognition of different geographical elements in a map. Providing the system in advance with all the knowledge required for its various application domains is simply illusory, especially in the case of wide-ranging projects such as those set up by governmental agencies.

A solution to these problems can be found in the application of data mining techniques. For instance, an INGENS user can train the system to recognize cells with systems of farms, by performing the SDMOQL query in Fig. 4. The interpreter analyzes the query and verifies its syntactic and semantic correctness. Then the Map

Descriptor generates a symbolic description for each specified cell (see Fig. 3) and the expert associates each symbolic description with a concept, in order to define the training set. Association is made by *binding* variable terms of one of the four concepts to be learned to the constants terms in the descriptions of map cells. This step is necessary to create the training set of positive and negative examples for the learning system ATRE [18], which is used in INGENS for classification tasks. The user marks 5, 11, 34 as instances of the class "system of farms", the cells 8, 16, 17 as instances of the class "fluvial landscape", the cells 15, 27, 28 as instances of the class "royal cattle track" and the cells 83, 84, 89 as instances of the class "system of Cliffs". This binding function is supported by INGENS GUI. The training set obtained is input to ATRE, which returns the classification rules. With reference to the above query, ATRE generates the following clauses:

class(X1) = system_of_farms
contain(X1,X2)= true,
area_parcel(X2) in [102.787..249.525],
density_parcel(X2) = high,
font_to_parcel(X3,X2) = outside,

"A cell is an example of a *system of farms* if it contains a parcel (X2) that has an area between 102,787 and 249,525 square meters and a high vegetation density, and a font (X3) that is outside the parcel."

class(X1) = .fluvial_landscape
contain(X1,X2) = true,
extension_road(X3) in [234.0..440.0],
canal_to_road(X2,X3) = almost_parallel,
distance_canal_to_road(X2,X3) in [42.0..300.0].

"A cell is an example of a *fluvial landscape* if it contains a canal (X2) and a street (X3). The street has an extension between 234.0 and 440.0 meters and is almost parallel to the canal. In particular, the distance between the canal and the street is between 42.0 and 300.0 meters."

class(X1) = royal_cattle_track
contain(X1,X2) = true,
extension_road(X2) in [1002.0..1162.0],
subtype_of(X2) = main_road.

"A cell is an example of a *royal cattle track* if it contains a street (X2) that is a main road and has an extension between 1002.0 and 1162.0 meters."

class(X1)= system_of_cliffs
contain(X1,X2) = true,
distance_contour_slope_to_contour_slope(X2,X3)in [2.0..74.0],
extension_contour_slope(X2) in [79.0..307.0].

"A cell is an example of a *system of cliffs* if it contains two contour slopes (X2, X3), such that the distance between them is between 79.0 and 307.0 meters. One contour slope (X2) has an extension between 2,0 and 74,0 meters."

Whether the induced theory is "correct", that is, whether it classifies correctly all other examples of map cells not in the training set is beyond the scope of this work. However, it is noteworthy that these rules are coherent with the definitions given by town planners for the four morphological concepts of interest [19].

Operational definitions like those reported above can be used either to retrieve new instances of the learned concepts from the Map Repository or to facilitate the formulation of a query involving geographical objects not present in map legends. For instance, by submitting the following query:

SELECT C
FROM M in Map, C in Cell, R in Road
WHERE M->name = "Canosa" AND C->map = M AND R->log_incell = C
AND R->type_road="main_road" AND class(C) = fluvial_landscape

the user asks INGENS to find all cells in the Canosa map that are classified as fluvial landscape and contain a main road. To check the condition defined by the predicate *class(C)=fluvial_landscape*, the *Query Interpreter* generates the symbolic description of each cell in the map and asks the Query Engine of the Deductive Database to prove the goal *class(C)=fluvial_landscape* given the logic program above.

6 Related Work

Several data mining query languages have been proposed in the literature. MSQL is a rule query language proposed by Imielinski and Virmani [12] for relational databases. It satisfies the closure property, that is, the result of a query is a relation that can be queried further. Moreover, a cross-over between data and rules is supported, which means that there are primitives in the language that can map generated rules back to the source data, and vice versa. The combined result of these two properties is that a data mining query can be nested within a regular relational query. SDMOQL do not allow users to formulate nested queries, however, as pointed out at the end of the previous section, it supports some form of cross-over between data and mined rules. This is obtained by integrating deductive inferences for extracted rules with data selection queries expressed in OQL.

Another data mining query language for relational databases is DMQL [8]. Its design is based on five primitives, namely the set of data relevant to a data mining task, the kind of knowledge to be mined, the background knowledge to be used in the discovery process, the concept hierarchies, the interestingness measures and thresholds for pattern evaluation. As explained in Section 4.1, the design of SDMOQL is based on a different set of principles. In particular, the specification of data relevant to a data mining task involves a separate specification for the geographical objects of interest for the application, for the set of automatically generated (primitive) descriptors and for the set of descriptors used to specify the patterns. An additional design principle is that of visualization, since in spatial data mining it is important to specify whether results have to be visualized or presented in a textual form.

GMQL is based on DMQL and allows the user to specify the set of relevant data for the mining process, the type of knowledge to be discovered, the thresholds to filter out interesting rules, and the concept hierarchies as the background knowledge [16].

In the process of selecting data relevant to the mining task, the user has to specify (1) the relevant tables, (2) the conditions that are satisfied by the relevant objects and (3) the properties of the objects which the mining process is based on. Conditions may involve spatial predicates on topological relations, distance relations and direction relations. Although data can be selected from several tables, mining is performed only on a single table which result from an SQL query (single table assumption). GMQL queries can generate different types of knowledge, namely characteristic rules, comparison rules, clustering rules and classification rules. In Koperski's thesis, an extension to association rules was also proposed but not implemented. Differently from GMQL, SDMOQL separate the physical representation of geographical objects from their logical meaning. Moreover, all observations reported above for DMQL applies to GMQL as well.

Finally, it is noteworthy that some object-oriented extension of DMQL, named ODMQL, has also been proposed [4]. The design of ODMQL is based on the same primitives used for DMQL, so the main innovation is that each primitive is in an OQL-like syntax. Path expressions are supported in ODMQL, while more advanced features of object-oriented query languages, such as the use of collections and methods, are not mentioned. An interesting aspect of ODMQL, which will be taken into account in further developments of SDMOQL, is that some concept hierarchies are automatically defined by the inheritance hierarchy of classes.

7 Conclusions

In this paper, a spatial data mining language for a prototypical GIS with knowledge discovery facilities has been partially presented. This language is based on a simplified OQL syntax and is defined in terms of the eight data mining primitives. For a given query, these primitives define the set of objects relevant to a data mining task, the kind of knowledge to be mined, the set of descriptors to be extracted from a digital map, the set of descriptors to be used for pattern description, the background knowledge to be used in the discovery process, the concept hierarchies, the interestingness measures and thresholds for pattern evaluation, and the expected representation for visualizing the discovered patterns. An interpreter of this language has been developed in the system INGENS. It interfaces a Map Descriptor module that can generate a first-order logic description of selected geographical objects. A full example of the query formulation and its results has been reported for a classification task used in the qualitative interpretation of topographic maps. An extension of this language to other spatial data mining tasks supporting quantitative interpretation of maps is planned for the near future.

Acknowledgments

This work is part of the MURST COFIN-2001 project on "Methods of knowledge discovery, validation and representation of the statistical information in decision tasks". Thanks to Lynn Rudd for her help in reading the paper.

References

1. DeMers, M.N.: Fundamentals of Geographic Information Systems. 2nd ed., John Wiley & Sons, (2000)
2. Dzeroski, S., Lavrac, N. (eds.): Relational Data Mining. Springer, Berlin, Germany, (2001)
3. Egenhofer, M.J.: Reasoning about Binary Topological Relations. In Proceedings of the Second Symposium on Large Spatial Databases, Zurich, Switzerland, (1991) 143–160
4. Elfeky M.G., Saad A.A., Fouad S.A.. ODMQL: Object Data Mining Query Language. In Proceedings of Symposium on Objects and Databases, Sophia Antipolis, France, (2001)
5. Ester, M., Frommelt, A., Kriegel, H. P., Sander, J.: Algorithms for characterization and trend detection in spatial databases. In Proceedings of the 4th Int. Conf. on Knowledge Discovery and Data Mining, New York City, NY, (1998) 44–50
6. Ester, M., Gundlach, S., Kriegel, H.P., Sander, J.: Database primitives for spatial data mining, In Proceedings of Int. Conf. on Database in Office, Engineering and Science, (BTW '99), Freiburg, Germany, (1999)
7. Güting, R.H.: An introduction to spatial database systems. VLDB Journal, 3,4 (1994) 357–399
8. Han, J., Fu, Y., Wang, W., Koperski, K., Zaïane, O. R.: DMQL: a data mining query language for relational databases. In Proceedings of the Workshop on Research Issues on Data Mining and Knowledge Discovery, Montreal, QB, (1996) 27–34
9. Han, J., Koperski, K., Stefanovic, N.: GeoMiner: A System Prototype for Spatial Data Mining. In Peckham, J. (ed.): SIGMOD 1997, Proceedings of the ACM-SIGMOD International Conference on Management of Data. SIGMOD Record 26, 2 (1997) 553–556
10. Han, J., Kamber, M.: Data mining, Morgan Kaufmann Publishers (2000)
11. Han, J., Kamber, M., Tung, A.K.H.: Spatial clustering methods in data mining. In H. J. Miller & J. Han (eds.), Geographic Data Mining and Knowledge Discovery, Taylor and Francis, London, UK, (2001) 188-217
12. Imielinski T., Virmani A.: MSQL: A query language for database mining. Data Mining and Knowledge Discovery, 3(4) (1999) 373–408
13. Klosgen, W., May, M.: Spatial Subgroup Mining Integrated in an Object-Relational Spatial Database. In: Elomaa, T., Mannila, H., Toivonen, H. (eds.) Principles of Data Mining and Knowledge Discovery (PKDD), 6th European Conference, LNAI 2431, Springer-Verlag, Berlin, (2002) 275–286
14. Koperski, K., Han, J.: Discovery of spatial association rules in geographic information database. In Advances in Spatial Database, Proceedings of 4th Symposium, SSD '95. (Aug. 6–9. Portland, Maine), Springer–Verlag, Berlin., (1995) 47–66
15. Koperski, K., Adhikary, J., Han, J.: Knowledge discovery in spatial databases: progress and challenges, Proc. SIGMOD Workshop on Research Issues in Data Mining and Knowledge Discovery, (1996)
16. Koperski, K.: A progressive refinement approach to spatial data mining. Ph.D. thesis, Computing Science, Simon Fraser University, (1999)
17. Lanza, A., Malerba, D., Lisi, L.F., Appice, A., Ceci, M.: Generating Logic Descriptions for the Automated Interpretation of Topographic Maps. In D. Blostein and Y.-B. Kwon (eds.) Graphics Recognition: Algorithms and Applications, Lecture Notes in Computer Science, 2390, Springer, Berlin, Germany, (2002) 200–210
18. Malerba, D., Esposito, F., Lisi, F.A.: Learning recursive theories with ATRE. In: Prade, H. (ed.): Proc. 13th European Conference on Artificial Intelligence, John Wiley & Sons, Chichester, England, (1998) 435–439
19. Malerba, D., Esposito, F., Lanza, A., Lisi, F.A, Appice, A.: Empowering a GIS with Inductive Learning Capabilities: The Case of INGENS. Journal of Computers, Environment and Urban Systems, Elsevier Science (in press).

20. Malerba, D., Esposito, F., Lanza, A., Lisi, F.A.: Machine learning for information extraction from topographic maps. In H. J. Miller & J. Han (eds.), Geographic Data Mining and Knowledge Discovery, Taylor and Francis, London, UK, (2001) 291–314
21. Malerba, D., Lisi, F.A.: An ILP method for spatial association rule mining. Working notes of the First Workshop on Multi-Relational Data Mining, Freiburg, Germany (2001) 18–29
22. Preparata, F., Shamos, M.: Computational Geometry: An Introduction. Springer-Verlag, New York (1985)
23. Sagonas, K. F., Swift, T., & Warren, D. S.: XSB as an Efficient Deductive Database Engine. In R. T. Snodgrass, & M. Winslett (Eds.): Proceedings of the 1994 ACM SIGMOD International Conference on Management of Data, Minneapolis, Minnesota, SIGMOD Record 23(2), (1994) 442–453
24. Sander J., Ester M., Kriegel H.-P., Xu X.: Density-Based Clustering in Spatial Databases: A New Algorithm and its Applications. Data Mining and Knowledge Discovery, Kluwer Academic Publishers, 2(2) (1998) 169–194

Towards Query Evaluation in Inductive Databases Using Version Spaces

Luc De Raedt

Institut für Informatik
Albert-Ludwig-University
Georges Koehler Allee 79
D-79110 Freiburg, Germany
deraedt@informatik.uni-freiburg.de

Abstract. An inductive query specifies a set of constraints that patterns should satisfy. We study a novel type of inductive query that consists of arbitrary boolean expressions over monotonic and anti-monotonic primitives. One such query asks for all patterns that have a frequency of at least 50 on the positive examples and of at most 3 on the negative examples.

We investigate the properties of the solution spaces of boolean inductive queries. More specifically, we show that the solution space w.r.t. a conjunctive query is a version space, which can be represented by its border sets, and that the solution space w.r.t. an arbitrary boolean inductive query corresponds to a union of version spaces. We then discuss the role of operations on version spaces (and their border sets) in computing the solution space w.r.t. a given query. We conclude by formulating some thoughts on query optimization.

Keywords: inductive databases, inductive querying, constraint based mining, version spaces, convex spaces.

1 Introduction

The concept of inductive databases was introduced by Imielinski and Mannila in [23]. Inductive databases are databases that contain both patterns and data. In addition, they provide an inductive query language in which the user cannot only query the data that resides in the database but also mine the patterns of interest that hold in the data. The long-term goal of inductive database research is to put data mining on the same methodological grounds as databases. Despite the introduction of a number of interesting query languages for inductive databases, cf. [10,12,14,15,20,18,6,27], we are still far away from a general theory of inductive databases.

The problem of inductive querying can be described as follows. Given is an inductive database $\mathcal{D}$ (a collection of data sets), a language of patterns $\mathcal{L}$, and an inductive query $q(\phi)$ expressed in an inductive query language. The result of the inductive query is the set of patterns $Th(q(\tau), \mathcal{L}, \mathcal{D}) = \{\tau \in \mathcal{L} \mid$

R. Meo et al. (Eds.): Database Support for Data Mining Applications, LNAI 2682, pp. 117–134, 2004.

$q(\tau)$ is true in database $\mathcal{D}$}. It contains all patterns τ in the language $\mathcal{L}$ that satisfy the inductive query q in the database $\mathcal{D}$[1]. Boolean inductive queries are arbitrary boolean expressions over monotonic or anti-monotonic constraints on a single pattern variable τ. An example query could ask for all patterns τ whose frequency is larger than 100 in data set D_1 and whose frequency in data set D_2 or D_3 is lower than 20, where the data sets D_i belong the inductive database $\mathcal{D}$. This type of query is motivated by our earlier MolFea system for molecular feature mining [18]. However, in MolFea, only conjunctive queries were handled. To the best of the author's knowledge, boolean inductive queries are the most general form of query that have been considered so far in the data mining literature (but see also [9,12,10]).

We investigate the properties of the solution space $Th(q, \mathcal{L}, \mathcal{D})$. More specifically, for conjunctive queries q the solution space is a version space, which can be compactly represented using its border sets S and G, cf. [18,8,22,21,24,25,29]. For arbitrary boolean inductive queries, $Th(q, \mathcal{L}, \mathcal{D})$ is a union of version spaces. Furthermore, solution spaces and their border sets can be manipulated using operations on version spaces, such as those studied by Hirsh [25] and Gunter et al. [13]. These operations are used in an algorithm for computing the solution spaces w.r.t. an inductive query. The efficiency (and the borders that characterize the outcome) of this algorithm depend not only on the semantics of the query but also on its syntactic form. This raises the issue of query optimization, which is concerned with transforming the original query into an efficient and semantically equivalent form. This situation is akin to that in traditional databases and we formulate some initial ideas in this direction, cf. also [9].

This paper is organized as follows. In Section 2, we briefly review the MolFea system, which forms the motivation for much of this work, in Section 3, we introduce some basic terminology, in Section 4, we discuss boundary sets, in Section 5, we present an algorithm to evaluate queries, in Section 6, we discuss query optimization, and finally, in Section 7, we conclude.

2 A Motivating Example

MolFea is a domain specific inductive database for mining features of interest in sets of molecules. The examples in MolFea are thus molecules, and the patterns are molecular fragments. More specifically, in [18] we employed the 2D structure of molecules, and linear sequences of atoms and bonds as fragments. An example molecule named AZT, a commonly used drug against HIV, is illustrated in Figure 1. Two interesting molecular fragments discovered using MolFea are:

```
'N=N=N-C-C-C-n:c:c:c=O'
'N=N=N-C-C-C-n:c:n:c=O'
```

[1] This notation is adapted from that introduced by Mannila and Toivonen [28].

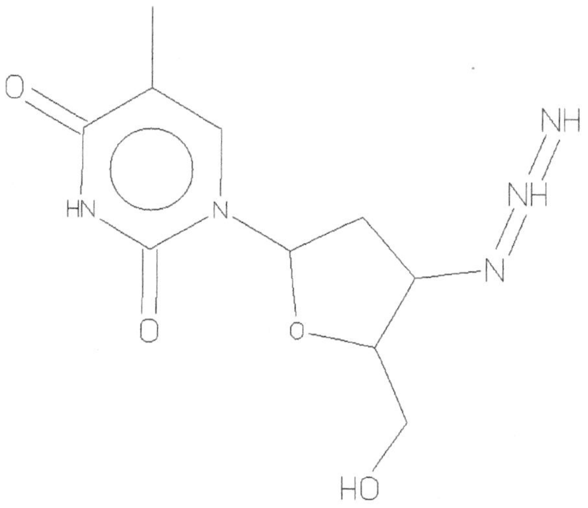

Fig. 1. Chemical structure of azidothymidine

In these fragments, ‘C’, ‘N’, ‘Cl’, etc. denote elements[2] , and ‘-’ denotes a single bond, ‘=’ a double bond, ‘#’ a triple bond, and ‘:’ an aromatic bond. The two fragments occur in AZT because there exist labelled paths in AZT that corresponds to these fragments.

These two patterns have been discovered using MolFea in a database containing over 40 000 molecules that have been tested in vitro for activity against HIV, cf. [18] for more details. On the basis of these tests, the molecules have been divided into three categories: confirmed active CA, moderately active MA and inactive I. The above patterns are solutions to a query of the form $freq(\tau, CA) \geq x \wedge freq(\tau, I) \leq y$ where x, y are thresholds. So, we were interested in finding those fragments that occur frequently in the data sets CA (consisting of about 400 substances) and that are infrequent in I (consisting of more than 40 000 substances). In order to answer such conjunctive queries, MolFea employs the level wise version space algorithm, cf. also Section 5.1 and [8]. It should be noted that MolFea has also been employed on a wide range of other molecular data sets, cf. [17]. In such applications, one is often interested in classification. When this is the case, one can proceed by first deriving interesting patterns using the type of query specified above and then using the discovered patterns as boolean attributes in a predictive data mining system, such as a decision tree or a support vector machine. As shown in [17], effective predictors can often be obtained using this method. Even though MolFea has proven to be quite effective on various real-life applications, it is limited in that it only processes conjunctive queries. In this paper, we study how this restriction can be lifted and consider arbitrary boolean inductive queries. The resulting framework can directly be incorporated in MolFea.

---

[2] Elements involved in aromatic bonds are written in lower-case.

3 Formalization

A useful abstraction of MolFea patterns and examples, which we will employ throughout this paper for illustration purposes, is the pattern domain of strings. This pattern domain should also be useful for other applications, e.g., concerning DNA/RNA, proteins or other bioinformatics applications. In the string pattern domain, examples as well as patterns are strings expressed in a language $\mathcal{L}_\Sigma = \Sigma^*$ over an alphabet Σ. Furthermore, a pattern p matches or covers an example e if and only if p is a substring of e, i.e., the symbols of p occur at consecutive positions in e. An inductive database may consist of different data sets. These data sets may correspond to different classes of data. E.g., in MolFea, we have analyzed classes of confirmed active, moderately active and inactive molecules w.r.t. HIV-activity [18].

Example 1. A toy database that we will be using throughout this paper is

- $e_1 = aabbcc$; $e_2 = abbc$; $e_3 = bb$; $e_4 = abc$; $e_5 = bc$; $e_6 = cc$
- $D_1 = \{e_1, e_2, e_3\} = \{aabbcc, abbc, bb\}$;
  $D_2 = \{e_4, e_5, e_6\} = \{abc, bc, cc\}$;
  $D_3 = D_1 \cup D_2$
- $p_1 = abb$; $p_2 = bb$; $p_3 = cc$
- $\mathcal{D} = \{D_1, D_2, D_3\}$

In addition to using the pattern domain of strings, we will employ the data miner's favorite item sets [1]. Then, if $\mathcal{I}$ is the set of items considered, examples e as well as patterns p are subsets of $\mathcal{I}$ and the language of all patterns $\mathcal{L} = 2^{\mathcal{I}}$. Furthermore, the database may then consist of various data sets, i.e., sets of item sets.

Throughout this paper, we will allow for inductive queries $q(\tau)$ that contain exactly one pattern variable τ. These queries can be interpreted as sets of patterns $Th(q(\tau), \mathcal{L}, \mathcal{D}) = \{\tau \in \mathcal{L} \mid q(\tau) \textit{ is true in } \mathcal{D}\}$, and for compactness reasons we will often write $q(\tau)$ to denote this solution set. Inductive queries are boolean expressions over atomic queries. An *atomic query* $q(\tau)$, or *atom* for short, is a logical atom $p(t_1, ..., t_n)$ over a predicate p of arity n where $n-1$ of the arguments t_i are specified (i.e. ground) and one of the arguments t_i is the pattern variable τ. Let us now - by means of illustration - review MolFea's atomic queries.

- Let g and s be strings. Then g *is more general than* s, notation $g \preccurlyeq s$, if and only if g is a substring of s. E.g., on our earlier example, $p_2 \preccurlyeq p_1$ evaluates to true, and $p_3 \preccurlyeq p_1$ to false. This predicate – defined in the context of strings – applies to virtually any pattern domain. It corresponds to the well-known generality relation in concept-learning, cf., [22].
- The $\preccurlyeq$ relation can now be used in atomic queries of the type $\tau \preccurlyeq p$, $\neg(\tau \preccurlyeq p)$, $p \preccurlyeq \tau$, and $\neg(p \preccurlyeq \tau)$, where τ is a pattern variable denoting the target pattern and p a specific pattern. E.g., $\tau \preccurlyeq abc$ yields as solutions the set of substrings of abc.

- Let p be a pattern and D a data set, i.e. a set of examples. Then $freq(p, D) = card\{e \in D \mid p \preccurlyeq e\}$, where $card(S)$ denotes the cardinality of the set S. So, $freq(p, D)$ denotes the number of instances in D covered by p, i.e. the frequency of p in D. E.g., $freq(p_3, D_2) = 1$.
- The $freq$ construct can now be used in atoms of the following form: $freq(\tau, D) \geq t$ and $freq(\tau,D) \leq t$ where t is a numerical threshold, τ is the queried pattern or string, and D is a given data set. E.g., $freq(\tau, D_2) \geq 2$ yields the set of all substrings of bc.

Throughout this paper, as in most other works on constraint based mining, we will assume that all atoms are either monotonic or anti-monotonic. These notions are defined w.r.t. the generality relation $\preccurlyeq$ among patterns. Formally speaking, the generality relation is a quasi-ordering (i.e., a relation satisfying the reflexivity and transitivity relations).

Definition 1. *An atom p is* monotonic *if and only if* $\forall x \in \mathcal{L} : (x \preccurlyeq y) \wedge p(x) \rightarrow p(y)$.

Definition 2. *An atom p is* anti-monotonic *if and only if* $\forall x \in \mathcal{L} : (x \preccurlyeq y) \wedge p(y) \rightarrow p(x)$.

Let us illustrate these definitions on the domain of item sets.

Example 2. Consider the lattice consisting of the item sets over $\{a, b, c, d\}$ graphically illustrated in Figure 2. The frequencies of the item sets on data sets d_1 and d_2 are specified between brackets.

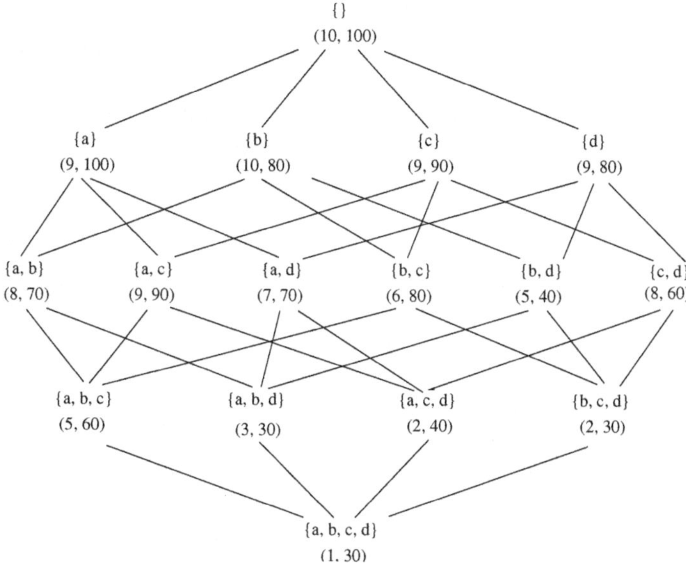

Fig. 2. An item set lattice with frequencies in data sets d_1 and d_2

Examples of anti-monotonic atoms include $freq(\tau, d_1) > 7$ for the item set domain and $\tau \preccurlyeq acd$ for the string domain. Monotonic ones include $freq(\tau, d_2) < 30$ and $\tau \succcurlyeq \{a, c, d\}$ for the item set example.

By now we can define boolean inductive queries.

Definition 3. *Any monotonic or anti-monotonic atom of the form $p(\tau)$ is a query. Furthermore, if $q_1(\tau)$ and $q_2(\tau)$ are queries over the same pattern variable τ, then $\neg q_1(\tau)$ and $q_1(\tau) \vee q_2(\tau)$ and $q_1(\tau) \wedge q_2(\tau)$ are also queries.*

The definition of the solution sets $Th(q, \mathcal{L}, \mathcal{D})$ can now be extended to boolean inductive queries in the natural way, i.e.

- $Th(q_1(\tau) \vee q_2(\tau), \mathcal{L}, \mathcal{D}) = Th(q_1(\tau), \mathcal{L}, \mathcal{D}) \cup Th(q_2(\tau), \mathcal{L}, \mathcal{D}) = q_1(\tau) \cup q_2(\tau)$;
- $Th(q_1(\tau) \wedge q_2(\tau), \mathcal{L}, \mathcal{D}) = Th(q_1(\tau), \mathcal{L}, \mathcal{D}) \cap Th(q_2(\tau), \mathcal{L}, \mathcal{D}) = q_1(\tau) \cap q_2(\tau)$;
- $Th(\neg q(\tau), \mathcal{L}, \mathcal{D}) = \mathcal{L} - Th(q(\tau), \mathcal{L}, \mathcal{D}) = \mathcal{L} - q(\tau)$.

This definition implies that $((freq(\tau, d_1) > 5) \vee (freq(\tau, d_2) < 90)) \wedge (\tau \preccurlyeq \{a, b\})$ is a query and that its solution set can be computed as $((freq(\tau, d_1) > 5) \cup (freq(\tau, d_2) < 90)) \cap (\tau \preccurlyeq \{a, b\})$. Moreover, the solution set for this query w.r.t. Figure 2 is $\{\{b\}, \{a, b\}\}$.

Using boolean inductive queries over the predicates available in MolFea, one can formulate the following type of queries:

- Traditional minimal frequency queries can be performed using $freq(\tau, D_1) \geq 2)$.
- Complex queries such as $(freq(\tau, D_{pos}) \geq n) \wedge (freq(\tau, D_{neg}) \leq m)$ ask for the set of patterns that are frequent on the positive examples D_{pos} and infrequent on the negatives in D_{neg}.
- Further syntactic constraints could be added. E.g., if we are interested only in patterns that are a substring of *abababab* and a superstring of *ab* we could refine the above query to $(freq(\tau, D_{pos}) \geq n) \wedge (freq(\tau, D_{neg}) \preccurlyeq m) \wedge (\tau \preccurlyeq abababab) \wedge (ab \preccurlyeq \tau)$.

In MolFea, inductive queries were required to be conjunctive. The techniques studied in this paper allow us to lift that restriction and to extend MolFea in order to answer any boolean inductive queries.

Since one can consider inductive queries as atoms over a single pattern variable, we will sometimes be talking about anti-monotonic and monotonic queries. The set of all monotonic queries will be denoted as $\mathcal{M}$ and the set of all anti-monotonic ones by $\mathcal{A}$. Some well-known properties of $\mathcal{A}$ and $\mathcal{M}$ are summarized in the following property.

Property 1. *If $a, a_1, a_2 \in \mathcal{A}$ and $m, m_1, m_2 \in \mathcal{M}$, then $\neg a \in \mathcal{M}$; $\neg m \in \mathcal{A}$; $(a_1 \vee a_2) \in \mathcal{A}$; $(a_1 \wedge a_2) \in \mathcal{A}$; $(m_1 \vee m_2) \in \mathcal{M}$ and $(m_1 \wedge m_2) \in \mathcal{M}$.*

4 Boundary Sets

Within the fields of data mining and machine learning it is well-known that the space of solutions $Th(q, \mathcal{L}, \mathcal{D})$ to specific types of inductive queries can be represented using boundary sets. E.g., Hirsh [24,25] has shown that sets that are convex and definite can be represented using their boundaries, Mellish [21] and Gunter et al. [13] also list conditions for the representability of sets using their borders in the context of version spaces, Mannila and Toivonen [28] have shown that the solution space for anti-monotonic queries can be represented using the set of minimally general elements. Furthermore, various algorithms exploit these properties for efficiently finding solutions to queries, cf. Bayardo's MaxMiner [3] and Mitchell's candidate-elimination algorithm [22] in the context of concept-learning. Most of the early work on version spaces was concerned with concept learning (i.e. queries of the form

$$freq(\tau, Pos) \geq \mid Pos \mid \wedge freq(\tau, Neg) \leq 0$$

where Pos is a set of positive and Neg of negative examples). In data mining, one typically investigates queries of the form $freq(\tau, D) \geq x$. One of the key contributions of the MolFea framework was the realization that inductive conjunctive queries can – when their solution set is finite – be represented using boundary sets. Furthermore, MolFea employed a novel algorithm, the so-called level-wise version space algorithm for computing these boundary sets (cf. also Section 5.1). Here, we will study how to extend this framework for dealing with arbitrary boolean queries.

The boundary sets, sometimes also called the borders, are the maximally specific (resp. general) patterns within the set. More formally, let P be a set of patterns. Then we denote the set of minimal (i.e. minimally specific) patterns within P as $min(P)$. Dually, $max(P)$ denotes the maximally specific patterns within P.

In the remainder of this paper, we will be largely following the terminology and notation introduced by Mitchell [22], Hirsh [24] and Mellish [21].

Definition 4. *The S-set $S(P)$ w.r.t. a set of patterns $P \subset \mathcal{L}$ is defined as $S(P) = max(P)$; and the G-set w.r.t. a set of patterns $P \subset \mathcal{L}$ is defined as $G(P) = min(P)$.*

We will use this definition to characterize the solution sets of certain types of queries.

Definition 5. *A set T is upper boundary set representable if and only if $T = \{t \in \mathcal{L} \mid \exists g \in min(T) : g \preccurlyeq t\}$; it is lower boundary set representable if and only if $T = \{t \in \mathcal{L} \mid \exists s \in max(T) : t \preccurlyeq s\}$; and it is boundary set representable if and only if $T = \{t \in \mathcal{L} \mid \exists g \in min(T), s \in max(T) : g \preccurlyeq t \preccurlyeq s\}$.*

Sets that are boundary set representable are sometimes also referred to as *version spaces*. Remark that the above definition allows the border sets to be infinitely large.

Example 3. Consider the constraint $freq(\tau, d_1) > 4$ over the item sets in Figure 2. Then $S(freq(\tau, d_1) > 4) = \{\{a,d\},\{b,d\},\{c,d\},\{a,b,c\}\}$; $G(\{a,b\} \preccurlyeq \tau) = \{\{a,b\}\}$, and finally, $S((freq(\tau, d_1) > 4) \wedge (\{a\} \preccurlyeq \tau)) = \{\{a,b,c\},\{a,d\}\}$ and $S((freq(\tau, d_1) > 4) \wedge (\{a,b\} \preccurlyeq \tau)) = \{\{a,b,c\}\}$.

Hirsh [25] has characterized version spaces using notions of convexity and definiteness as specified in Theorem 1.

Definition 6. *A set P is convex if and only if for all $p_1, p_2 \in P, p \in \mathcal{L} : p_1 \preccurlyeq p \preccurlyeq p_2$ implies that $p \in P$.*

Definition 7. *A set P is definite if and only if for all $p \in P, \exists s \in max(P), g \in min(P) : g \preccurlyeq p \preccurlyeq s$.*

Theorem 1 (Hirsh). *A set is boundary set representable if and only if it is convex and definite.*

Let us now investigate the implications of Hirsh's framework in the context of inductive queries.

Lemma 1. *Let $a \wedge m$ be a query such that $a \in \mathcal{A}$ and $m \in \mathcal{M}$. Then $Th(a \wedge m, \mathcal{L}, \mathcal{D})$ is convex.*

This property directly follows from the definitions of monotonicity and anti-monotonicity. If we want to employ border representation for representing solution sets we also need that $Th(a \wedge m, \mathcal{L}, \mathcal{D})$ is definite. According to Hirsh [25], any finite set is also definite.

Lemma 2. *If P is a convex and finite set, then it is boundary set representable.*

This implies that the solution space for queries of the form $a \wedge m$ over the pattern language of item sets, is boundary representable. What about infinite pattern languages such as Σ^*, the set of strings over the alphabet Σ? Consider the pattern set Σ^*. This set is not definite because $max(\Sigma^*)$ does not exist. Indeed, for every string in Σ^* there is a more specific one. One way to circumvent this problem is to introduce a special bottom element $\perp$ that is by definition more specific than any other string in Σ^*, cf. also [13,21], and to add it to the language of patterns $\mathcal{L}$. However, this is not a complete solution to the problem. Indeed, consider the language Σ^* over the alphabet $\Sigma = \{a, b\}$ and the set $As = a^*$ (where we employ the regular expression notation). This set is again not definite because $\perp \notin As$ and infinite strings (such as $a^\infty \notin \Sigma^*$) are not strings. Thus $max(As)$ does not exist. Furthermore, the set As is the solution to the inductive query $\neg(b \preccurlyeq \tau)$. Other problems arise because the boundary set may be infinitely large. Consider, e.g., the constraint $(a \preccurlyeq \tau) \wedge (b \preccurlyeq \tau)$ over the language Σ^*, where $\Sigma = \{a, b, c\}$. The G-set in this case would be $ac^*b \cup bc^*a$ because any string containing an a and a b must be more specific than a string of the form ac^nb or bc^na. Thus, it seems that not all queries result in boundary set representable version spaces. This motivates our definition of *safe* queries.

Definition 8. *A conjunctive query q is safe if and only if its solution set is boundary set representable and its borders are finite.*

This situation is akin to that in traditional databases, where some queries are also unsafe, e.g. queries for the set of tuples that do not belong to a given relation.

One immediate question that arises is which queries are safe for a particular type of pattern language $\mathcal{L}$ and constraint. Let us first remark, that all queries in the case of finite languages (such as e.g. item sets) are safe. On the other hand, for infinite pattern languages (such as strings) and the MolFea primitives, we have that

- $\tau \preccurlyeq s$ where $s \in \Sigma^*$ is safe because $S = \{s\}$
- $\neg(\tau \preccurlyeq s)$ with $s \in \Sigma^*$ is safe because $G = max\{g \in \Sigma \cup \Sigma^2 \cup ... \cup \Sigma^k \cup \Sigma^{k+1} \mid k = length(s) \; and \; \neg(g \preccurlyeq s)\}$. This is a finite set as all of its constituents are.
- $s \preccurlyeq \tau$ is safe because $G = \{s\}$.
- $\neg(s \preccurlyeq \tau)$ is not safe as shown above.
- $freq(\tau, D) \geq t$ is safe, because if $t \geq 1$ then patterns τ satisfying $freq(\tau, D) \geq t$ must be substrings of strings in D and there are only a finite number of them; on the other hand if $t = 0$ then $S = \{\bot\}$, so S always exist and is finite.
- $freq(\tau, D) \leq t$ is safe, because there are only a finite number of substrings of D that are excluded from considerations. So, $G = max\{g \in \Sigma \cup \Sigma^2 \cup ... \cup \Sigma^k \cup \Sigma^{k+1} \mid k = length(s) \; where \; s \; is \; the \; longest \; string \; in \; D \; and \; freq(g, D) \leq t\}$. This is a finite set.
- Furthermore, if query q is of the form $a \wedge m$ where $a \in \mathcal{A}$ and $m \in \mathcal{M}$ for which a is safe, then q will also be safe. The reason for this is that a finite S restricts the solution set to being finite (as there exist only a finite number of substrings of strings in S).

5 Query Evaluation

Let us now investigate how queries can be evaluated and how the boundary sets can be employed in this context.

5.1 Conjunctive Queries

The first observation is that - as argued above - the solution sets to safe conjunctive queries are completely characterized by their S and G-sets. Furthermore, there exist several algorithms that can be used to compute such border sets. First, it is well-known that the S-set or positive border for safe anti-monotonic queries can be computed using adaptations of the level-wise algorithm [19] such as the MaxMiner system by Bayardo [3]. Dual versions of these algorithms could be used to compute the G set w.r.t. monotonic safe queries, cf. [8,18]. Secondly, several algorithms exist that either compute the S and G set w.r.t. a safe conjunctive query, such as the level wise version space algorithm underlying MolFea

[8]. This algorithm combines the level-wise algorithm with principles of the candidate elimination algorithm studied by Mellish and Mitchell [21,22]. Furthermore, within MolFea, the level-wise version space algorithm has been used to deal with reasonably large data sets, such as the HIV data set discussed in Section 2. In addition, several algorithms exist that would compute the whole solution set of a conjunctive query over a finite pattern domain, cf. [4,5]. It should be possible to adapt these last algorithms in order to generate the borders only. In the remainder of this paper, we will not give any further details on how to compute a specific S and/or G-set as this problem has been studied thoroughly in the literature. Furthermore, we will assume that queries are safe unless stated otherwise.

5.2 Boolean Queries

Let us now address the real topic of this paper, which is how to compute and characterize the solution space w.r.t. general boolean queries. Before investigating these solution spaces more closely, let us mention that when q and q' are logically equivalent then their solution sets are identical. This is formalized below.

Property 2. *Let q and q' be two boolean queries that are logically equivalent. Then $Th(q, \mathcal{L}, \mathcal{D})= Th(q', \mathcal{L}, \mathcal{D})$ for any inductive database $\mathcal{D}$.*

This property is quite useful in showing that the solution space of a safe general boolean query can be specified as the union of various convex sets.

Let q be a boolean inductive query. Then $Th(q, \mathcal{L}, \mathcal{D})$ will consist of the union of a finite number of convex sets $\cup_i Th(q_i, \mathcal{L}, \mathcal{D})$. To see why this is the case, one can rewrite the query q into its disjunctive normal form $q = q_1 \vee ... \vee q_n$. In this form each of the q_i is a conjunctive query. Furthermore, if these sub-queries q_i are safe, their solution sets $Th(q_i, \mathcal{L}, \mathcal{D})$ form version spaces that can be represented by their border sets. This motivates the following definition and result.

Definition 9. *A boolean inductive query q is safe if and only if there exists an equivalent formula $q' = q_1 \vee ... \vee q_k$ that is in DNF and for which all q_i are safe.*

Notice that the fact that a query is safe does not imply that all possible rewrites in DNF will yield safe q_i. Indeed, let a (resp. b) denote the atom $(a \preccurlyeq \tau)$ (resp. $(b \preccurlyeq \tau)$). Then the query a is equivalent to $(a \wedge b) \vee (a \wedge \neg b)$ of which the first conjunction is not safe.

Lemma 3. *Let q be a safe boolean query. Then $Th(q, \mathcal{L}, \mathcal{D})$ is a union of version spaces, each of which can be represented by its boundary sets.*

The question of solving boolean queries is now reduced to that of computing the version spaces or their border sets.

5.3 Operations on Border Sets

We will now employ a simple algebra that will allow us to compute the boundary sets w.r.t. given boolean queries. The algebraic operations (such as intersection and union) are not new. They have been studied before by Haym Hirsh [25] and Gunter et al. [13]. However, Hirsh and Gunter studied version spaces in quite different contexts such as concept-learning and in the case of Gunter et al. also truth maintenance. Our context, i.e. inductive querying using version spaces, is new and so is the incorporation of the results about safety and the use of the pattern domain of strings.

We first show how the boundary sets of a safe query of the form $q_1 \Delta q_2$ (where Δ is conjunction or disjunction) can be computed in terms of the boundary sets of the queries q_i.

Lemma 4. *(Disjunction)*

- *If* $a_1, a_2 \in \mathcal{A}$ *and* a_1 *and* a_2 *are safe then* $a_1 \vee a_2$ *is safe and* $Th(a_1 \vee a_2, \mathcal{L}, \mathcal{D})$ *is lower boundary representable using* $S(a_1 \vee a_2) = max(S(a_1) \cup S(a_2))$.
- *If* $m_1, m_2 \in \mathcal{M}$ *and* m_1 *and* m_2 *are safe, then* $m_1 \vee m_2$ *is safe and* $Th(m_1 \vee m_2, \mathcal{L}, \mathcal{D})$ *is upper boundary set representable using* $G(m_1 \vee m_2) = min(G(m_1) \cup G(m_2))$.

Let us first define the necessary operations used when intersecting two version spaces. Here *mgg* and *mss* denote the minimally general generalizations and maximally specific specializations of the two elements[3].

Definition 10. *(mgg and mss)*

- $mgg(s_1, s_2) \equiv max\{s \in \mathcal{L} \mid s \preceq s_1 \wedge s \preceq s_2\}$.
- $mss(g_1, g_2) \equiv min\{g \in \mathcal{L} \mid g_1 \preceq g \wedge g_2 \preceq g\}$.

Lemma 5. *(Conjunction)*

- *If* $a_1, a_2 \in \mathcal{A}$ *and* a_1, a_2 *and* $a_1 \wedge a_2$ *are safe, then* $Th(a_1 \wedge a_2, \mathcal{L}, \mathcal{D})$ *is lower boundary representable using* $S(a_1 \wedge a_2) = max\{m \mid m \in mgg(s_1, s_2)$ *where* $s_i \in S(a_i)\}$.
- *If* $m_1, m_2 \in \mathcal{M}$ *and* m_1, m_2 *and* $m_1 \wedge m_2$ *are safe then* $Th(m_1 \wedge m_2, \mathcal{L}, \mathcal{D})$ *is upper boundary set representable using* $G(m_1 \wedge m_2) = min\{m \mid m \in mss(g_1, g_2)$ *where* $g_i \in G(m_i)\}$.

Lemma's 4 and 5 motivate the introduction of the following operations:

- $G(m_1) \vee G(m_2) = min(G(m_1) \cup G(m_2))$
- $S(a_1) \vee S(a_2) = max(S(a_1) \cup S(a_2))$
- $S(a_1) \wedge S(a_2) = max\{m \mid m \in mgg(s_1, s_2)$ *where* $s_i \in S(a_i)\}$
- $G(m_1) \wedge G(m_2) = min\{m \mid m \in mss(g_1, g_2)$ *where* $g_i \in G(m_i)\}$.

[3] Gunter et al. [13] call this the quasi-meet and quasi-join, Mellish [21] uses *aa* and *bb* to denote these operations.

Notice that the conjunctive operation on border sets is - in general - not safe. An illustration is the query $(a \preceq \tau) \wedge (b \preceq \tau)$ which we discussed earlier. The problem with this query is that even though both atoms are safe, their conjunction is not, due to an infinitely large G-set.

Lemma 6. *If q_1, q_2 and $q_1 \wedge q_2$ are safe, then $Th(q_1 \wedge q_2, \mathcal{L}, \mathcal{D})$ is boundary set representable using $S(q_1 \wedge q_2) = S(q_1) \wedge S(q_2)$ and $G(q_1 \wedge q_2) = G(q_1) \wedge G(q_2)$.*

This operation corresponds to Hirsh's version space intersection operation, which he employs in the context of concept-learning and explanation-based learning.

The corresponding property for disjunctive queries $q_1 \vee q_2$ does not hold in general, i.e., $q_1 \vee q_2$ is not necessarily a convex space.

Example 4. Consider that we have the version spaces VS_1 and VS_2 containing item-sets, and having the following boundary sets : $S_1 = \{\{a,b,c,d\}\}; G_1 = \{\{a\}\}$ and $S_2 = \{\{c,d,e,f\}\}; G_2 = \{\{d,e\}\}$. In this example, the set $VS_1 \cup VS_2$ is not boundary set representable. The reason is that the version space corresponding to these boundary sets would also include elements like $\{a,f\}$ which do neither belong to VS_1 nor to VS_2.

In general, if we have a query that is a disjunction $q_1 \vee q_2$, we will have to keep track of two version spaces, that is, we need to compute $S(q_1)$, $G(q_1)$ and $S(q_2)$ and $G(q_2)$. The semantics of such a union or a disjunction of version spaces is then that all elements that are in at least one of the version spaces are included.

Now, everything is in place to formulate a procedure that will generate the boundary sets $Th(q, \mathcal{L}, \mathcal{D})$ corresponding to a query q. In this procedure, we assume that all queries and subqueries are safe. Furthermore, (S, G) denotes the version space characterized by the border sets S and G, i.e. the set $\{\tau \in \mathcal{L} \mid \exists s \in S, g \in G : g \preceq \tau \preceq s\}$. Thus the notation $(S_1, G_1) \cup (S_2, G_2)$ denotes the union of the two version spaces characterized by (S_1, G_1) and (S_2, G_2). This notation can trivially be extended towards intersection. Furthermore, it is useful in combination with the operations on version spaces. Indeed, consider that

$$(S_1, G_1) \cap (S_2, G_2) = (S_1 \wedge S_2, G_1 \wedge G_2)$$

which denotes the fact that the intersection of the two version spaces is a version space obtained by performing the $\wedge$ operation on its corresponding border sets.

The symbols $\top$ and $\bot$ denote the sets of minimally, respectively maximally specific elements. These are assumed to exist. If they do not exist, as in the case of $\bot$ for strings, one has to introduce an artificial one.

By now we are able to formulate our algorithm, cf. Figure 3. It essentially computes the borders corresponding to the solution sets of the queries using the logical operations introduced earlier.

Various rewrites can be applied to simplify the final result. As an example consider $(S, G_1) \cup (S, G_2) = (S, G_1 \cup G_2)$ and $(S_1, G) \cup (S_2, G) = (S_1 \cup S_2, G)$.

Example 5. To illustrate the above algorithm, consider the query $(freq(\tau, d_1) \geq 5) \vee (freq(\tau, d_2) \leq 85)$. The algorithm would compute $(\{\{a,b,c\}, \{b,d\}, \{c,d\}\},$

function $vs(q)$ **returns** a set of boundary sets S_i, G_i
such that $Th(q, \mathcal{L}, \mathcal{D}) = \cup_i (S_i, G_i)$
case $q \in \mathcal{A}$ **do**
return $(S(q), \{\top\})$
case $q \in \mathcal{M}$ **do**
return $(\{\bot\}, G(q))$
case q of $a \wedge m$ with $a \in \mathcal{A}, m \in \mathcal{M}$ **do**
return $(S(a \wedge m), G(a \wedge m))$
case q of the form $q_1 \wedge q_2$ with $q_1, q_2 \notin \mathcal{A} \cup \mathcal{M}$ **do**
call $vs(q_1)$ and $vs(q_2)$
assume $vs(q_1)$ returns $(S_1, G_1) \cup ... \cup (S_n, G_n)$
assume $vs(q_2)$ returns $(S'_1, G'_1) \cup ... \cup (S'_m, G'_m)$
return $\bigcup_{1 \preccurlyeq i \preccurlyeq n, 1 \preccurlyeq j \preccurlyeq m} (min(S_i \wedge S'_j), max(G'_i \wedge G'_j))$
case q of the form $q_1 \vee q_2$ with $q_1, q_2 \notin \mathcal{A} \cup \mathcal{M}$ **do**
call $vs(q_1)$ and $vs(q_2)$
assume $vs(q_1)$ returns $(S_1, G_1) \cup ... \cup (S_n, G_n)$
assume $vs(q_2)$ returns $(S'_1, G'_1) \cup ... \cup (S'_m, G'_m)$
return $(S_1, G_1) \cup ... \cup (S_n, G_n) \cup (S'_1, G'_1) \cup ... \cup (S'_m, G'_m)$

Fig. 3. Computing the solution set w.r.t. an inductive query

$\{\{\}\})$ for the first atom and $(\{\{a, b, c, d\}\}, \{\{b\}, \{d\}\})$ for the second one and then take the union (cf. the last case in the algorithm).

6 Query Optimization

The results and efficiency of the previous algorithm strongly depend on the syntactic form of the inductive query. Indeed, consider two queries q_1 and q_2 that are logically equivalent but syntactically different. Even though the solution sets are the same in both cases, the output representations of the boundaries can greatly differ. As an example consider the boolean expression $q_1 = (a_1 \vee a_2) \wedge (m_1 \vee m_2)$ and its logically equivalent formulation $q_2 = (a_1 \wedge m_1) \vee (a_1 \wedge m_2) \vee (a_2 \wedge m_1) \vee (a_2 \wedge m_2)$. The result of q_1 will be represented as a single version space, because the query is of the form $a \wedge m$ (third case). On the other hand, the result of q_2 would be represented as union of four version spaces (last case). Secondly, if one applies the algorithm in a naive way, various boundary sets may be recomputed using q_2, e.g. the $S(a_i)$ and $G(m_i)$.

It is therefore advantageous to consider optimizing the queries. Optimization then consists of reformulating the query into a form that is 1) logically equivalent and 2) that is in a form that is more efficient to compute. When formulating query optimization in this form, there is a clear analogy with that of circuit design and optimization. Indeed, when designing logical circuits one also starts from a logical formula (or possibly a truth-table) and logically rewrites it in the desired optimized form. In this optimization process certain connectives are to be preferred and other ones are to be avoided. This is akin to the query optimization problem that we face here, where the different primitives and operations may have different costs. It is thus likely that the solutions that exist for circuit design can be adapted towards those for query optimization. The connection with circuit design remains however the topic of further work.

Below we discuss one form of optimization, due to De Raedt et al. [9], where the optimization criterion tries to minimize the number of calls to the level wise version space algorithm. In addition, some simple but powerful means of reasoning about queries are also discussed in Sections 6.2 and 6.3.

6.1 Minimizing the Number of Version Spaces

In the context of query optimization, one desirable form of query is $a \wedge m$ where $a \in \mathcal{A}$ and $m \in \mathcal{M}$. Let us call this type of query the version space normal form. If one can rewrite a query in the version space normal form, the result will be a single version space (provided that the query is safe). We can then also compute the boundary sets using known algorithms such as the level wise version space algorithm by [8] and that by Boulicaut [5]. Notice that it is not always possible to do this. Consider e.g. a query of the form $a \vee m$ where both a and m are atomic. A natural question that arises in this context is how many version spaces are – in the worst case[4] – needed to represent the solution space to an inductive query. This problem is tackled in a recent paper by De Raedt et al. [9]. More formally, De Raedt et al. study the problem of reformulating a boolean inductive query q into a logically equivalent query of the form $q_1 \wedge ... \wedge q_k$ where each $q_i = a_i \wedge m_i$ is in version space normal form (i.e. a_i is an anti-monotonic query and m_i a monotonic one) and such that k is minimal. The number k is called the dimension of the query q. This is useful for query optimization because using the reformulated query in the algorithm specified in Figure 3, would result in the minimum number k of calls to an algorithm for computing a version space. As an example, consider the queries $q_1 = (a_1 \vee a_2) \wedge (m_1 \vee m_2)$ and its logically equivalent formulation $q_2 = (a_1 \wedge m_1) \vee (a_1 \wedge m_2) \vee (a_2 \wedge m1) \vee (a_2 \wedge m_2)$. q_1 is in version space normal form, which shows that the dimension of both queries is 1. Now, De Raedt et al. provide a procedure for determining the dimension of a query q and for rewriting it into subqueries q_i that are in version space normal form. This procedure could be used to rewrite the query q_2 into q_1. More specifically, they show that how to obtain the a_i and m_i where each a_i is of the form $(a_{1,1} \wedge ... \wedge a_{1,n_1}) \vee ... \vee (a_{k,1} \wedge ... \wedge a_{k,n_k})$ and each m_i of the form $(m_{1,1} \wedge ... \wedge m_{1,n_1}) \vee ... \vee (m_{k,1} \wedge ... \wedge m_{k,n_k})$. Observe that each a_i is anti-monotonic because all the conjunctions in a_i are (a conjunction of anti-monotonic atoms is anti-monotonic) and a disjunction of anti-monotonic queries is also anti-monotonic. Similarly, each m_i is monotonic. A further contribution by De Raedt et al. is that they show that the dimension of a query $q(\mathcal{T})$ corresponds to the length of the longest alternating chain in its solution space. More formally,

[4] The best case corresponds to the situation where all atoms evaluate to the empty set. In this uninteresting case, the number of version spaces needed is 0. The worst case arises when all queries (that are not logically equivalent) have a different solution set.

an alternating chain of length k is a sequence $p_1 \preccurlyeq ... \preccurlyeq p_{2k-1}$ such that for all odd i, $p_i \in q(\tau)$ and for all even i, $p_i \notin q(\tau)$.

In the context of query optimization, two other points are worth mentioning. These concern the use of background knowledge and subsumption.

6.2 Subsumption

We previously explained that two logically equivalent queries have the same set of solutions. Often it is useful to also consider subsumption among different queries.

Lemma 7. *Let q and q' be two boolean queries. If q logically entails q', then $Th(q, \mathcal{L}, \mathcal{D}) \subseteq Th(q', \mathcal{L}, \mathcal{D})$. We say that q' subsumes q.*

This property can be used in a variety of contexts, cf. [2]. First, during one query answering session there may be several related queries. The user of the inductive database might first formulate the query q' and later refine it to q. If the results of the previous queries (or query parts) are stored, one could use them as the starting point for answering the query. Secondly, one might use it during the query optimization process. E.g., suppose one derives a $q \vee q'$ such that $q \rightarrow q'$. One could the simplify the query into q'.

As opposed to this logical or intentional subsumption test, it is also possible to perform an *extensional* subsumption test among version spaces.

Definition 11. *A version space (S_1, G_1) extensionally subsumes (S_2, G_2) if and only if $S_2 \cup G_2 \subseteq (S_1, G_1)$.*

Whenever q is subsumed by q', $(S(q), G(q))$ will be extensionally subsumed by $(S(q'), G(q'))$ though the reverse property does not always hold.

6.3 Background Knowledge

For many query languages it is useful to also formulate background knowledge. Such background knowledge then contains properties about the primitives sketched. We illustrate the idea on frequency. Let $freq(p, d)$ denote the frequency of pattern p on data set d. Atoms would typically impose a minimum (or maximum frequency) on a given data set. We can then incorporate the following logical sentences in our background theory KB.

$$\forall p, d, c_1, c_2 : (freq(p, d) > c_1) \wedge (c_1 > c_2) \rightarrow freq(p, d) > c_2$$

$$\forall p, d, c_1, c_2 : (freq(p, d) < c_2) \wedge (c_1 > c_2) \rightarrow freq(p, d) < c_1$$

Such background knowledge can be used to reason about queries. This motivates the following theorems.

Property 3. *Let q and q' be two boolean queries and let KB be the background theory about the constraint primitives. If $KB \models q \rightarrow q'$, then $Th(q, \mathcal{L}, \mathcal{D}) \subseteq Th(q', \mathcal{L}, \mathcal{D})$.*

In the light of the formulated background theory the first lemma would allow us to conclude that $freq(\tau, d) > 3$ subsumes $freq(\tau, d) > 5$.

7 Conclusion

We have introduced a novel and expressive framework for inductive databases in which queries correspond to boolean expressions. To the best of the author's knowledge this is the most general formulation of inductive querying so far, cf. also [12,9]. Secondly, we have shown that this formulation is interesting in that it allows us to represent the solution space of queries using border sets and also that it provides us with a logical framework for reasoning about queries and their execution. This provides hope that the framework could be useful as a theory of inductive databases.

The results in the paper are related to the topic of version spaces, which have been studied for many years in the field of machine learning, cf. e.g. [24,25,22,26,29,8]. The novelty here is that the framework is applied to that of inductive querying of boolean expressions and that this requires the use of disjunctive version spaces (cf. also [26,30]).

The work is also related to constraint based data mining, cf. e.g. [15,18]. However, as argued before, the form of query considered here is more general. The use of border sets for inductive querying has recently also been considered by [11].

Acknowledgements

This work was partly supported by the ESPRIT FET project cInQ. The author is grateful to the cInQ partners for several interesting and inspiring discussions on this topic, in particular to Jean-Francois Boulicaut, Manfred Jaeger, Stefan Kramer, Sau Dan Lee and Heikki Mannila.

References

1. R. Agrawal, T. Imielinski, A. Swami. Mining association rules between sets of items in large databases. In *Proceedings of ACM SIGMOD Conference on Management of Data*, pp. 207–216, 1993.
2. E. Baralis, G. Psaila. Incremental Refinement of Mining Queries. In Mukesh K. Mohania, A. Min Tjoa (Eds.) *Data Warehousing and Knowledge Discovery, First International Conference DaWaK '99* Proceedings. Lecture Notes in Computer Science, Vol.1676, Springer Verlag, pp. 173–182, 1999.
3. R. Bayardo. Efficiently mining long patterns from databases. In *Proceedings of ACM SIGMOD Conference on Management of Data*, 1998.
4. C. Bucila, J. Gehrke, D. Kifer, W. White. DualMiner: A dual pruning algorithm for item sets with constraints. In *Proc. of SIGKDD*, 2002.
5. Jean Francois Boulicaut, habilitation thesis, INSA-Lyon, France, 2001.
6. Jean-Francois Boulicaut, Mika Klemettinen, Heikki Mannila: Querying Inductive Databases: A Case Study on the MINE RULE Operator. In *Proceedings of PKDD-98*, Lecture Notes in Computer Science, Vol. 1510, Springer Verlag, pp. 194–202, 1998.

7. L. De Raedt, A Logical Database Mining Query Language. In *Proceedings of the 10th Inductive Logic Programming Conference*, Lecture Notes in Artificial Intelligence, Vol. 1866, Springer Verlag, pp. 78–92, 2000.
8. L. De Raedt, S. Kramer, The level wise version space algorithm and its application to molecular fragment finding, in *Proceedings of the Seventeenth International Joint Conference on Artificial Intelligence*, Morgan Kaufmann, in press, 2001.
9. L. De Raedt, M. Jaeger, S.D. Lee, H. Mannila. A theory of inductive query answering. In *Proceedings of the 2nd IEEE Conference on Data Mining*, Maebashi, Japan, 2002.
10. L. De Raedt. A perspective on inductive databases. SIGKDD Explorations, ACM, Vol. 4(2):69–77, 2002.
11. A. Giacometti, D. Laurent and C. Diop. Condensed Representations for Sets of Mining Queries. In *Proceedings of the ECML-PKDD Workshop on Knowledge Discovery in Inductive Databases*, Helsinki, 2002.
12. B. Goethals, J. Van den Bussche. On supporting interactive association rule mining. In *Proc. DAWAK*, LNCS Vol. 1874, Springer Verlag, 2000.
13. C. Gunter, T. Ngair, D. Subramanian. The common order-theoretic properties of version spaces and ATMSs, *Artificial Intelligence*, 95, 357–407, 1997.
14. J. Han, Y. Fu, K. Koperski, W. Wang, and O. Zaiane, DMQL: A Data Mining Query Language for Relational Databases, in SIGMOD'96 Workshop on Research Issues on Data Mining and Knowledge Discovery, Montreal, Canada, June 1996.
15. J. Han, L. V. S. Lakshmanan, and R. T. Ng, Constraint-Based, Multidimensional Data Mining, *Computer*, Vol. 32(8), pp. 46–50, 1999.
16. D. Gunopulos, H. Mannila, S. Saluja: Discovering All Most Specific Sentences by Randomized Algorithms. In Foto N. Afrati, Phokion Kolaitis (Eds.): *Database Theory - ICDT '97, 6th International Conference*, Lecture Notes in Computer Science, Vol. 1186, Springer Verlag, pp. 41–55, 1997.
17. S. Kramer and L. De Raedt. Feature construction with version spaces for biochemical applications. In *Proceedings of the 18th International Conference on Machine Learning*, Morgan Kaufmann, 2001.
18. S. Kramer, L. De Raedt, C. Helma. Molecular Feature Mining in HIV Data, in *Proceedings of the Seventh ACM SIGKDD International Conference on Knowledge Discovery and Data Mining*, ACM Press, in press, 2001.
19. H. Mannila and H. Toivonen, Levelwise search and borders of theories in knowledge discovery, *Data Mining and Knowledge Discovery*, Vol. 1, 1997.
20. R. Meo, G. Psaila and S. Ceri, An extension to SQL for mining association rules. *Data Mining and Knowledge Discovery*, Vol. 2 (2), pp. 195–224, 1998.
21. C. Mellish. The description identification algorithm. *Artificial Intelligence*, Vol. 52 (2), pp,. 151–168, 1990.
22. T. Mitchell. Generalization as Search, *Artificial Intelligence*, Vol. 18 (2), pp. 203–226, 1980.
23. T. Imielinski and H. Mannila. A database perspective on knowledge discovery. *Communications of the ACM*, 39(11):58–64, 1996.
24. H. Hirsh. Generalizing Version Spaces. *Machine Learning*, Vol. 17(1): 5–46 (1994).
25. H. Hirsh. Theoretical underpinnings of version spaces. Proc. IJCAI 1991.
26. T. Lau, S. Wolfman, P. Domingos, D.S. Weld, Programming by demonstration using version space algebra, *Machine Learning*, to appear.
27. F. Giannotti, G. Manco: Querying Inductive Databases via Logic-Based User-Defined Aggregates. In *Proceedings of PKDD 99*, Lecture Notes in Artificial Intelligence, Springer, 1999.

28. H. Mannila and H. Toivonen, Levelwise search and borders of theories in knowledge discovery, *Data Mining and Knowledge Discovery*, Vol. 1, 1997.
29. G. Sablon, L. De Raedt, and Maurice Bruynooghe. Iterative Versionspaces. *Artificial Intelligence*, Vol. 69(1–2), pp. 393–409, 1994.
30. G. Sablon. Iterative and Disjunctive Versionspaces. Ph.D. Thesis. Katholieke Universiteit Leuven, 1995.

The GUHA Method, Data Preprocessing and Mining

Petr Hájek[1], Jan Rauch[2], David Coufal[1], and Tomáš Feglar

[1]Institute of Computer Science, Academy of Sciences of the Czech Republic, Prague
[2]University of Economics, Prague

Abstract. The paper surveys basic principles and foundations of the GUHA method, relation to some well-known data mining systems, main publications, existing implementations and future plans.

1 Introduction: Basic Principles

GUHA (General Unary Hypotheses Automaton) is a method originated in Prague (in Czechoslovak Academy of Sciences) in mid-sixties. Its main *principle* is *to let the computer generate and evaluate all hypotheses that may be interesting from the point of view of the given data and the studied problem.* This principle has led both to a specific theory and to several software implementations. Whereas the latter became quickly obsolete, the theory elaborated in the mean time has its standing value. Typically hypotheses have the form "Many A's are B's" (B is highly frequented in A) of "A,S are mutually positively dependent". (Note that what is now called "association rules" in data mining occurs already in the first 1966 paper [14] on GUHA, see below.) A second feature, very important for GUHA, is its *explicit logical and statistical foundations.*

Logical foundations include *observational calculi* (a kind of predicate calculi with only finite models and with generalized quantifiers, serving to express relations among the attributes valid in data) and *theoretical calculi* (a kind of modal predicate calculi serving to express probabilistic or other dependencies among the attributes, meaningful in the universe of discourse). Statistical foundations include principles of statistical hypotheses testing and other topics of *exploratory data analysis.* Statistical hypothesis testing is described as a sort of inference in logical sense. But note that GUHA is not bound to generation of *statistical* hypotheses; the logical theory of observational calculi is just logic of *patterns* (associations, dependencies etc.) contained (true) in the data.

The monograph [15] contains detailed exposition of fundamentals of this theory. The underlying logical calculi are analyzed and several basic facts for corresponding algorithms are proved. Special attention is paid to deduction rules serving for optimization of knowledge representation and of intelligent search[1].

[1] This book [15] has been not more obtainable since several years ago. We are happy to announce that its publisher, Springer-Verlag, reverted the copyright to the authors which has made possible to *put the text of the book on web for free copying* as a report of the Institute of Computer Science [16].

R. Meo et al. (Eds.): Database Support for Data Mining Applications, LNAI 2682, pp. 135–153, 2004.

The reader is hoped to realize that the aim of this paper is not mere stressing of antiquity of GUHA; on the contrary, we want to show the reader that GUHA (even if old) presents an approach to data mining (knowledge discovery) that is valuable and useful *now*. We summarize and survey here the underlying notions and theory (Sect. 2 and 3), one contemporary implementation and its performance evaluation (Sect. 4), as well as an example of particular application to financial data (Sect. 5); furthermore, we comment on the relation of GUHA to current data mining and data base systems (Sect. 6 and 7). We conclude with some open problems.

2 Hypotheses Alias Rules

For simplicity imagine the data processed by GUHA as a rectangular matrix of zeros and ones, the rows corresponding to objects and columns to some attributes. (Needless to say, much more general data can be processed.) In the terminology of [1], columns correspond to *items* and rows describe itemsets corresponding to *transactions*. In logical terminology one works with predicates $P_1, \ldots, P_n$ (names of attributes), *negated predicates* $\neg P_1, \ldots \neg P_n$, elementary conjunctions (e.g. $P_1 \& \neg P_3 \& P_7$) and possibly elementary disjunctions $(P_1 \vee \neg P_3 \vee P_7)$.

Needless to say, an object *satisfies* the formulas $P_1 \& \neg P_3 \& P_7$ if its row in the data matrix has 1 in the fields no. 1 and 7 and has 0 in the field no. 3; etc.

A *hypothesis* (association rule, observational statement) has the form

$$\varphi \sim \psi$$

where φ, ψ are elementary conjunctions and $\sim$ is the sign of association. Logically speaking it is a *quantifier*; in the present context just understand the word "quantifier" as synonymous with "a notion of association". The formulas φ, ψ determine four frequencies a, b, c, d (the number of objects in the data matrix satisfying $\varphi \& \psi, \varphi \& \neg\psi, \neg\varphi \& \psi, \neg\varphi \& \neg\psi$ respectively). They are often presented as a *four-fold table:*

	ψ	$\neg\psi$
φ	a	b
$\neg\varphi$	c	d

The semantics of $\sim$ is given by a function $tr_\sim$ assigning to each four-fold table a, b, c, d the number 1 (true – the formulas with this table are associated) or 0 (false – not associated). Clearly there may be many such functions, thus many quantifiers. The intuitive notion of (positive) association is: the numbers a, d (counting objects for which both φ, ψ are true or both are false – coincidences) somehow dominates the numbers b, c (differences). Precisely this leads to the following natural condition on the truth function $tr_\sim$ of $\sim$: if (a, b, c, d) and (a', b', c', d') are two four-fold tables and $a' \geq a, b' \leq b, c' \leq c, d' \geq d$

(coincidences increased, differences diminished) then $tr_{\sim}(a,b,c,d) = 1$ implies $tr_{\sim}(a',b',c',d') = 1$. Such quantifiers are called *associational* (see [14]).

Let us shortly discuss three classes of associational quantifiers.

(1) *Implicational.* Example: The quantifier $\Rightarrow_{p,s}$ of *founded implication:* hypothesis true if $a/(a+b) \geq p$ and $a \geq s$ (see [14]). This is almost the semantics of Agrawal [1] (only instead giving a lower bound for the absolute frequency a he gives a lower bound *minsup* for the relative frequency a/m where m is the number of transactions; this is indeed a very unessential difference.)

This quantifier does not depend on c, d (the second line of the four-fold table. Thus it satisfies:

if $a' \geq a, b' \leq b$ and $tr_{\Rightarrow^*}(a,b,c,d) = 1$ then $tr_{\Rightarrow^*}(a',b',c',d') = 1$.

Quantifiers with this property are called *implicational.* They are very useful; $\varphi \Rightarrow^* \psi$ expresses the intuitive property that *many objects satisfying* φ *satisfy* ψ. For statistically motivated implicational quantifiers see [14]. But they have one weakness: they say nothing on how many objects satisfying $\neg\varphi$ satisfy ψ : also many? or few? Both things can happen. The former case means that in the whole data many objects have ψ. Then $\varphi \Rightarrow^* \psi$ does not say much. This leads to our second class:

(2) *Comparative.* Example: The quantifier $\sim$ of simple deviation: hypothesis true if $ad > bc$ (equivalently, if $\frac{a}{a+b} > \frac{c}{c+d}$; in words ψ is more frequent among objects satisfying φ than among those satisfying $\neg\varphi$).

Note that this does not say that *many* objects having φ have ψ; e.g. among objects having $\neg\varphi$, 10% have ψ, but among those having φ, 30% have ψ. (Trivial example: smoking increases the occurence of cancer.) An associational quantifier is *comparative* if $tr_{\sim}(a,b,c,d) = 1$ implies $ad > bc$. We may make the simple quantifier above dependent on a parameter, and define $tr_{\sim}(a,b,c,d) = 1$ if $ad > K.bc$ (K a constant) etc. For statistically motivated comparative quantifiers (Fisher, chi-square) see again [14].

(3) *Combined.* We may combine these quantifiers in various ways, getting new quantifiers. Here are some examples. Let $\Rightarrow^*$ be an implicational quantifier and $\sim^*$ a comparative associational quantifier.

Define $\varphi \Leftrightarrow^* \psi$ iff $\varphi \Rightarrow^* \psi$ and $\psi \Rightarrow^* \varphi$. Thus

$tr_{\Leftrightarrow^*}(a,b,c,d) = 1$ iff $tr_{\Rightarrow^*}(a,b,c,d) = 1$ and $tr_{\Rightarrow^*}(a,b,c,d) = 1$.

This quantifier depends only on a, b, c (and it is not comparative). It is a *pure doubly implicational quantifier* [36]. Pure doubly implicational quantifiers are a special case of doubly implicational quantifiers [37].

Define $\varphi \equiv^* \psi$ iff $\varphi \Rightarrow^* \psi$ and $\neg\varphi \Rightarrow^* \neg\psi$. Then

$tr_{\equiv^*}(a,b,c,d) = 1$ iff $tr_{\Rightarrow^*}(a,b,c,d) = 1$ and $tr_{\Rightarrow^*}(d,c,b,a) = 1$.

This is a *pure equivalence quantifier* [36].

Exercise: Show for $\Rightarrow^*_{p,s}$ that if $p \geq 0.5$ then the corresponding quantifier $\equiv^*_{p,s}$ is comparative.

Last example: define $\varphi \Rightarrow^{\sharp} \psi$ if $\varphi \Rightarrow^{*} \psi$ and $\varphi \sim \psi$. Thus

$$tr_{\Rightarrow^{\sharp}}(a,b,c,d) = 1 \text{ iff } tr_{\Rightarrow^{*}}(a,b,c,d) = tr_{\sim}(a,b,c,d) = 1.$$

Now $\varphi \Rightarrow^{*} \psi$ says: many objects having φ have ψ and not so many objects having $\neg\varphi$ have ψ.

Various implementations of GUHA work with various choices of such quantifiers, either directly (generating true sentences e.g. with $\Leftrightarrow^{*}$) or indirectly: during the interpretation of results with $\Rightarrow^{*}$, say, one can sort our those some sentences $\varphi \Rightarrow^{*} \psi$ satisfying also $\varphi \sim \psi$.

3 More on the Underlying Theory

Here we survey some notions and aspects of theoretical foundations of GUHA, mentioned in the introduction. The section may be skipped at first reading.

Observational logical calculi. The *symbols* used are: unary predicate $P_1, \ldots, P_n$, a unique object variable x (that may be omitted in all occurrences), logical connectives (say, $\&, \vee, \rightarrow, \neg$ – conjunction, disjunction, implication, negation), one or more quantifiers (and brackets). Atomic formulas have the form $P_i(x)$ (or just P_i); open formulas result from atomic ones using connectives (e.g. $(P_1 \& P_3) \vee \neg P_7$ etc.) If $P_1, \ldots, P_n$ name the columns of a given data matrix it should be clear what it means that an object *satisfies* an open formula in the data matrix.

Each quantifier has *dimension* 1 (is applied to one open formula) or 2 (applied to two formulas). (Higher dimensions possible.) *Classical* quantifiers $\forall, \exists$ (for all, there is) have dimension 1. If φ is open formula then $(\forall x)\varphi$ and $(\exists x)\varphi$ are formulas (in brief, $\forall\varphi, \exists\varphi$). Other examples: $(Many\, x)\varphi, (Based\, x)\varphi$. Let a be the number of objects satisfying φ, b the number of objects satisfying $\neg\varphi$. The semantics of a quantifier q is given by a truth function $Tr_q(a,b) \in \{0,1\}$. For example, $Tr_{\forall}(a,b) = 1$ iff $b = 0$; $Tr_{\exists}(a,b) = 1$ iff $a > 0$. Given $0 < p \leq 1$, $Tr_{Many}(a,b) = 1$ iff $a/(a+b) \geq p$; given a natural number $\tau > 0$, $Tr_{Based}(a,b) = 1$ iff $a \geq \tau$. Examples of quantifiers of dimension 2 and their truth functions were discussed above. A formula $(qx)\varphi$ is *true* in given data of $Tr_q(a,b) = 1$ where a, b are the frequencies as above. Formulas of the form $(qx)\varphi$ (q one-dimensional quantifier) or $(\sim x)(\varphi, \psi)$ (written also $\varphi \sim \psi$, two-dimensional quantifier) are called *prenex closed formulas.* Other closed formulas result from prenex ones using connectives, eg. $\forall\varphi \rightarrow Many(\varphi)$.

A *deduction rule* has some assumptions and a conclusion. It is *sound* if for each data set D, whenever the assumptions are true in D, then the conclusion is true in D.

Examples: Let φ, ψ, χ be open formulas.

(1) For each associational quantifier $\sim$, the following is a sound rule:

$$\frac{\varphi \sim \psi}{\varphi \sim (\varphi \& \psi)}$$

(2) For each implicational quantifier $\Rightarrow^*$, the following are sound rules:

$$\frac{\varphi \Rightarrow^* \psi}{\varphi \Rightarrow^* (\psi \vee \chi)} \qquad \frac{(\varphi \& \chi) \Rightarrow^* \psi}{\varphi \Rightarrow^* (\psi \vee \neg\chi)}$$

(3) A quantifier $\sim$ is σ-based (σ natural) if $\frac{\varphi \sim \psi}{Based_\sigma(\varphi \& \psi)}$ is a sound rule, i.e. whenever $\varphi \sim \psi$ is true in D then the frequency $a = fr(\varphi \& \psi)$ satisfies $a \geq \sigma$. Then the following rules are sound:

$$\frac{\neg Based_\sigma(\varphi \& \psi)}{\neg(\varphi \sim \psi)} \qquad \frac{\neg Based_\sigma(\varphi \& \psi)}{\neg Based_\sigma(\varphi \& \psi \& \chi)}.$$

This is very useful for pruning the tree of all hypotheses $\varphi \sim \psi$: if we meet φ, ψ such that for them $a \leq \sigma$ then all hypotheses $(\varphi \& \chi) \sim \psi$ and $\varphi \sim (\psi \& \chi)$ may be deleted from consideration.

(4) Some quantifiers (e.g. our SIMPLE, FISHER, CHISQ) make the following rules (of commutativity and negation) sound:

$$\frac{\varphi \sim \psi}{\psi \sim \varphi}, \qquad \frac{\varphi \sim \psi}{\neg\varphi \sim \neg\psi}.$$

This is also useful for the generation of hypotheses.

The general question if the deduction rule of the form $\frac{\varphi \sim \psi}{\varphi' \sim \varphi'}$ is sound was also studied [33], [37]. It was shown that there is relatively simple condition equivalent to the fact that the deduction rule $\frac{\varphi \sim \psi}{\varphi' \sim \varphi'}$ is sound. The condition depends on the class of quantifiers the quantifier $\sim$ belongs to. (There are classes of implicational, Σ - double implicational and Σ - equivalency quantifers.) This condition concerns several propositional formulas of the form $\phi(\varphi, \psi, \varphi', \varphi')$ derived from φ ψ, φ' and φ'. It is crucial if these formulas are tautologies of propositional calculus or not.

Observe that instead of saying that a rule, $\frac{\Phi}{\Psi}$ say, is sound (Φ, Ψ closed formulas) we may say that the formula $\Phi \rightarrow \Psi$ (where $\rightarrow$ is the *connective* of implication) is a tautology – is true in all data sets. Note that computational complexity of associational and implicational tautologies (i.e. formulas being tautologies for each associational/implicational quantifier) has been studied in [11] and [9].

To close this section, let us briefly comment on our *theoretical logical calculi.* They serve as statistical foundations for the case that observational hypotheses use quantifiers defined by some test statistics in statistical hypothesis testing. In this case one has, besides the observational language, its theoretical counterpart, enabling to express hypotheses on probabilities of open formulas in the unknown universe, from which our data form a sample, for example $P(\psi/\varphi) > P(\psi)$ (the conditional probability of ψ given φ is bigger than the unconditional probability of ψ). Let $THyp(\varphi, \psi)$ be such a theoretical hypothesis and let $\varphi \sim \psi$ be an observational hypothesis. This formula is a *test* for $THyp(\varphi, \psi)$ if under some "frame assumption" $Frame$, one can prove that if $THyp(\varphi\psi)$ were false then the probability that $\varphi \sim \psi$ is true in a data sample would be very small:

$$\frac{Frame, \neg THyp(\varphi, \psi)}{P(\varphi \sim \psi \text{ true in data }) \leq \alpha}$$

where α is a significance level. (This is how a logician understands what statisticians are doing.)

Note that the above concern just *one* pair $THyp(\varphi, \psi), \varphi \sim \psi$. What can one say if we find *several* hypotheses $\varphi \sim \psi$ true in the data? This is the problem of *global interpretation* of GUHA results, discussed also in [15].

A generalization of the above for testing fuzzy hypotheses in elaborated in [26], [27]. Further theoretical results concerning namely new deduction rules, fast verification of statistically motivated hypotheses and dealing with missing information were achieved in [33] and later in [36]. Some of these results are published in [35], [37] and [38]. These results are applied in the GUHA procedure 4ft-Miner.

4 The GUHA Procedure 4ft-Miner

4.1 4ft-Miner Overview

The GUHA method is realised by GUHA-procedures. A GUHA-procedure is a computer program, the input of which consists of the analysed data and of a few parameters defining a possibly very large set of potentially interesting hypotheses. The GUHA procedure automatically generates particular hypotheses from the given set and tests if they are supported by analysed data. This is done in an optimized way using several techniques of eliminating as many hypotheses as possible from the verification since their truth/falsity follows from the facts obtained in the given moment of verification. The output of the procedure consists of all prime hypotheses. The hypothesis is prime if it is true in (supported by) the given data and if it does not follow from the other output hypotheses.

Several GUHA procedures were implemented since 1966, see Section 6. One of the current implemetations is the *GUHA procedure 4ft-Miner.* It is a part of the *academic software system LISp-Miner.* The purpose of this system is to support teaching and research in the field of KDD (knowledge discovery in databases). The system consists of several procedures for data mining, machine learning and for data exploration and transformations. It is developed by a group of teachers and students of the University of Economics Prague see [44]. The whole LISp-Miner system can be freely downloaded, see [47].

The GUHA procedure 4ft-Miner deals with data matrices with nominal values. An example of such data matrix $\mathcal{M}$ is in figure 1.

Rows of data matrix $\mathcal{M}$ correspond to observed objects, columns of data matrix correspond to attributes $V_1, \ldots, V_K$. We suppose that $v_{1,1}$ is the value of attribute V_1 for the first object, $v_{n,K}$ is the value of attribute V_K for the last n-th object etc.

V_1	V_2	$\dots$	V_K
$v_{1,1}$	$v_{1,2}$	$\dots$	$v_{1,K}$
$\vdots$	$\vdots$	$\vdots\vdots\vdots$	$\vdots$
$v_{n,1}$	$v_{n,2}$	$\dots$	$v_{n,K}$

Fig. 1. Data matrix $\mathcal{M}$

The 4ft-Miner mines for hypotheses of the form $\varphi \approx \psi$ and for the conditional hypotheses $\varphi \approx \psi/\chi$. Here φ, ψ and χ are conjunctions of literals. A literal can be positive or negative. A positive literal is an expression of the form $V(\alpha)$ where V is an attribute and α is a proper subset of the set of all categories (i.e. possible values) of V. A negative literal is the negation of a positive literal. The set α is the coefficient of the literal $V(\alpha)$.

The literal $V(\alpha)$ is true in the row of the given data matrix if and only if the value v in this row and in the column corresponding to the attribute V is an element of α. Let {A,B,C,D,E,F} be a set of all possible values of the attribute V. Then V(A) and V(A,B,F) are examples of literals derived from the attribute V. Here {A} is the coefficient of V(A) and {A,B,F} is the coefficient of V(A,B,F).

A conditional hypothesis $\varphi \approx \psi/\chi$ is true in the data matrix $\mathcal{M}$ iff the hypothesis $\varphi \approx \psi$ is true in the data matrix $\mathcal{M}/\chi$. The data matrix $\mathcal{M}/\chi$ consists of all rows of the data matrix satisfying the boolean attribute χ (we suppose there is such a row).

The GUHA procedure 4ft-Miner automatically generates and verifies the set of hypotheses in the given data matrix. The set of hypotheses to be generated and tested is defined by

- definition of all antecedents,
- definition of all succedents,
- definition of all conditions,
- definition of the quantifier and its parameters (there are 17 types of various quantifiers, see http://lispminer.vse.cz/overview/4ft_quantifier.html).

The antecedent is the conjunction of literals automatically generated from the given set of antecedent attributes. (There cannot be two literals created from the same attribute in one hypothesis.) The set of all antecedents is given by

- a list of attributes - some of them are marked as basic (each antecedent must contain at least one basic attribute),
- a minimal and maximal number of attributes to be used in antecedent,
- a definition of the set of all literals to be generated from each attribute.

The set of all literals to be generated for particular attribute is given by:

- a type of coefficient (see below),
- a minimal and maximal number of categories in the coefficient,
- positive/negative literal option:

- only positive literals will be generated,
- only negative literals will be generated,
- both positive and negative literals will be generated.

There are six types of coefficients: subsets, intervals, left cuts, right cuts and cuts. 4ft-Miner generates all possible coefficients of the given type with respect to the given minimal and maximal number of categories in the coeficient. An example of the coefficient of the type *subset* is the coefficient {A,B,F} of the literal V(A,B,F).

Further coefficients can be defined for attributes with ordinal values only. Let us suppose that the attribute O has the categories 1, 2, ..., 8, 9. Then the coefficient of the literal O(4,5,6,7,8) is of the type *interval* - it corresponds to the interval of integer numbers $\langle 4, 8 \rangle$. The coefficient of the literal O(4,6,9) is not of the type interval. A *left cut* is an interval that begins at the first category of the attribute. The coefficient of the literal O(1,2,3) is of the type left cut and the coefficient of the literal O(4,5,6,) is not of the type left cut. Analogously a *right cut* is an interval that ends at the last category of the attribute. A *cut* is either a left cut or a right cut.

The output of the procedure 4ft-Miner consists of all prime hypotheses. There are also various possibilities for sorting and filtering of the output hypotheses.

The GUHA procedure 4ft-Miner was several times applied e.g. in medicine (see e.g. [4]), sociology and traffic. Modular architecture of the LISp-Miner system makes it possible to create a "tailor-made" interface customized for specialists in various fields. The only notions familiar to the field specialists are used in such interface. An experiment with on-line data mining can b found at http://euromise.vse.cz/stulong-en/online/. A very simple interface to the procedure 4ft-Miner is here used. Let us remark that there are some activities related to the conversion of hypotheses produced by 4ft-Miner into natural language [45].

4.2 A Performance Analysis

The usual A-priori algorithm [1] is not applicable for 4ft Miner. Instead, it the fast-mining-bit-string approach based on representation of analysed data by suitable strings of bits. The core of this approach is representation of each possible value (i.e. category) of each attribute by a string of bits [31]. This representation makes possible to use simple data structures to compute bit-string representations of each antecedent and of each succedent. Bit-string representations of antecedent and succedent are used to compute necessary four-fold tables in a very fast way [42]. The resulting algorithm is linearly dependent on the number of rows of the analysed data matrix. Let us note that this approach makes possible to effectively mine for conditional asociation rules and for rules with literals with coefficients with more than one value, e.g. V(A,B,F) see above.

We describe the solution of the same task on three data matrices with the same structure but with various numbers of rows to demonstrate linearity of this algorithm. Our task was derived from a database of the fictitious bank

Id	Age	Sex	Salary	District	Amount	Repayment	Loan
1	41–50	M	high	Prague	20–50	1–2	good
2	31–40	F	average	Pilsen	> 500	9–10	bad
...	...	...	...	...	...	...	...
6181	41–50	F	high	Brod	250–500	2–3	good

Fig. 2. Data matrix LoanDetail

BARBORA (see http://lisp.vse.cz/pkdd99/). We start with data matrix LoanDetail, see Fig. 2.

Each row of data matrix LoanDetail corresponds to a loan, there are 6181 loans. Columns Age, Sex, Salary and District describe clients, columns Amount, Payment and Loan correspond to loans. The first row describes a loan given to a man with the age 41–50. He has a high salary and he lives in Prague. Further, he borrowed amount of money in the interval 20–50 thousands of Czech crowns, he repays 1–2 thousands of Czech crowns and the quality of his loan is good from the point of view of the bank.

There are five possible values 21–30, ..., 61–70 of the attribute Age, six values < 20, 20–50, 50–100, 100–250, 250–50, > 500 of the attribute Amount and ten posible values ≤ 1, 1–2, , ..., 9–10 of the attribute Repayment.

We are interested in all pairs ⟨*segment of clients, type of loan*⟩ such that the assertions *to be a member of segment of clients* and *to have a bad loan of certain type* are in some sense equivalent. This can expressed by a conditional association rule

$$SEGMENT \Leftrightarrow_{0.95,20} Loan(bad) \; / \, TYPE$$

where *SEGMENT* is Boolean attribute defining a *segment of clients*, *Loan(bad)* is Boolean attribute that is true iff the corresponding loan is bad and *TYPE* is Boolean attribute defining *type of loan.*

The above rule $SEGMENT \Leftrightarrow_{0.95,20} Loan(bad) \, / \, TYPE$ is true in the data matrix LoanDetail if the rule $SEGMENT \Leftrightarrow_{0.95,20} Loan(bad)$ is true in the data matrix LoanDetail/*TYPE* (i.e. in the data matrix consisting of all rows of data matrix LoanDetail sastisfying the Boolean attribute *TYPE*) describing type of loan. The association rule $SEGMENT \Leftrightarrow_{0.95,20} Loan(bad)$ is true in data matrix LoanDetail/*TYPE* if the condition $\frac{a}{a+b+c} \geq 0.95 \wedge a \geq 20$ is satisfied where a, b, c are frequencies from four-fold table of Boolean attributes *SEGMENT* and *Loan(bad)* in data matrix LoanDetail/*TYPE*:

	Loan(bad)	¬ *Loan(bad)*
SEGMENT	a	b
¬*SEGMENT*	c	d

Let us remark that the rule $SEGMENT \Leftrightarrow_{0.95,20} Loan(bad) \, / \, TYPE$ says: If we consider only loans of the type *TYPE* then there are at least 95 per cent of loans satisfying both *SEGMENT* and *Loan(bad)* among loans satisfying *SEGMENT* or *Loan(bad)*.

The Boolean attribute SEGMENT is derived from columns Age, Sex, Salary and District of data matrix LoanDetail. An example is a segment of all men 21–30 years old defined by SEGMENT = Sex(M) ∧ Age(21–30). The Boolean attribute TYPE is defined similarly by columns Amount and Repayment. An example is Boolean attribute Amount(< 20)∧ Repayment(1–2) defining a type of loans: borrowed less than 20000 Czech crowns and repaid between 1000 and 2000 Czech crowns each month.

Our task can be solved by the 4ft-Miner procedure where all antecedents to be automatically generated are specified as conjunctions of one to four literals of the form Age(?), Sex(?), Salary(?) and either District(?) or District(?,?) (we consider single district or pairs of districts) where each ? can replaced by one possible value of the corresponding attribute. There are 6 literals of the form Age(?), 2 literals of the form Sex(?), 3 literals of the form Salary(?) and $3003 = \binom{77}{1} + \binom{77}{2}$ literals of the form either District(?) or District(?,?).

If we use terminology of the 4ft-Miner input we can say that the minimal length of antecedent is 1, maximal length of antecedent is 4, positive literals - subsets will be generated for all attributes Age, Sex, Salary and District. The minimal number of categories in all coefficients will be 1, the maximal number of categories in the coefficients for Age, Sex and Salary will be 1, the maximal number of categories in the coefficients for District will be 2. This defines a set of more than 250 000 antecedents.

Similarly we can ask 4ft-Miner to use the Booolean attribute Loan(bad) as a succedent and further to automatically generate all conditions as conjunctions of one or two literals of the form Amount(?) or Repayment(?). There are 76 such conditions. Together there are more than $19 * 10^6$ rules of the form $SEGMENT \Leftrightarrow_{0.95,20} Loan(bad)\ /\ TYPE$ defined this way.

The task of automatical generation and verification of this set of rules is solved in 56 seconds (PC with Pentium II processor, 256 MB RAM). Let us emphasise that only 80 549 rules had actually to be fully processed due to various optimisations used in the 4ft-Miner procedure (pruning of the tree of hypotheses). The result of the run of the 4ft-Miner consists of 10 true association rules. An example is the rule

$$\text{Salary(average)} \wedge \text{District(Olomouc, Tabor)} \Leftrightarrow_{1.0,27} \text{Loan(bad)}\ /$$

$$/\text{Amount(20–50)} \wedge \text{Repayment(2–3)}\ .$$

The four-fold table of this rule is

	Loan(bad)	¬ Loan(bad)
Salary(average) ∧ District(Olomouc, Tabor)	27	0
¬ Salary(average) ∧ District(Olomouc, Tabor)	0	144

It concerns the data (sub)matrix LoanDetail/(Amount(20–50)∧Repayment(2–3)) with 171 rows.

We solved the same task on the data matrices LoanDetail_10 and LoanDetail_20 on the same computer. Data matrix LoanDetail_10 has ten times more rows than data matrix LoanDetail. Each row of data matrix LoanDetail is used ten times in data matrix LoanDetail_10. Analogously, each row of data matrix LoanDetail is used twenty times in data matrix LoanDetail_20. Data matrix LoanDetail_20 has twenty times more rows than data matrix LoanDetail.

We use the quantifier $\Leftrightarrow_{0.95,200}$ in the data matrix LoanDetail_10 instead of the quantifier $\Leftrightarrow_{0.95,20}$ to ensure the same behaviour of the optimised algorithm. Similarly we use the quantifier $\Leftrightarrow_{0.95,400}$ in the data matrix LoanDetail_20. The solution times necessary to solve the same task on all data matrices are compared in the following table.

Data matrix	LoanDetail	LoanDetail_10	LoanDetail_20
number of rows	6181	61810	123 620
time total in seconds	56	513	1066
quantifier	$\Leftrightarrow_{0.95,20}$	$\Leftrightarrow_{0.95,200}$	$\Leftrightarrow_{0.95,400}$
fully processed rules	80 549	80 549	80 549
from them true	10	10	10

We can conclude that the solution time is approximately linearly dependent on the number of rows of the analysed data matrix.

5 An Example of Application of GUHA

This section gives a short report on the GUHA approach to the discovery challenge for a financial data set, which was a part of the 3rd European Conference on Principles and Practice of Knowledge Discovery in Databases (PKDD'99), held in Prague, September 15–18, 1999.

The goal of challenge was to characterize clients of the BARBORA fictitious bank (see the preceding section), particularly those having problems with loan payments, using data mining methods. Note that the run described here is different from that described in the preceding section (and was done using the GUHA+- implementation [46]).

The financial data were stored in eight tables of relational database. Each table consisted of several attributes characterizing different properties of an account. Hence the object of investigation was an account and attributes its properties. An account had static and dynamic characteristics. Static characteristics were given by tables *account, client, disposition, permanent order, loan, credit card* and *demographic data.* Dynamic ones were given by table *transaction.*

5.1 Data Description and Preprocessing

Within the preprocessing stage, new attributes were computed on base of original ones. The following list consists of attributes computed from static characteristics of account:

ACCOUNT_YEAR; ACCOUNT_FREQ; ACCOUNT_DIST%
OWNER_SEX; OWNER_AGE; OWNER_DIST%; USER
ORDER_INSURANCE; ORDER_HOUSEHOLD; ORDER_LEASING;
ORDER_LOAN; ORDER_SUM; ORDER_OTHER;
LOAN_STATUS; LOAN_AMOUNT; LOAN_DURATION;
LOAN_PAYMENT; LOAN_YEAR
CARD_TYPE; CARD_YEAR.

Since lack of space for detail explanation of attributes we do this in the next section only for attributes presented in found hypotheses. On aggregate, 48 attributes describing static characteristics of account were computed. Note that ACCOUNT_DIST% and OWNER_DIST% are shortcuts each for 15 attributes related to properties of district an account is held or an owner lives.

Dynamic characteristics of account were given by table *transaction*. On base of them these new attributes were computed:

AvgM_AMOUNT_SIGN; MinM_AMOUNT_SIGN; MaxM_AMOUNT_SIGN;
AvgM_VOLUME; MinM_VOLUME; MaxM_VOLUME;
AvgM_IBALANCE; MinM_IBALANCE; MaxM_IBALANCE;
AvgM_WITHDRAWAL_CARD; AvgM_CREDIT_CASH;
AvgM_COLLECTION; AvgM_WITHDRAWAL_CASH;
AvgM_REMITTANCE; AvgM_INSURANCE; AvgM_STATEMENT;
AvgM_CREDITED_INTEREST; AvgM_SANCTION_INTEREST;
AvgM_HOUSEHOLD; AvgM_PENSION; AvgM_LOAN;
AvgM_TRANSACTION_#.

Attributes give average, minimal or maximal month cash flow with respect to different types of transactions. Attribute AvgM_TRANSACTION_# refers to averaged number of realized transactions per a month. On aggregate, 22 attributes describing dynamic behavior of an account were defined.

The total number of attributes characterizing an account used for GUHA method was 70. Attributes were either nominal (range was finite set of features, typically presence or absence of some fact) or ordinal (range was set of reals typically giving amount of money in some transaction). Attributes were categorized into several categories, (a category is actually Boolean attribute) at a medium into 5 for each one, to obtain basic matrix of zero and ones suitable for GUHA processing.

5.2 Discovered Knowledge

We had this natural point of view of good and bad clients. If a loan for client is granted then client is good if there have been no problems with loan payments, and clearly client is bad when there have been problems with payment. In our discovery work we concentrated on exploring appropriate characterizations of this notion of good or bad client. The consequence is that we aimed only on data of clients with granted loan.

We assume that eminent interest of every bank is to reveal in advance if a client asking for a loan is good or bad according to our definition. This prediction has to be based on information the bank knows in time of asking for a loan. That is why we carried out our exploration from data (namely data from table *transaction*) that were older than date when the loan was granted.

In the first phase we aimed on hypotheses with single antecedents and characterization of good clients. These clients have accounts satisfying category LOAN_STATUS:GOOD. By proceeding preprocessed data by GUHA+– software we have explored hypotheses summarized in Table 1. We have used Fisher quantifier with the significance level alpha=0.05 and restricted ourselves only on hypotheses supported at least by 15 objects ($a \geq 15$). In Table 1 presented statistic Prob is given by fraction $a/(a+b)$ and characterizes hypotheses in sense of an implication.

Table 1. Characterization of good clients - succedent LOAN_STATUS:GOOD

#	antecedent	ff-table		Fisher	Prob
1	AvgM_SANCTION_INTEREST:NO	603	50	6.12e-024	0.92
		3	26		
2	ORDER_HOUSEHOLD:YES	421	20	5.03e-013	0.95
		185	56		
3	USER:YES	145	0	3.76e-009	1.00
		461	76		
4	CARD_TYPE:CARD_YES	165	5	1.39e-005	0.97
		441	71		

Hypotheses are sorted increasingly according to value of Fisher statistic. Hypothesis #1 says that there is strong association between absence of sanction interest payment and good loan payments. Hypothesis #2 says that there are no problems with loan payments if household permanent order is issued on the account. Hypotheses #3 and #4 ties good loan payments policy with a presence of another client who can manipulate with account (typically, client's wife or husband) or presence of card issued to the account, respectively.

Presented Table 1 can be used for characterization of bad clients, satisfying LOAN_STATUS:BAD category, as well. Due to symmetry of Fisher quantifier we can read each hypothesis as "if a client do not satisfy antecedent of hypothesis then there is greater probability than on average that the loan will not be paid." For example for hypothesis #3, if only owner can manipulate with the account then it is more risky to grant him by a loan than if there would be another person who can manipulate with the account as well.

In the second phase of our exploration we aimed on hypotheses with compound antecedents of length 2. We used Fisher quantifier with significance level alpha=0.001, restricted on hypotheses supported at least by 15 objects ($a \geq 15$). We applied other restriction to choose only 100% implication hypotheses (Prob=1.00).

These hypotheses are 100% implications as it can be seen from the four fold tables. In antecedents of hypotheses of Table 2 there are combinations of other properties of account that (in data) gives 100% warranty that loan will be paid

Table 2. Characterization of good clients - succedent LOAN_STATUS:GOOD

#	antecedent	ff-table		Fisher	Prob
1	ORDER_OTHER:NO	208	0	1.31e-013	1.00
	ORDER_HOUSEHOLD:YES	398	76		
2	ACC_DIST_A7: > 3	139	0	9.33e-009	1.00
	ORDER_HOUSEHOLD: > 5000	467	76		

back. The category ORDER_OTHER:NO says that there is no other kind of permanent order issued to the account besides for insurance, household, leasing or loan payments. Category ORDER_ HOUSEHOLD:> 5000 clearly denotes clients with household permanent order payment greater than 5000. Category ACC_DIST_A7:> 3 says that bank address is in the district with more than 3 municipalities with number of inhabitants from 2000 to 10000.

Form of antecedents for LOAN_STATUS:BAD succedent reveals that critical property for bad loan payments is presence of sanction interest, but even in this case the loan can be paid back. However, if there is sanction interest issued on account together with a property from set given by first members of antecedents of hypotheses in Table 3, there were always problems with payments. The categories in the table have the following meaning:

Table 3. Characterization of bad clients - succedent LOAN_STATUS:BAD

#	antecedent	ff-table		Fisher	Prob
1	OWNER_DIST_A13 : > 3	21	0	6.27e-022	1.00
	AvgM_SANCT._INTEREST:YES	55	606		
2	OWNER_DIST_A7: ≤ 6	20	0	7.41e-021	1.00
	AvgM_SANCT._INTEREST:YES	56	606		
3	OWNER_DIST_A6: ≤ 27	17	0	1.11e-017	1.00
	AvgM_SANCT._INTEREST:YES	59	606		

OWNER_DIST_A13:> 3 - unemployment rate in the year 1996 and in the district where the client lives was greater than 3%;

OWNER_DIST_A7:≤ 6 - the client's address is in a district with maximally 6 municipalities with number of inhabitants from 2000 to 10000;

OWNER_DIST_A6:≤ 27 - the address of the client is in a district with maximally 27 municipalities with number of inhabitants from 500 to 2000.

In this section we presented a short example of GUHA method's employment in data mining process. To full review of coping with financial data set challenge by GUHA see [2].

6 Relation to (Relational) Data Mining and Discovery Science

Both data mining and discovery science are terms that emerged recently and have aims similar to each other as well as to the main ideas of GUHA declared from its beginning: to develop methods of discovering (mining) knowledge form data (usually large data). Relations of GUHA and its ancestors to data mining and discovery science were analyzed in [35], [17], [26], [12].

In particular, our hypotheses described above are more general than Agrawal's association rules particularly by (1) explicit use of negations and (2) choice from a variety of quantifiers, not just FIMPL. On the other hand, Agrawal's data mining is particularly developed for processing extremely huge data, which influence the choice of techniques for pruning the system (tree) of hypotheses = rules. These aspects and possibilities of mutual influence of GUHA and Agrawal's approach are analyzed especially in [12].

The fact that Agrawal's notion of an association rule [1] occurs in fact in [14] has remained unnoticed for long; but note that it is explicitly stated e.g. in [29]. Let us stress that priority questions are by far not the most important thing; what is valuable is possible mutual influence. (This is discussed e.g. in [12]).

A relatively new direction of data mining is *relational data mining* [5]. It means that the input of the mining procedure is not one single database table but several tables. GUHA procedures dealing with several database tables were also defined and studied see [33], [34], [40]. Such GUHA procedure mines for *multi-relational association rules.* GUHA hypotheses as described above are formulae of observational calculi that can be obtained by a modification of classical predicate calculi. Multi-relational hypotheses are formulae of many-sorted observational calculi with possibly binary, ternary,.. predicates that can be obtained by a modification of classical many-sorted predicate calculi. Some theoretical results concerning e.g. decidability of many-sorted observational calculi were achieved [33]. It can be useful to compare GUHA approach to further techniques of relational data mining.

7 The Relation of the GUHA Method to Modern Database Systems

The GUHA implementations were used in various research domains (e.g. in medicine [22,28,43], pharmacology [23,24], banking [2,30] or in meteorology [3]), but admittedly, they never got a broad use.

During the rather long history of the method there were created several implementations of the method reflecting the rapid development in information technologies (IT). The first implementations were realized on MINSK22 computer in 60's and on mainframes in 70's. In the 80's implementations were transferred on IBM PC platform which gave the PC-GUHA implementation for MS DOS operation system. There are two present implementations named GUHA +– [46] and 4FT-Miner [39]. The latter one was described above in Section 3.

In the first implementations of the method data were stored in stand alone plain files. In these files data were already dichotomized (the oldest implementations) or in a raw form, consequently dichotomized in an automatic way on base of scripts determined by user (PC-GUHA). However, such a way of access to raw data was untenable in a light of developments in a database software industry. Therefore present implementations (GUHA+–, 4FT-Miner) employ universal ODBC interface to access raw data typically stored in form of tables in MS Access or MS Excel software. But progress continues.

Note that a GUHA-DBS database system was proposed in early 1980's by Pokorný and Rauch [32], see also [34]. This is now rather obsolete and this development was neglected in GUHA for long. But the development of the 4ft-Miner brought important progress in this direction [44]. See also [6].

Contemporary trends in data mining area are driven by requirements for processing huge data sets which brings new research and implementation problems for GUHA and its logical and statistical theory. Standard sources of huge databases - created typically in a dynamic way - are hypermarkets, banks, internet applications, etc. These data sets are enormous and professional databases as Oracle or MS SQL Server have to be used to manage them. To cope with these modern trends there was established a research group formed around COST Action 274 aiming on a new implementation of the GUHA method enabling to work efficiently with large data sets.

Actually, there are generally two ways possible of raw data access in a new GUHA method's implementation.

The first way is to transform respective data objects of modern database systems (e.g., data cubes of MS SQL Server 2000) into standard tables and then access data by old algorithms and their generalisations in the spirit of relational data mining. The second approach is to homogeneously interconnect GUHA core algorithms with data access algorithms offered by modern data base systems issuing into a qualitatively new (faster) processing of huge data sets. Detailed description of work on this task can be found in the report [7].

The other direction of research we are interested in is an effort for incorporation of GUHA method into modern databases as a support tool for an intellectual workflow related to information classification and structuring, analytical processing and decision support modeling [6] see also experiments with incorporating GUHA procedure 4ft-Miner into medical data mining process at http://euromise.vse.cz/stulong-en/.

Rapid progress in the area of data base management systems (DBMS) sufficiently simplified information classification and structuring processes. Powerful database engines developed in last few years also improved analytical processing capability of a DBMS. The most significant gap in the automation of an intellectual workflow lies between analytical processing and decision support modeling (these two processes influence each other within decision driven loop [7].)

To understand more precisely what the gap is we developed new model "IT preferences for a decision making" [8]. This model was applied for an evaluation of three different discovery engines currently existing (MDS Engine, GUHA Engine and DEX-HINT engine). This model enables us to specify a new combination of features required for a successful GUHA implementation (called GUHA Virtual Machine) in the near future. The target implementation framework of the GUHA Virtual Machine relates to complex analytical and decision problems like radioactive waste management [21].

8 Conclusion

Research tasks include: Further comparison of GUHA theory with the approaches of data mining for mutual benefits. Systematic development of the theory in relation to fuzzy logic (in the style of Hájek's monograph [10]). Development of observational calculi for temporal hypotheses (reflecting time). Systematic development of the database aspects, in particular improving existing methods and elaborating new methods of data pre-processing and post-processing of GUHA results. Design of a new GUHA-style system based on data received from distributed network resources, and construction of a model of the customer decision processes.

Acknowledgement

Theoretical and practical development of the GUHA method is the subject of Czech participation in the EU COST project 274 (TARSKI).

References

1. Agrawal R., Manilla H., Sukent R., Toivonen A., Verkamo A.: Fast discovery of Association rules. Advance in Knowledge Discovery and Data Mining, AAA Press 1996, pp. 307–328.
2. Coufal D., Holeňa M., Sochorová A.: Coping with Discovery Challenge by GUHA. In: Discovery Challenge. A Collaborative Effort in Knowledge. Discovery from Databases. - Prague, University of Economics 1999, pp. 7–16, PKDD'99 European Conference on Principles and Practice of Knowledge Discovery in Databases /3./, Prague, Czech Rep.
3. Coufal D.: GUHA Analysis of Air Pollution Data. In: Artificial Neural Nets and Genetic Algorithms. Proceedings of the International conference ICANNGA'2001 Prague (Ed.: Kůrková V., Steele N.C., Neruda R.,Kárný M.) Springer Wien 2001, pp. 465–468).
4. Dolejší, P. - Lín, V.- Rauch, J. - Šebek, M.: System of KDD Tasks and Results within the STULONG Project In: (Berka ed.). Discovery Challenge Workshop Notes. University Helsinki, 2002.
5. Džeroski S., Lavrač N.(Eds): Relational Data Mining. Springer 2001, 398 pp.
6. Feglar T.: The GUHA architecture. Proc. Relmics 6, Tilburg (The Netherlands), pp. 358–364.
7. Feglar T.: The GUHA Virtual Machine - Frameworks and Key Concept - Frameworks and Key Concept, Research Report COST 274, Year 2001.
8. Feglar T.: Modeling of a Engine Based Approach to the Decision Support, Proceedings of ADBIS 2002 Conference, Vol. 2. Research Communications, Bratislava, September 2002, pp. 98–107.
9. Hájek P.: On generalized quantifiers, finite sets and data mining. Accepted for presentation on the conference IIS 2003 (Intelligent Information Systems) Zakopane, Poland June 2–5 2003
10. Hájek P.: Metamathematics of Fuzzy Logic, Kluwer 1998.
11. Hájek P.: Relations in GUHA-style data mining. In H. de Swart, ed.: Relational methods in Computer Science, LNCS 2561, Springer Verlag 2002, p. 81 pp.

12. Hájek P.: The GUHA method and mining association rules. Proc. CIMA'2001 (Bangor, Wales) 533–539
13. Hájek P.: The new version of the GUHA procedure ASSOC, COMPSTAT 1984, pp. 360–365.
14. Hájek P., Havel I., Chytil M.: The GUHA method of automatic hypotheses determination, Computing 1(1966) 293–308.
15. Hájek P., Havránek T.: Mechanizing Hypothesis Formation (Mathematical Foundations for a General Theory), Springer-Verlag 1978, 396 pp.
16. Hájek P., Havránek T.: Mechanizing Hypothesis Formation (Mathematical Foundations for a General Theory). Internet edition. http://www.cs.cas.cz/~hajek/guhabook/
17. Hájek P., Holeňa M.: Formal logics of discovery and hypothesis formation by machine. To appear in Theoretical Computer Science.
18. Hájek P. (guest editor): International Journal of Man-Machine Studies, vol. 10, No 1 (special issue on GUHA). Introductory paper of the volume is Hájek, Havránek: The GUHA method - its aims and techniques. Int. J. Man-Machine Studies 10(1977) 3–22.
19. Hájek P. (guest editor): International Journal for Man-Machine Studies, vol. 15, No 3 (second special issue on GUHA)
20. Hájek P., Sochorová A., Zvárová J.: GUHA for personal computers, Comp. Stat., Data Arch. 19, pp. 149–153.
21. Hálová J., Feglar T.: Systematic approach to the choice of optimum variant of radioactive waste management, Proceedings of 6th ISAHP 2001 Conference, Berne, Switzeland, August 2001, pp. 139–144.
22. Hálová J., Žák P.: Coping Discovery challenge of mutagenes discovery with GUHA+/– for windows. In: The Fourth Pacific-Asia Conference on Knowledge Discovery and Data Mining. Workshop KDD Challenge 2000. International Workshop on KDD Challenge on Real-world Data. - Kyoto, - 2000, pp. 55–60, Pacific-Asia Conference on Knowledge Discovery and Data Mining /4./, Kyoto, Japan
23. Hálová J., Žák P.: Drug Tailoring by GUHA +/– for Windows. In: Challenges for MCDM in the New Millenium. Abstracts. - Ankara, Middle East Technical University 2000, pp. 50, International Conference MCDM /15./, Ankara, Turkey
24. Hálová J., Žák P.: Fingerprint Descriptors in Tailoring New Drugs Using GUHA Method. In: 51th Meeting of the European Working Group Multicriteria Aid for Decisions. Program and Abstracts. - Madrid, - 2000, pp. 25, Meeting of European Working Group Multicriteria Aid for Decisions /51./, Madrid, Spain, 00.03.30-00.03.3
25. Havránek T.: The statistical modification ond interpretation of GUHA method, Kybernetika 7(1971) 13–21.
26. Holeňa M.: Fuzzy hypotheses for GUHA implications, Fuzzy Sets and Systems 98 (1998), 101–125.
27. Holeňa M.: Exploratory data processing using a fuzzy generalization of the GUHA approach, Fuzzy Logic, Baldwin et al., ed. Willey et Sons, New York, 1996, pp. 213–229.
28. Holubec L., jr., Topolcan O., Pikner R., Pecen L., Holubec L., sen., Fínek J., Ludvíková M.: Discriminative Level of Tumor Markers after Primary Therapy in Colorectal Carcinoma Patients. In: ISOBM Meeting. Abstract Book. - Barcelona, - 2001, pp. 173,

29. Lin W., Alvarez S. A., Ruiz C.: Collaborative recommendation via adaptive association rule mining. Web-KDD 2000.
30. Pecen L., Pelikán E., Beran H., and Pivka D.: Short-term fx market analysis and prediction. In Neural Networks in Financial Engeneering (1996), pp.189–196
31. Rauch J.: Some Remarks on Computer Realisations of GUHA Procedures. International Journal of Man-Machine Studies, 10, (1978) 23–28
32. Pokorný D., Rauch J.: The GUHA-DBS Data Base System. Int. Journ. Math. Machine Studies 15 (1981), pp. 289–298.
33. Rauch J.: Logical foundations of mechanizing hypothesis formation from databases (in Czech). PhD. thesis, Mathematical Institute of the Czechoslovak academy of Sciences, 1986
34. Rauch J.: Logical problems of statistical data analysis in databases. Proc. Eleventh Int. Seminar on Database Management Systems (1988), pp. 53–63.
35. Rauch J.: Logical Calculi for Knowledge Discovery. Red. Komorowski, J. - Zytkow, J. Berlin, Springer Verlag 1997, pp. 47–57.
36. Rauch, J.: Contribution to Logical Foundations of KDD (in Czech). Assoc. Prof. Thesis, Faculty of Informatics and Statistics, University of Economics Prague, 1998
37. Rauch, J.: Classes of Four-Fold Table Quantifiers. In Principles of Data Mining and Knowledge Discovery, (J. Zytkow, M. Quafafou, eds.), Springer-Verlag, pp. 203–211, 1998.
38. Rauch, J.: Four-Fold Table Calculi and Missing Information. In JCIS'98 Proceedings, (Paul P. Wang, editor), Association for Intelligent Machinery, pp. 375–378, 1998.
39. Rauch J., Šimůnek M.: Mining for 4ft association rules. Proc. Discovery Science 2000 Kyoto, Springer Verlag 2000, pp. 268–272
40. Rauch, J.: Interesting Association Rules and Multi-relational Association Rules. In Communications of Institute of Information and Computing Machinery, Taiwan. Vol. 5., No. 2, May 2002, pp. 77–82.
41. Rauch, J.: Mining for Scientific Hypotheses. In Meij, J.(Editor): Dealing with the data flood. Mining Data, Text and Multimedia. STT/Beweton, The Hague. 2002. pp. 73–84
42. Rauch J., Šimůnek M.: Alternative approach to Mining Association Rules. In IEEE ICDM02 Workshop Proceedings The Foundation of Data Mining and Knowledge Discovery (T.Y. Lin and Setsuo Ohsuga Eds,) pp. 157–162
43. Šebesta V., Straka L.: Determination of Suitable Markers by the GUHA Method for the Prediction of Bleeding at Patients with Chronic Lymphoblastic Leukemia. In: Medicon 98, Mediterranean Conference on Medical and Biological Engineering and Computing /8./, Lemesos, Cyprus
44. Šimůnek M.: Academic KDD Project LISp-Miner. Accepted for publication at Intelligent Systems Design and Applications (ISDA'03), Tulsa, Oklahoma
45. Strossa,P.- Rauch,J.: Converting Association Rules into Natural Language - an Attempt.Accepted for the presentation at the conference IIS 2003 (Intelligent Information Systems) Zakopane, Poland June 2–5 2003
46. GUHA+- project web site http://www.cs.cas.cz/ics/software.html
47. LISp-Miner system web site http://lispminer.vse.cz

Constraint Based Mining of First Order Sequences in SeqLog

Sau Dan Lee and Luc De Raedt

Institut für Informatik
Albert-Ludwigs-Universität Freiburg
Germany
{danlee,deraedt}@informatik.uni-freiburg.de

Abstract. A logical language, SeqLog, for mining and querying sequential data and databases is presented. In SeqLog, data takes the form of a sequence of logical atoms, background knowledge can be specified using Datalog style clauses and sequential queries or patterns correspond to subsequences of logical atoms. SeqLog is then used as the representation language for the inductive database mining system MineSeqLog. Inductive queries in MineSeqLog take the form of a conjunction of a monotonic and an anti-monotonic constraint on sequential patterns. Given such an inductive query, MineSeqLog computes the borders of the solution space. MineSeqLog uses variants of the famous level-wise algorithm together with ideas from version spaces to realize this. Finally, we report on a number of experiments in the domains of user-modelling that validate the approach.

1 Introduction

Data mining has received a lot of attention recently, and the mining of knowledge from data of various models has been studied. One popular data model that has been studied concerns sequential data [1,2,3,4,5,6]. Many of these approaches are extensions of the classical level-wise itemset discovery algorithm "Apriori" [7]. However, the data models that have been used so far for modelling sequential patterns are not very expressive and are often based on some form of propositional logic. The need for more expressive kind of patterns arises, e.g., when modelling Unix-users [8]. As an example, the command sequence

1. `ls`
2. `vi paper.tex`
3. `latex paper.tex`
4. `dvips paper.dvi`
5. `lpr paper.ps`

can be represented as a sequence of first-order terms: "`ls vi(paper.tex) latex(paper.tex) dvips(paper.dvi) lpr(paper.ps)`". With such a represen-

R. Meo et al. (Eds.): Database Support for Data Mining Applications, LNAI 2682, pp. 154–173, 2004.

tation model, it is possible to discover first-order rules such as "`vi(X) latex(X)` is frequent".[1]

Researchers such as Mannila and Toivonen [9] have realized the need for such more expressive frameworks. They have introduced a simple data model that allows one to use binary predicates as well as variables. In doing so, they have taken a significant step into the direction of multi-relational data mining and inductive logic programming. However, when comparing their data model to those traditionally employed in inductive logic programming, such as Datalog, the model is more limited and less expressive. Indeed, traditional inductive logic programming languages would allow the user to encode background knowledge (in the form of view predicates or clauses). They would possess a fix-point semantics as well as an entailment relation.

In this chapter, we first introduce a simple logical data model for mining sequences, called SeqLog. It is in a sense the sequence equivalent of the Datalog language for deductive databases. Moreover, we provide a formal semantics, study the entailment and subsumption relations and provide clause like mechanisms to define view predicates. Through the introduction of the SeqLog language, we put sequential data mining on the same methodological grounds as inductive logic programming. This framework may be useful also for other mining or learning tasks in inductive logic programming. In this context, our lab has also developed an approach for analysing SeqLog type sequences based on Hidden Markov Models, cf. [10].

Next, we will see how the MineSeqLog system uses SeqLog to mine for SeqLog patterns of interest in sequential data. MineSeqLog combines principles of the level-wise search algorithm with version spaces in order to find all patterns that satisfy a constraint of the form $a \wedge m$ where a is an anti-monotonic constraint (such as "the frequency of the target patterns on the positives is at least 10%") and a monotonic constraint m (such as the "frequency on the negatives is at most 1%"). While most attention in the data mining community goes to handling an anti-monotonic constraint (e.g. minimum frequency), we argue that it is also useful to consider monotonic constraints (e.g. maximum frequency), especially in conjunction with an anti-monotonic constraint. Our group came across such a need when we developed the molecular feature miner (MolFea) [11]. With a conjunction of both kinds of constraints, we are able to instruct MineSeqLog and MolFea to find rules and patterns that are frequent in a certain data subset, but infrequent in another subset. Such rules are useful for distinguishing patterns exhibited by the two subsets. Using the Unix user modelling example, this dual-constraint approach allows us to find out command sequences that are frequently used by novice users, but seldom used by advanced users. The key design issue in MineSeqLog was the development of an optimal refinement operator for SeqLog. We will go through this operator in detail. Finally, we validate the approach using some experiments in the domain of user modelling.

---

[1] We follow the Prolog notation here in which identifiers for variables start with capital letters.

This chapter is organized as follows: in Sect. 2, we introduce SeqLog and define its semantics; in Sect. 3, we define the mining task addressed in SeqLog and provide the main algorithms; in Sect. 4, we define an optimal refinement operator for SeqLog (without functors); in Sect. 6, we present some preliminary experiments, and finally, in Sect. 7, we conclude and touch upon related work.

2 SeqLog: SEQuential LOGic

In this section, we introduce a representational framework for sequences called SeqLog. The framework is grounded in first order logic and akin to Datalog except that SeqLog represents sequences rather than relations or predicates. This also motivates the use of the traditional logical terminology, which we briefly review here.

An *atom* $p(t_1, ..., t_n)$ consists of a relation symbol p of arity n followed by n terms t_i. A *term* is either a constant or a variable[2]. A *substitution* θ is a set of the form $\{ v_1 \leftarrow t_1, ..., v_n \leftarrow t_n \}$ where the v_i are variables and the t_i terms. One can *apply a substitution* θ on an expression e yielding the expression $e\theta$ which is obtained by simultaneously replacing all variables v_i by their corresponding terms t_i.

We can now introduce sequences and sequential databases.

Simple Sequence. A simple sequence is a possibly empty ordered list of atoms. The empty list is denoted by ω. A few examples of simple sequences are:
1. `latex(kdd,tex) xdvi(kdd,dvi) dvips(kdd,dvi) lpr(hpml,kdd)`
2. `latex(FileName,tex)   xdvi(FileName,tex)`
3. `helix('A', h(right,alpha), 7)   strand('SA','A',1,0,6)`

Simple sequences are used to represent data in a sequential database.

Complex Sequence. A complex sequence is a possibly empty sequence of atoms separated by operators l_0 op_1 l_1 op_2 l_2 $op_3 ... op_n$ l_n. The two operators that are employed are $\lhd$ (which we often omit for readability reasons) and $<$. The former, i.e. $\lhd$, denotes the 'direct successor' operator, the latter, i.e. $<$, encodes the transitive closure of $\lhd$. An example of a complex sequence, is `latex(FileName,tex)` $<$ `dvips(FileName,dvi)`. It states that the atom `dvips(FileName,dvi)` occurs somewhere after `latex(FileName,tex)`. Complex sequences are used to represent sequential patterns. In this chapter, we use the term "query" interchangeably with "pattern".

Heads and Tails of Queries. The *head*(q) of a complex sequence q denotes the maximal prefix of q that does not contain the operator $<$. The remainder of the query will be referred to using *tail*(q). E.g. if $q = a\ b\ c < d\ e$ then *head*$(q) = a\ b\ c$ and *tail*$(q) = d\ e$.

Given two sequences q and s we can define a notion of subsumption.

[2] In principle, we could also allow for functors. However, our optimal refinement operator (Sect. 4) is optimized for the case without functors.

Simple Subsumption. A simple sequence $l_0 l_1 l_2 \ldots l_n$ s-subsumes a sequence $s_0\ op_1\ ...\ op_m\ s_m$ if and only if there exists a substitution θ such that $(l_0\ l_1\ l_2\ ...\ l_n)\theta$ is a subsequence of $s_0 ... s_m$, i.e. $\exists i : l_0\theta = s_i \wedge ... \wedge l_n\theta = s_{i+n}$ and $op_{i+1} = ... = op_{i+n} = \lhd$.

Subsumption. A complex sequence q subsumes a sequence $s_0\ op_1\ ...\ op_n\ s_n$ if and only if there exists a substitution θ and an integer i such that $head(q)\theta$ s-subsumes $s_0\ op_1\ ...\ op_i\ s_i$ and $tail(q)\theta$ subsumes $s_{i+1} ... s_n$.

Under this definition, a simple sequence s can only subsume sequence c if c has a simple subsequence c' such that $s \sqsubseteq c'$.

For example, the sequence $s_1 = b\ c(X)$ s-subsumes the sequence $s_2 = a\ b\ c(p)\ d(q)$ with the substitution $\{X \mapsto p\}$. The sequence $s_3 = a\ b < c(X)$ subsumes s_2 and also $s_4 = a\ b < c(p)\ d(q)$ with the same substitution. However, s_3 does not subsume $s_5 = a < b\ c(p)\ d(q)$, as $head(s_3) = a\ b$ does not s-subsume any fragment of s_5.

When a sequence s_1 subsumes another sequence s_2 we will write $s_1 \sqsubseteq s_2$. The introduced notions of subsumption will be useful to query sequential databases (cf. below) and also to reason about the generality of patterns.

Sequences in SeqLog correspond to base predicates in Datalog. Akin to Datalog, we allow the user also to specify view predicates in terms of queries.

A Sequential Clause is an expression of the form $h \leftarrow q$ where h is a literal and q is a (possibly complex) sequence. Predicates appearing in the conclusion part of clauses will be called view predicates.

Notice that sequential clauses can be recursive.

A Sequential Database D consists of a set of sequential clauses, $clauses(D)$, and a set of sequences, $sequences(D)$.

By now, we have everything available to define the semantics of SeqLog. This is realized by analogy to the well-known T_p operator in computational logic.

Definition 1. *Let Q be sequential database. Then*

$$T_Q(S) = S \cup \{s_0\ op_1\ ...\ op_i\ s_i op_{i+1} h\theta\ op_j\ s_j\ ...\ s_m \mid s_0\ op_1\ ...\ op_m\ s_m \in S$$

$$\wedge \exists (h \leftarrow l_1 ... l_k) \in clauses(Q), \theta : l_1\theta = s_{i+1} \wedge l_k\theta = s_{j-1} \wedge$$

$$(l_1 ... l_k)\theta \sqsubseteq s_{i+1}\ op_{i+2}\ ...\ op_{j-1}\ s_{j-1}\}$$

We can then inductively define $T_Q^n(S) = T_Q(T_Q^{n-1}(S))$. The fix-point is then $T_Q^\infty(S)$. It defines the meaning of the database S.

Definition 2. *A sequence s is logically entailed by a sequential database Q, notation $Q \models s\theta$ if and only if $\exists s' \in T_Q^\infty(sequences(Q))$ such that s subsumes s' with substitution θ.*

Sequences of the form s can be regarded as queries in the above definition.

Consider for example the following sequential database: { "$b\ a\ a$", "$p \leftarrow a$", "$p \leftarrow a\ p$"}. Then $T_Q^{\infty}(\{b\ a\ a\}) = \{b\ a\ a, b\ p\ a, b\ a\ p, b\ p\ p, b\ p\}$. We would therefore have that $Q \models b < p$ and $Q \models b\ a < p$.

At this point, we wish to stress that resolution type mechanisms can be employed in order to obtain algorithms for reasoning about entailment. Also, s-subsumption can be decided in time polynomial in the length of the sequence, whereas testing for subsumption is NP-complete. Though we will not elaborate much further on the use of SeqLog outside the context of constraint based mining, the reader familiar with inductive logic programming should notice that the entailment relation could be employed as a coverage relation. This, in turn, allows us to adapt the traditional inductive logic programming settings for use with SeqLog instead of Prolog. In this way it is straightforward to define a concept-learning task for sequential data in SeqLog.

3 Constraint Based Mining in SeqLog

3.1 Conjunctive Constraints

Now that we have defined our SeqLog formalism, it becomes possible to mine sequential data. The data mining task addressed uses a background theory KB (a set of clauses in SeqLog) as well as a set of sequences D (possibly divided into subsets $D_1, ..., D_n$). The aim is then to find all sequences satisfying the specified constraints with regard to KB and D. A variety of constraints can be used. The only requirement is that the overall constraint $a \wedge m$ can be written as the conjunction of a monotonic m and an anti-monotonic component a.

Definition 3. *A constraint p is* anti-monotonic *if and only if $\forall$ sequences x : $(x \sqsubseteq y) \wedge p(y) \rightarrow p(x)$.*

Definition 4. *A constraint p is* monotonic *if and only if $\forall$ sequences x : $(x \sqsubseteq y) \wedge p(x) \rightarrow p(y)$.*

For example, the "minimum frequency" constraint commonly used in data mining is an anti-monotonic one. In particular, since any sequence that satisfies a pattern y must also satisfy all patterns $x \sqsubseteq y$, the frequency for y must be no higher than that for x. Thus, if the minimum frequency constraint is satisfied for pattern y, it must also be satisfied by all $x \sqsubseteq y$. On the other hand, a "maximum frequency" constraint is monotonic.

Notice that this is a quite expressive framework given that one can compose complex anti-monotonic (resp. monotonic) constraints on the basis of simpler ones. This is because the conjunction and disjunction of a set of anti-monotonic (resp. monotonic) constraints is still anti-monotonic (resp. monotonic), whereas the negation of an anti-monotonic (resp. monotonic) constraint is monotonic (resp. anti-monotonic). Within MineSeqLog, the following primitives are directly supported. They are inspired by similar primitives for simple sequences in the molecular feature miner MolFea [11].

- $T \sqsubseteq p$, $p \sqsubseteq T$, $\neg(T \sqsubseteq p)$ and $\neg(p \sqsubseteq T)$: where T is the unknown target query and p is a logical sequence; this type of primitive constraint denotes that T should (resp. should not) subsume the sequence p; e.g., the constraint $a\ b\ c \sqsubseteq T$ specifies that the target pattern T should be subsumed by $a\ b\ c$.
- $freq(T, E)$ denotes the frequency of a pattern T on a set of sequences E; the frequency of a pattern T on a data-set E is defined as the number of sequences in E that T matches (possibly taking into account clauses in the background knowledge B). More formally, we have

$$freq(T, E) = |\{e \in E \mid B \cup \{e\} \models T\}|$$

  E.g., the frequency of $a < b$ on the data set $E = \{a\ b\ c, a\ c\ b, a\ c\}$ is 2.
- $freq(T, E_1) \leq c_1$, $freq(T, E_2) \geq c_2$ where the c_i are positive integers and E_1 and E_2 are sets of sequences; this constraint denotes that the frequency of T on the data-set E_i should be larger than (resp. smaller than) or equal to c_i; e.g., the constraint $freq(T, Pos) \geq 100$ denotes that the target patterns T should have a minimum frequency of 100 on the set of positive sequences Pos.

These primitive constraints can now conjunctively be combined in order to declaratively specify the target queries of interest. Note that the conjunction may specify constraints w.r.t. any number of data-sets, e.g., imposing a minimum frequency on a set of positive sequences, and a maximum one on a set of negative ones. E.g., we can express the questions "What Unix command sequences are typically used by experts only?" with $a =$ "$freq(s, \text{expert}) > \text{threshold}_1$" and $m =$ "$freq(s, \text{novice}) < \text{threshold}_2$", with databases "expert" and "novice" denoting sequences of Unix commands used by expert users and novice users, respectively.

3.2 Characterizing the Solution Space

It is well-known that the space of solutions $Sol(q)$ to certain types of constraints can be represented using boundary sets. E.g., Hirsh [12] has shown that sets that are convex and definite can be represented using their boundaries; Mannila and Toivonen [13] have shown that the solution space for anti-monotonic queries can be represented using the set of minimally general elements. Furthermore, various algorithms exploit these properties for efficiently finding solutions to queries, cf. Bayardo's MaxMiner [14] and Mitchell's candidate-elimination algorithm [15] as well as our MolFea system [11].

The boundary sets, sometimes also called the borders, are the most specific (resp. the most general) patterns within (or just outside) the set. More formally, let P be a set of patterns. Then we denote the set of minimal (i.e. minimally specific) patterns within P as $\min(P)$. Dually, $\max(P)$ denotes the maximally specific patterns within P. In Mannila and Toivonen's terminology this corresponds to the positive border $BD^+(P)$.

In the remainder of this paper, we will be largely following Mitchell and Hirsh's notation and terminology. We will also assume that the cardinality of

$\mathcal{L}$ is finite (finiteness implies that all subsets of $\mathcal{L}$ are definite, one of the two requirements for having boundary set representability, cf. [12])[3].

Definition 5. *The S-set w.r.t. a query $q \in Q$ is defined as $S(q) = \max(Sol(q)) = \{s \in Sol(q) | \nexists s' \in Sol(q) \wedge s \sqsubseteq s' \wedge s' \not\sqsubseteq s\}$. The G-set w.r.t. a query $q \in Q$ is defined as $G(q) = \min(Sol(q)) = \{s \in Sol(q) | \nexists s' \in Sol(q) \wedge s' \sqsubseteq s \wedge s \not\sqsubseteq s'\}$.*

Definition 6. *A set T is upper boundary set representable if and only if $T = \{t \in \mathcal{L} \mid \exists s \in \max(T) : t \sqsubseteq s\}$; it is lower boundary set representable if and only if $T = \{t \in \mathcal{L} \mid \exists g \in \min(T) : g \sqsubseteq t\}$; and a set T is boundary set representable if and only if $T = \{t \in \mathcal{L} \mid \exists g \in \min(T), s \in \max(T) : g \sqsubseteq t \sqsubseteq s\}$.*

Sets that are boundary set representable are sometimes also referred to as *version spaces*. It is well-known that finite solution spaces of constraints of the form $a \wedge m$ with a an anti-monotonic and m a monotonic are boundary set representable, cf. [12,16,11].

The key problem that is remaining now, is how to compute the version space for a MineSeqLog query of the form $a \wedge m$. One approach would be to employ the level wise version space algorithm employed in MolFea [16,11]. This algorithm combines the well-known level-wise algorithm with the description identification algorithm of [17]. In this paper, we present another approach, the MineSeqLog algorithm (Algorithm 1). This algorithm reuses a level-wise algorithm for discovering patterns under anti-monotonic constraints *only*, such as Apriori [7] or MaxMiner [14]. Reusing these algorithms has the advantage that they are already well studied, with many optimization and implementation techniques available. Further improvements to these algorithms automatically applies to our approach, too. We present our version of MaxMiner, FindMaximalPatterns, in Sect. 5 (Algorithm 2).

FindMaximalPatterns is called twice, as sketched in Algorithm 1. In the first invocation, FindMaximalPatterns finds the set of maximal patterns for constraint a. The second invocation finds the set of maximal patterns satisfying $\neg m$, which is also a set of minimal patterns just *not* satisfying m.

It is easy to see that

$$Sol(a \wedge m) = \{p \mid \exists u \in U : p \sqsubseteq u \text{ and } \neg\exists l \in L : p \sqsubseteq l\}$$

This directly follows from the observation that the negation $\neg m$ of a monotonic constraint m is anti-monotonic. Furthermore, we want to find those patterns p that satisfy the anti-monotonic constraint a (hence $\exists u \in U : p \sqsubseteq u$) and do not satisfy m (hence $\neg\exists l \in L : p \sqsubseteq l$). Not all the elements of U and L found in lines 8 and 9 are useful, however. For any $u \in U$ such that we can find an $l \in L$ satisfying $u \sqsubseteq l$, we observe that if there is any pattern $p \sqsubseteq u$, we have automatically $p \sqsubseteq u \sqsubseteq l$ and hence $p \notin Sol(a \wedge m)$. So, such a u is redundant. It is thus pruned in line 11. On the other hand, an $l \in L$ is redundant if for

[3] Finiteness can be guaranteed by imposing a finite upperbound on the length of the sequences in $\mathcal{L}$.

all patterns $p \sqsubseteq l$, there is no $u \in U$ such that $p \sqsubseteq u$. This is equivalent to $mgg(l, u) = \{\omega\}\ \forall u \in U$[4]. So, these elements of L are pruned in line 12 above.

Regardless of whether one applies the level-wise version space algorithm or the bi-directional MaxMiner approach, it is crucial for efficiency reasons, to employ a so-called optimal refinement operator. This is elaborated in the next section.

Algorithm 1 MineSeqLog

1: /* *Input:* */
2: /*   *a = an anti-monotonic constraint* */
3: /*   *m = a monotonic constraint* */
4: /* *Output:* */
5: /*   *U = the set of maximal patterns satisfying $a \wedge m$* */
6: /*   *L = the set of minimal patterns not satisfying $a \wedge m$* */
7:
8: compute $U = S(a)$, the set of maximally specific patterns satisfying a using FindMaximalPatterns.
9: compute $L = S(\neg m)$ using FindMaximalPatterns; the set of maximally specific patterns not satisfying m.
10: /* *prune U and L :* */
11: Prune away all $u \in U$ satisfying: $\exists l \in L, u \sqsubseteq l$.
12: Prune away all $l \in L$ satisfying: $\forall u \in U, mgg(l, u) = \{\omega\}$ where $mgg(x, y)$ signifies the set of the minimally general generalizations of two patterns.

4 An Optimal Refinement Operator for SeqLog

4.1 Optimality

A refinement operator is an operator ρ that maps each pattern p to a set of specializations of it, i.e. $\rho(p) \subseteq \{p' \in \mathcal{L} \mid p \sqsubseteq p'\}$. With such an operator, we can then employ e.g. level-wise algorithms to generate and test patterns that satisfy the anti-monotonic constraints.

For *optimality*, we further require that:

Completeness. Applying the operator ρ on p (possibly repeatedly), it is possible to generate all other patterns that p subsumes. In other words, $\cup_{r=0}^{\infty}\rho^r(p) = \{p' \in \mathcal{L} \mid p \sqsubseteq p'\}$. This requirement guarantees that we will not miss any patterns that may satisfy the constraints.

Single Path. Given pattern p, there should exist exactly one sequence of patterns $p_0 = \omega, p_1, \ldots, p_n = p$ such that $p_{i+1} \in \rho(p_i)$ for all i. This requirement helps ensuring that no query is generated more than once, i.e., there are no duplicates.

[4] Here, we assume that $\omega \notin Sol(a \wedge m)$. If $\omega \in Sol(a \wedge m)$, then we would have $a(\omega) \wedge m(\omega) \implies m(\omega) \implies m(\varphi)\ \forall \varphi \in \mathcal{L}$ by the monotonicity. This would mean that m is a trivial predicate that is always true and hence $Sol(a \wedge m)$ would degenerate to $Sol(a)$, for which L is simply the empty set.

When working with propositional patterns, such an optimal refinement operator is rather straightforward to devise. However, when working with first order expressions, optimal refinements operators may not always exist [18]. Furthermore, the use of naïve (non)-optimal refinement operators may lead to severe efficiency problems, cf. the work by Nijssen and Kok [19], who elegantly solved many of the efficiency problems with the Warmr system [20].

Therefore, in the remainder of this section, we elaborate on the optimal refinement operator that we have developed for *functor-free* SeqLog.

4.2 Basic Ideas

Four operations are identified for refining a certain SeqLog query Q_g into a more specific one Q_s. More precisely, $Q_g \sqsubseteq Q_s$. In other words, the Q_s so generated is the most general specialization of Q_g. We will use the query $Q_g = a(X) < b(Y)$ as an example.

Lengthening. Add a new atom to the query, thus increasing the length of the query by one literal. The new term can be inserted into any position with any operator. E.g. $Q_s = a(X) < b(Y) < c(Z)$ or $Q_s = a(X)\ c(Z) < b(Y)$.

Promotion. If the query has a $<$ operator, replace it with $\triangleleft$. For example, $Q_s = a(X)\ b(Y)$.

UniVar. Pick any two variables from the query and unify them. For instance, $Q_s = a(X) < b(X)$.

Instantiation. Pick any variable from the query and replace it with a term. E.g. $Q_s = a(f(U, V)) < b(Y)$.

Note that for Lengthening and Instantiation, we need to introduce new terms to form the refined query Q_s. Where shall we draw these from? In practice, we restrict the set of all possible SeqLog queries by specifying a list of predicate names, together with their arities, which may appear in the generated sequences. Terms for predicates not appearing in this list are not generated by our refinement operator.

Furthermore, the operator UniVar in the above form causes problems in practice. With n variables in Q_g, there are $\binom{n}{2} = O(n^2)$ possible pairs to unify. With a large number of possible queries, this becomes intolerable. Therefore, in our MineSeqLog system, we allow the user to specify *types* for the arguments of the terms as in many ILP systems, e.g. Warmr [20]. Arguments of different types will never be unified. This helps preventing the generation of nonsense queries such as $Q_s =$"cook(Person$_1$, Food) $<$ eat(Food, Food)" formed by unifying variables "Food" with "Person$_2$" from $Q_g =$"cook(Person$_1$, Food) $<$ eat(Person$_2$, Food)". Types can also be used in MineSeqLog to restrict the terms that are used for the Instantiation of variables, so that only meaningful values will be used for the substitution.

With these four basic operations, we can create a refinement operator that takes a query Q_g and applies each operation (in multiple possible ways) to generate new queries Q_s. Such a refinement operator will satisfy the completeness criterion. However, it generates a lot of duplicates and is hence not satisfying the single path criterion. For instance, the specialization $a < b < c$ may be generated from $a < b$, $a < c$ or $b < c$.

We define a measure, the refinement level vector, on each query. The measure helps us define restrictions on the above refinement operations to ensure that no duplicates are generated.

4.3 Refinement Level Vector

Given any functor-free SeqLog query q , the *refinement level vector* $\boldsymbol{v}(q)$ is defined as a 4-dimensional vector of the form (l, p, u, i), where:

- l is the number of predicates in q.
- p is the number of $\lhd$ operators in q.
- u is the number of arguments in q minus the number of distinct variables (but not constant arguments) in q minus the number of constant arguments in q.
- i is the number of constant arguments in q.

Furthermore, we call the first norm of this vector, $\|\boldsymbol{v}(q)\|_1 = l + p + u + i$ the *refinement level* of q. Our goal is to make sure that given a query q, any of the four refinement operations on q will always generate a new query q' so that $\|\boldsymbol{v}(q')\|_1 = \|\boldsymbol{v}(q)\|_1 + 1$.

The refinement level vector $\boldsymbol{v}(q)$ for any given query q has the following properties:

- $\boldsymbol{v}(q) = (0, 0, 0, 0)$ if and only if q is the empty query.
- All the 4 subordinate values of $\boldsymbol{v}(q)$ are non-negative integers, for all valid queries q.

4.4 Duplicate Avoidance

In order to avoid the generation of duplicates and to satisfy the single-path criterion, we need to add restrictions to our 4 refinement operations. Because of space limitations, we can only briefly present the restrictions in this paper. Below, q denotes the query being refined and q' denotes a refined query. The symbols l, p, i, u denote any non-negative integers.

Lengthening. Apply only if $\boldsymbol{v}(q) = (l, 0, 0, 0)$. Moreover, new atoms are only added at the end with the $<$ operator. Result: $\boldsymbol{v}(q') = (l + 1, 0, 0, 0)$.

Promotion. Apply only if $\boldsymbol{v}(q) = (l, p, 0, 0)$ When promoting an operator from $<$ to $\lhd$, only do so when this operator is followed by no other $\lhd$ operators. This is required to avoid duplication. For example, "a⊲b⊲c" could be generated from both "a⊲b <c" and "a<b ⊲c" by Promotion. However, the above restriction forbids the latter, and only allows "a⊲b⊲c" to be promoted from "a⊲b <c". Result: $\boldsymbol{v}(q') = (l, p + 1, 0, 0)$.

UniVar. Apply for all $\boldsymbol{v}(q) = (l, p, u, i)$. Of the two variables chosen to be unified, one of them must be not yet unified with any other variables. Moreover, this variable must not be followed by any other already unified variables.

These constraints are needed to avoid duplicates as illustrated in Fig. 1, which illustrates the possible paths through which a query $f(X, X', Y, Y')$ can be refined to $f(X, X, X, X)$ using UniVar. The above restrictions prune away all paths in dashed lines, leaving a unique spanning tree rooted at query $f(X, X', Y, Y')$. Result: $\boldsymbol{v}(q') = (l, p, u+1, i)$.

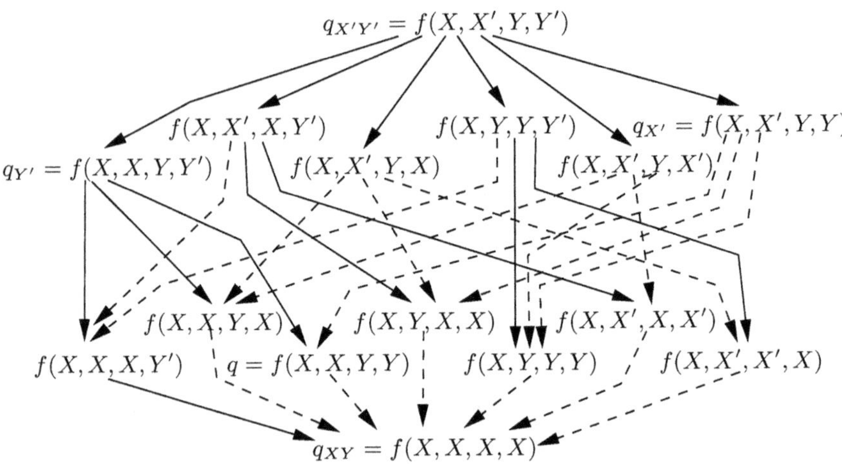

Fig. 1. Example of duplicate avoidance for UniVar

Instantiation. Apply only if $\boldsymbol{v}(q) = (l, p, 0, i)$. Moreover, in q no other arguments to the right of the variable to be instantiated should be the result of a previous Instantiation, i.e. successive instantiations are performed from left to right. This restriction is required to avoid duplicates as illustrated in Fig. 2. The figure illustrates the different possible paths to refine $f(A, B, C)$ to $f(\mathsf{a}, \mathsf{b}, \mathsf{c})$. The restriction prunes away the paths shown with dashed lines, leaving a spanning tree rooted at $f(A, B, C)$. Result: $\boldsymbol{v}(q') = (l, p, 0, i+1)$.

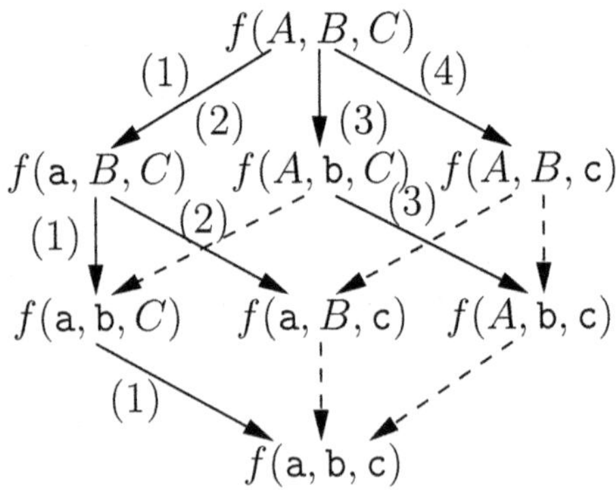

Fig. 2. Example of duplicate avoidance for Instantiation (DataLog case)

Note that each of these four operations increases exactly one ordinate of the refinement level vector by one. Each of these operations has a corresponding inverse operation, which decreases the corresponding ordinate in the refinement level vector by one. Given any valid sequential query, we can apply the inverse of UniVar successively to generalize it, decreasing its u value by one at a time, until the u value reaches zero. We can then apply the inverse of Instantiation to decrease the value of i until it becomes zero. Similarly, we can decrease the p value to zero by the inverse of Promotion and the value of l to zero by the inverse of Lengthening. This sequence of inverse operations corresponds to a reversed sequence of refinement operations that refines the original query from the empty query. So, the original query can be generated by refining the empty query. This is true for any valid sequential query. Therefore, any valid sequential query can be refined from the empty query. Our refinement operation is thus complete, despite the additional restrictions.

Moreover, the sequence of refinements described above to obtain any query from the empty query is also unique, because of the restrictions added to the 4 refinement operations. So, the single-path criterion is also met. Thus, our refinement operator is optimal. As an example, consider the query "$q_7 = a(X)\ b(Y,g) < c(h,X)$", with a refinement level vector $\boldsymbol{v}(q_7) = (3,1,2,1)$. The only way to get this query using our refinement operators is from "$q_6 = a(X_1)\ b(Y,g) < c(h,X_2)$" by Unification, where $\boldsymbol{v}(q_6) = (3,1,2,0)$. This in turn must have come from 2 Instantiations, via "$q_5 = a(X_1)\ b(Y,g) < c(H,X_2)$" and "$q_4 = a(X_1)\ b(Y,G) < c(H,X_2)$", where $\boldsymbol{v}(q_5) = (3,1,1,0)$ and $\boldsymbol{v}(q_4) = (3,1,0,0)$. The only way to obtain q_4 is a Promotion from "$q_3 = a(X_1) < b(Y,G) < c(H,X_2)$", with $\boldsymbol{v}(q_3) = (3,0,0,0)$. This must have come from Lengthening, via "$q_2 = a(X_1) < b(Y,G)$" (where $\boldsymbol{v}(q_2) = (2,0,0,0)$), and "$q_1 = a(X_1)$" (where $\boldsymbol{v}(q_1) = (1,0,0,0)$). Eventually, we get q_1 by Lengthening the empty query $q_0 = \omega$, which has $\boldsymbol{v}(q_0) = (0,0,0,0)$). It is left as an exercise for the reader to check that $q_0 = \omega, q_1, \ldots, q_7$ is the only way to obtain q_7 from q_0 using our optimal refinement operator repeatedly.

4.5 Further Optimizations

Using the idea from the Apriori and the apriori_gen function, the Lengthening operation can be further optimized when dealing with anti-monotonic constraints. We can perform Lengthening on the set Q of all q with $\boldsymbol{v}(q) = l$ in batch and generate the set Q' of all q' with $\boldsymbol{v}(q') = l+1$. We first sort the elements of Q in lexicographical order. Then, for every pair of $q \in Q$ queries sharing a common length $l-1$ prefix, join them to form 4 length $l+1$ candidate queries. e.g. the pair "a <b <c" and "a <b <d" produces "a <b <c <c", "a <b <c <d", "a <b <d <c" and "a <b <d <d". All such generated queries, are collected and inserted into the set Q'. The monotonic property guarantees that all suitable SeqLog queries of length $l+1$ are found inside Q'. Moreover, this generation method never generates any query q' with $\boldsymbol{v}(q') = (l+1,0,0,0)$ more than once.

Another optimization when dealing with only anti-monotonic constraints is to consider only those atoms which, when treated as singleton sequences, satisfy the constraints. This is because sequences containing any other atoms will not satisfy the constraints. (If they do, then so will the singleton sequences formed from these atoms because of the anti-monotonicity—a contradiction.) E.g., when finding frequent itemsets, only frequent items need to be added when generating candidate itemsets of larger sizes.

5 The MineSeqLog Algorithm

Having an optimal refinement operator, we can now devise an efficient algorithm for solving the data mining problem specified in Sect. 3. We call this algorithm MineSeqLog.

The inputs to the algorithm are two constraints: a and m, where a is anti-monotonic and m is monotonic. These constraints are predicates on SeqLog queries against a given database D (or a particular subset of it). The algorithm then discovers all SeqLog queries that satisfy the given constraints. The algorithm also takes a set F as input. This specifies the set of predicate names (with the specified arities) to be used to generate the terms in the SeqLog patterns.

The MineSeqLog algorithm is already given in Sect. 3.2, and is thus not repeated here. There, we mentioned a FindMaximalPatterns algorithm, which is listed in Algorithm 2. Most of the work of MineSeqLog is delegated to this algorithm. This algorithm requires that the input constraint a be anti-monotonic. It is an iterative level-wise discovery algorithm based on the same framework as the well-known Apriori algorithm [7]. In iteration k, FindMaximalPatterns generates a candidate set C_k of potential patterns and then tests them against the given anti-monotonic constraint a, possibly scanning the involved database subsets. Those that satisfy the constraint are added to L_k. C_1 is the set of interesting atoms generated from F. In subsequent iterations, C_k is generated by applying the GenerateCandidates algorithm described later. This latter algorithm takes a set of patterns of the same refinement level and uses the optimal refinement operator described in Sect. 4 to generate the patterns of the next refinement level. For efficiency reasons, we keep the refinement level vector $\boldsymbol{v}(q)$ together with each pattern q in C_k as well as L_k. We will later see that we can compute the refinement level vectors easily without calculating them from scratch in each iteration.

The GenerateCandidates algorithm (Algorithm 3) takes as input a set of interesting patterns of the same refinement level (i.e. the sum of the coordinates in the refinement level vector, see Sect. 4.3) and computes their optimal refinements. The refined patterns are all of a refinement level one higher than that of the input patterns. For efficiency, the input and output sets are actually ordered pairs containing the pattern as well as its refinement level vector. GenerateCandidates can compute the refinement level vectors of the newly generated Candidates immediately, without having to use the definition given in Sect. 4.3.

The GenerateCandidates algorithm given here does not apply the Lengthening operation (Sect. 4.2) directly. Rather, it employs the optimization described in Sect. 4.5. This uses the join operation in Apriori to efficiently compute the lengthened candidates.

Algorithm 2 FindMaximalPatterns

```
/* Input: */
/*  a = an anti-monotonic constraint */
/*  D = the database to be mined */
/*  F = a set of predicate names with arities */
/* Output: S = the set of maximal patterns satisfying a */

/* First Iteration: */
C_1 ← {(f(X_1, X_2, ..., X_n), (1,0,0,0)) | (f, n) ∈ F} /* (1,0,0,0) is the refinement
level vector */
L_1 ← {(x, v) ∈ C_1 | pattern x satisfies constraint a}

k ← 2
loop /* k-th Iteration: */
  C_k ← GenerateCandidates(L_{k-1})
  if C_k = ∅ then
    exit loop
  end if
  L_k ← {(x, v) ∈ C_k | pattern x satisfies constraint a}
end loop

/* Results: */
return S ← max(⋃_{l=1}^{k} L_k)
```

Algorithm 3 GenerateCandidates

```
/* Input: */
/*  L_{k-1} = the set of interesting patterns discovered at the previous level */
/* Output: C_k = the set of new candidates */

F_{k-1} ← {x|(x, v) ∈ L_{k-1} ∧ v = (k − 1, 0, 0, 0)}
F_k ← Join(F_{k-1}) /* Using the idea of Apriori's join */
C_k ← {(x, (k, 0, 0, 0)|x ∈ F_k}
for all (x, v) ∈ L_{k-1} do
  if v has the form (l, p, 0, 0) then /* candidate for promotion */
    apply the Promotion operation on x to get y
    if successful, C_k ← C_k ∪ (y, (l, p + 1, 0, 0))
  end if
  if v has the form (l, p, i, 0) then /* candidate for instantiation */
    apply the Instantiation operation on x to get y
    if successful, C_k ← C_k ∪ (y, (l, p, i + 1, 0))
  end if
  if v has the form (l, p, i, u) then /* candidate for univar */
    apply the UniVar operation on x to get y
    if successful, C_k ← C_k ∪ (y, (l, p, i, u + 1))
  end if
end for
return C_k.
```

6 Experiments

We have implemented the algorithms in Sect. 5 in SICSTUS Prolog and performed some preliminary experiments on a Unix command history database [21,8] to test out our ideas. The implementation uses the refinement operator in Sect. 4 and hence discovers functor-free SeqLog queries. The database is a set of simple (i.e. without the < operator) SeqLog facts, although the patterns discovered are complex sequences (which may contain the < operator).

6.1 The Unix Command History Data

The database is obtained from 168 users of Unix `csh`. The Unix commands that they used over a period of timer were recorded. These are represented as a set of simple SeqLog facts. Each command is represented as an atom with exactly 2 parameters: the command arguments an the directory from which the command is invoked. The sequence of commands used by a single user in a single login session is represented as a SeqLog sentence. For example, if user Mary logs in and invokes the commands `cd myprog`, `ls *.c`, `mail`, `exit`, this login session is represented by the SeqLog sequence:

```
cd('myprog','/home/mary')
ls('*.c','/home/mary/myprog')
mail('','/home/mary/myprog')
exit('','/home/mary/myprog')
```

A large number of login sessions of different users are gathered. These users are divided into four groups: novice programmers (nov), experienced programmers (exp), non-programmers (non) as well as computer scientists (sci). This allows us to partition the database into 4 disjoint subsets. Table 1 shows a summary of some measures on the data.

Table 1. Summary statistics of the data

Subset (D)	users	sequences	threshold (θ)	time (s)	patterns	$\vert M\vert$
nov	55	5164	80	676	796	38
exp	36	3859	100	604	834	61
non	25	1906	200	890	533	100
sci	52	7751	250	2345	208	74

6.2 The Mining Task

The anti-monotonic constraint we used in the experiment is $\mathsf{frequent}(p, D; \theta)$, where p represents the pattern, i.e. SeqLog query, in question and D represents the database subset (one of {nov, exp, non, sci}). It evaluates to *true* if the frequency of pattern p in the database subset D is no fewer than θ (see Sect. 3.1). For simplicity, the monotonic constraints that we used are simply the negation of "frequent". Moreover, we have assigned a threshold value to each data subset and always used that threshold for it. These values are shown in table 1. Many different threshold values have been attempted and we present here those that yield a moderate amount of patterns. The database is preprocessed to obtain the 150 most frequent commands. These, with an arity of 2, form the set F. Thus, the mining algorithm will only generate patterns out of the atoms corresponding to these commands. The results of applying FindMaximalPatterns to these data sets are also shown in table 1. The "time" column shows the wall-clock time taken for the algorithm to complete on a Pentium III (600MHz) PC running Linux kernel 2.4. The "patterns" column shows the number of patterns found to satisfy the given constraints. The "$|M|$" column shows the number of patterns in the set of maximal patterns.

We then used our MineSeqLog program to discover interesting patterns from the database. For each pair of distinct database subsets D_1 and D_2, we denote by $D_1 : D_2$ the constraint obtained from the conjunction:

$$\mathsf{frequent}(p, D_1; \theta_{D_1}) \wedge \neg\mathsf{frequent}(p, D_2; \theta_{D_2})$$

e.g. when D_1 is 'nov' and D_2 is 'exp', this constraint will cause MineSeqLog to discover command sequences that are often used by novice programmers but only seldom used by experienced programmers. Each such constraint is then specified to MineSeqLog, which then discovers all interesting patterns satisfying the constraint. The results are summarized in table 2. In each case, the table shows the number of patterns found and the sizes of the sets U and L (see Sect. 3.2) before and after pruning. The length of the longest sequence was 4 in all cases. At this point the reader may have noticed that the size of $U \cup L$ can be larger than the number of solutions. For version spaces, the size of $S \cup G$ can be at most twice the number of solutions. At present we are still investigating this issue for U and L as well as whether more elements from U and L could be pruned.

Table 2. Results of running MineSeqLog

Constraint	number of patterns found	before pruning			after pruning		
		$\|U\|$	$\|L\|$	$\|U\|+\|L\|$	$\|U\|$	$\|L\|$	$\|U\|+\|L\|$
nov:non	679	38	100	138	37	29	66
nov:exp	401	38	61	99	27	41	68
nov:sci	716	38	74	112	33	62	95
exp:nov	609	61	38	99	35	36	71
exp:non	686	61	100	161	58	100	158
exp:sci	756	61	74	135	51	68	119
non:nov	489	100	38	138	93	13	106
non:exp	479	100	61	161	93	27	120
non:sci	501	100	74	174	98	61	159
sci:nov	161	74	38	112	68	30	98
sci:non	193	74	100	174	71	95	166
sci:exp	160	74	61	135	63	49	112

6.3 Experimental Results

Some interesting example patterns in the results include:

- The sequences discovered in non:exp formed a subset of those in non:sci and non:nov. (E.g. The sequence "`nroff`(A,B) < `wordproc`(C,D)" is common to all these result sets.) The sequences discovered in nov:exp formed a subset of those in nov:non. (E.g. "`more`(A,B) `more`(C,D) < `more`(E,F) < `more`(G,H)" is common to both.) The result of sci:exp is also a subset of sci:non. (e.g. They both contain "`more`(A,B) < `more`(C,D) < `more`(E,D)".) Perhaps, experienced programmers tend to use a larger set of command sequences, so that there are fewer sequences that are infrequent in their data set.
- The sequences in non:sci and sci:non are disjoint. (E.g. In non:sci, the following pattern is found: "`emacs`(A,B) < `wordproc`(C,B) < `emacs`(D,B) < `wordproc`(E,B)". It does not occur in sci:non. However, sci:non contains "`mailnews`(A,B) < `mailnews`(A,B) < `mailnews`(A,B)", which is not found in non:sci.) This suggests that non-programmers use command sequences very different from those of computer scientists.
- The sequences in nov:exp and exp:nov are also disjoint. (E.g. "`script`(A,B) < `script`(A,C) < `script`(A,B) < `script`(A,B)" is found in nov:exp but not in exp:nov. On the other hand, "`make`(A,B) < `make`(C,B) < `make`(C,B) < `make`(C,B)" is found in exp:nov but not nov:exp.) Maybe, novice and experienced programmers tend to use very different command sequences. (e.g. experienced programmers tend to master the `make` utility.)

7 Related Work

There is plenty of related work to MineSeqLog. First, there are the already mentioned propositional approaches to mining sequential data. Secondly, there is the approach by Mannila and Toivonen [9] that considers a form of relational sequential patterns in which there is only one type of event, no background knowledge, the order must be expressed with a total order relation on a special time attribute and instead of unification, equality must be expressed with an explicit relation. Thirdly, there is the work by [22], who also investigate logical sequential patterns. However, Masson et al. extend the Spirit system by [4], which employs a grammar as a syntactic bias mechanism to mine *propositional* sequences. Masson's model cannot express the equivalent of our < operator, cannot express background knowledge in the form of clauses, and is limited to minimum frequency constraints only. Fourthly, there is the traditional work on frequent query discovery in multi-relational data mining and inductive logic programming, e.g. [20,23]. Although one could in principle also use these techniques to find sequential patterns, it would be less efficient to do so. Indeed, the typical inductive logic programming approach to representing sequential data consists of adding to each fact a number (that denotes the element in the sequence). If one then wants to express that one fact comes after the other, one must add predicates of the type *successor* or *after* together with the relevant arguments. This leads to a larger search space and also does not exploit possible optimizations for matching (e.g. s-subsumption is polynomial). Finally, the work is also related to constraint based data mining. In a sense it was motivated by our earlier work on mining for linear (or sequential) fragments within molecules in the MolFea system [11]. The present approach extends the MolFea framework with more expressive fragments but retains the powerful primitives that were present in MolFea.

8 Conclusions

We have introduced a logical language for representing and reasoning about sequential data. Other researchers (such as Bonner [24]) have also introduced languages for processing sequences based on computational logic. However, the main contribution of SeqLog is that the notions of subsumption, entailment and a fix point semantics are given. These act as direct analogues of corresponding notions in Prolog. This is important because subsumption and entailment are central to inductive logic programming. As a consequence, mining SeqLog sequences will be analogous to inductive logic programming.

The optimal refinement operator (see Sect. 4) discussed in this paper is restricted to functor-free SeqLog. One future direction is to generalize this refinement operator to work for general SeqLog, by allowing Instantiation with general terms instead of just constant terms. Another future direction is concerned with the use of SeqLog as a representation model for other mining and learning tasks. In this context, our lab has already explored the upgrading of Hidden Markov Models for use with SeqLog, cf. [10], and we are now interested in predicate learning and distance based learning within the SeqLog language. It is our hope that the SeqLog framework will inspire further research in databases, data mining and inductive logic programming.

Acknowledgements

This work was supported by the EU IST FET project cInQ, contract number IST-2000-26469. The authors are especially grateful to Maurice Bruynooghe, Tamas Horvath, Stefan Kramer, Heikki Mannila and Esko Ukkonen for interesting discussions and input on this work. We would also like to thank Nico Jacobs for providing us the data for the experiments.

References

1. Agrawal, R., Srikant, R.: Mining sequential patterns. In Yu, P.S., Chen, A.L.P., eds.: Proc. 11th Int. Conf. Data Engineering, ICDE, IEEE Press (1995) 3–14
2. Srikant, R., Agrawal, R.: Mining sequential patterns: Generalizations and performance improvements. In Apers, P.M.G., Bouzeghoub, M., Gardarin, G., eds.: Proc. 5th Int. Conf. Extending Database Technology, EDBT. Volume 1057., Springer-Verlag (1996) 3–17

3. Mannila, H., Toivonen, H., Verkamo, A.I.: Discovering frequent episodes in sequences. In Fayyad, U.M., Uthurusamy, R., eds.: First International Conference on Knowledge Discovery and Data Mining (KDD-95). (1995)
4. Garofalakis, M.N., Rastogi, R., Shim, K.: SPIRIT: Sequential pattern mining with regular expression constraints. In: Proceedings of the 25th International Conference on Very Large Data Bases (VLDB '99), San Francisco, Morgan Kaufmann (1999) 223–234
5. Wang, K.: Discovering patterns from large and dynamic sequential data. Journal of Intelligent Information Systems **9** (1997) 33–56
6. Zaki, M.J.: Fast mining of sequential patterns in very large databases. Technical Report 668, Computer Science, University of Rochester, PO Box 270226, Rochester, NY 14627, U.S.A. (1997)
7. Agrawal, R., Srikant, R.: Fast algorithms for mining association rules. In Bocca, J.B., Jarke, M., Zaniolo, C., eds.: Proc. 20th Int. Conf. Very Large Data Bases, VLDB, Morgan Kaufmann (1994) 487–499
8. Jacobs, N., Blockeel, H.: From shell logs to shell scripts. In Rouveirol, C., ed.: Proc. 11th International Conference, ILP 2001 (Lecture notes in computer science, 2157: Lecture notes in artificial intelligence), Strasbourg, France, Springer Verlag (2001) 80–90
9. Mannila, H., Toivonen, H.: Discovering generalized episodes using minimal occurrences. In Simoudis, E., Han, J.W., Fayyad, U., eds.: Proceedings of the Second International Conference on Knowledge Discovery and Data Mining (KDD-96), AAAI Press (1996) 146
10. Kersting, K., Raiko, T., Kramer, S., De Raedt, L.: Towards discovering structural signatures of protein folds based on logical hidden markov models. In: Proceedings of the Pacific Symposium on Biocomputing (PSB-2003), Kauai, Hawaii, U.S.A. (2003)
11. Kramer, S., De Raedt, L., Helma, C.: Molecular feature mining in hiv data. In: KDD-2001: The Seventh ACM SIGKDD International Conference on Knowledge Discovery and Data Mining, Association for Computing Machinery (2001) ISBN: 158113391X.
12. Hirsh, H.: Theoretical underpinnings of version spaces. In: Proceedings of the Twelfth International Joint Conference on Artificial Intelligence (IJCAI91), Morgan Kaufmann Publishers (1991) 665–670
13. Mannila, H., Toivonen, H.: Levelwise search and borders of theories in knowledge discovery. Data Mining and Knowledge Discovery **1** (1997) 241–258
14. Bayardo, R.: Efficiently mining long patterns from databases. In: Proceedings of ACM SIGMOD Conference on Management of Data. (1998)
15. Mitchell, T.: Generalization as search. Artificial Intelligence **18** (1980) 203–226
16. De Raedt, L., Kramer, S.: The levelwise version space algorithm and its application to molecular fragment finding. In: IJCAI01: Seventeenth International Joint Conference on Artificial Intelligence. (2001)
17. Mellish, C.: The description identification algorithm. Artificial Intelligence (1990)
18. Nienhuys-Cheng, S.H., de Wolf, R.: Foundations of Inductive Logic Programming (Lecture Notes in Computer Science, 1228). Springer Verlag (1997)
19. Nijssen, S., Kok, J.N.: Faster association rules for multiple relations. In: IJCAI. (2001) 891–896
20. Dehaspe, L., Toivonen, H.: Discovery of frequent datalog patterns. Data Mining and Knowledge Discovery Journal **3** (1999)

21. Greenberg, S.: Using unix: Collected traces of 168 users. Research Report 88/333/45, Department of Computer Science, University of Calgary, Calgary, Canada. (1988)
22. Masson, C., Jacquenet, F.: Mining frequent logical sequences with spirit-log. In: Proc. 12th International Conference on Inductive Logic Programming, ILP 2002, Sydney, Australia (2002)
23. De Raedt, L.: A logical database mining query language. In: Proceedings ILP'00. Volume 1866 of LNCS., springer (2000) 78–92
24. Bonner, A.J., Mecca, G.: Sequence datalog: Declarative string manipulation in databases. In: Logic in Databases. (1996) 399–413

Interactivity, Scalability and Resource Control for Efficient KDD Support in DBMS

Matthias Gimbel, Michael Klein, and P.C. Lockemann

Universität Karlsruhe, Fakultät für Informatik,
Am Fasanengarten 5, 76128 Karlsruhe, GERMANY

Abstract. The conflict between resource consumption and query performance in the data mining context often has no satisfactory solution. This is in sharp contrast to the needs of the analysts for interactive response times and has rendered the seamless integration of data mining operators into common multiuser database systems a difficult and (so far) not very successful task. This paper describes an approach that allows to combine preprocessing and data mining operators into one common KDD-aware implementation algebra such that interactivity, scalability and resource efficiency can simultaneously be achieved. The basic idea of our framework is pipelining. However, since there is a danger of blocking pipelines, we introduce controlled ordering-, cardinality- and special-value-properties of the data stream across the whole query tree up to the complex data mining operators. The framework builds on a spezialized index that is basically an extension of the UB-Tree and efficiently provides various data orderings. These orderings and the remaining properties are then exploited by the KDD-algebra operators to release results and internal data structures early enough to allow pipelined, resource-efficient query processing with interactive response times. This paper describes the framework and demonstrates its benefits in preprocessing and in the parallel and interactive detection of outliers.

1 Introduction

Decision support applications and especially data mining tasks are inconceivable without database support. In order to achieve the necessary scalability with today's massive datasets, database systems split query processing into successive tasks, process one operator after the other and swap intermediate results to disk. Performance optimization is achieved by clever organization of the data on disk, proper design of algorithms and operator ordering. However, these techniques are insufficient in a KDD environment which is explorative in nature and where the analyst needs continuously to reevaluate the results and refine the query.

Techniques such as the exploitation of access patterns in order to support the most common queries, e.g., by one-dimensional indexes delivering a special sort order efficiently, or data partitioning in order to allow for data parallel processing

R. Meo et al. (Eds.): Database Support for Data Mining Applications, LNAI 2682, pp. 174–193, 2004.

in a parallel environment, appear problematic: In explorative processes it is not known in advance which index will be useful or which partitioning scheme will allow for well-balanced data parallelism. At the very least, optimized KDD support needs parallelization schemes and index structures that are less vulnerable to changing access patterns.

Further, analysts in data mining must take decisions on how to proceed while the query is still being processed. They cannot wait until the last operator has finished but need to interact in a more timely fashion, perhaps to start a new iteration cycle as soon as first signs of wrong choice of parameters has been detected. The conclusion from this observation is that optimized KDD support needs an operator scheduling strategy that achieves scalability under the condition of interactivity.

In this paper, we develop techniques that make database support for large-scale query processing in KDD resource-efficient, scalable and interactive. The central idea is to provide interactivity through the extensive use of pipelining. To achieve blocking-free and balanced pipelining, we split the operators into primitives and eliminate blocking obstacles by exploiting ordering properties and cardinality information of the data stream flowing through the query tree. To establish these properties efficiently, we provide a multidimensional index structure providing several data orderings that can be used by each operator of our KDD query algebra to compute final results on partial data, and to release these to the next operator and finally, at the end of the operator chain, to the human analyst. At the same time the internal data structures of the operators can be cleared so that resource limits can be met without the need for swapping data to peripheral storage. This allows for interactive query processing with controllable resources and especially for non-blocking pipelining in parallel environments.

The paper is organized as follows. After a survey of related work, we analyze in Section 3 common implementations of typical KDD preprocessing operators to determine which properties of the processed data stream are useful to achieve our goals. In Section 4, we will describe the concept of data stream quality, which formalizes these properties. Section 5 shows how such qualities can be provided efficiently. After a short demonstration of the benefits that our framework provides to query processing, we describe in Section 6 how interactive data mining algorithms can benefit from this framework to allow for parallel and interactive query processing in KDD. A summary, a survey of open problems and suggestions for further work conclude the paper.

2 Related Work

In recent years, efforts towards database support for KDD concentrated on the development of special languages to express data mining queries. Starting with the query language DMQL [5] and continuing all the way to the formulation of a sophisticated abstract data mining algebra in [9], there have been many attempts to design languages for data mining queries. What is still missing is the integration of KDD with database system kernels at the implementation level. Despite the lack of integration there have been several attempts to improve interactivity by delivering early results in query processing. We will now study these approaches and discuss how well they combine interactivity with the pipelining requirements needed to achieve scalability and resource efficiency within the complex analysis chains in KDD.

From the early beginnings of System R, the idea of improving performance by pipelining and achieving it through the exploitation of ordering properties within the data stream has been incorporated into database optimizers [16]. Its success rests on the existence of frequently used sort orders which are supported by suitable one-dimensional indices, an issue quite difficult to enforce in KDD as mentioned before.

Early results without requiring special data stream properties at the input can only be obtained by more complex operators. The first representative of this type was the pipelined hash join [20]. The work in [7] extends it with out-of-core techniques to limit its massive resource consumption. The paper proposes to swap parts of the hash tables to disk and defines two processing stages: a regular stage that works similar to the original algorithm, and a clean-up stage where the portions of the hash table swapped to disk are processed. The work in [18] extends this with a third reactive phase, which is activated when regular input is blocked. Overall these various phases lead to complex pipeline scheduling problems solved by the specialized scheduling scheme in [19]. A similar goal to provide early join results is described in [3] for sort-based joins: The authors add early joins for the data being in memory during the initial sort and merge steps in order to deliver early results. The overall execution time grows slightly because of the more complex internal join processing and, in contrast to the original sort-merge-join, the data being delivered has no specific order. What is more, the rate in which results are delivered vastly varies during runtime.

The single-operator solutions described so far have in common that they do not guarantee specific qualities on the data delivered so far although these could be useful for subsequent operators as, e.g., GROUPBY, and produce results in vastly varying rates making well-balanced pipelines difficult to maintain.

A different approach to achieve interactivity in the presence of incomplete results is online query processing, where interactivity is traded for accuracy. The goal is to provide early results that reflect the trends within the data, and do so with an increasing level of confidence as query processing progresses. The most prominent operator in this context is the ripple join [4], which provides the possibility to adjust the rate at which the inputs are processed to allow the user to favor the "most interesting" input. [15] can support this approach by an efficient best-effort-reordering in specific cases (without guarantees that all relevant data elements have been provided) but this does not solve the scalability problem. What is more, long sequences of (inaccurate) preprocessing operators like reorderers, joins and aggregations in an analysis chain cannot provide enough accuracy for subsequent data mining algorithms.

In a parallel environment, where classical data parallelism leads to scalability problems in case of data skew, and limits interactivity by splitting query processing into several tasks, pipelining seems to be the only solution in order to achieve interactivity. Because of its bandwidth requirements, pipelining was seen mainly as a technique for shared memory (SMP) machines, and even there the problems were regarded as particularly serious [2]. Nevertheless, there has been substantial work in the area of pipelining implementations of complex operators [20], cost models [21,17] and optimization strategies [11]. But the problem that especially the complex operators block the pipeline and therefore destroy interactivity and scalability still remains unsolved.

3 The Source of the Problems

Our main thesis at this point is that if we wish to truly integrate data mining and database systems, we should integrate the KDD-specific operators into the query algebra of database systems. To do so our first task is to analyze the common preprocessing operators used in KDD to determine the reasons for resource consumption and for the lack of interactivity due to blocking. As seen in Figure 1, this includes standard database operators like select, join, etc. as well as KDD specific operators (sample, history etc.). For reasons of space, we limit the paper to the following four operators:

- `GROUP`, which calculates an aggregation function f of tuples coinciding on the values of attributes in some fixed list T_U (like in the known SQL construct)
- `JOIN`, which joins two relations using the join attributes T_U and T_V
- `SAMPLE`, which provides a randomly chosen sample of g tuples from the relation
- `NORMALIZE`, which does a linear rescaling of attribute A_{R_i} so that the values fit into a given interval $[a, b]$.

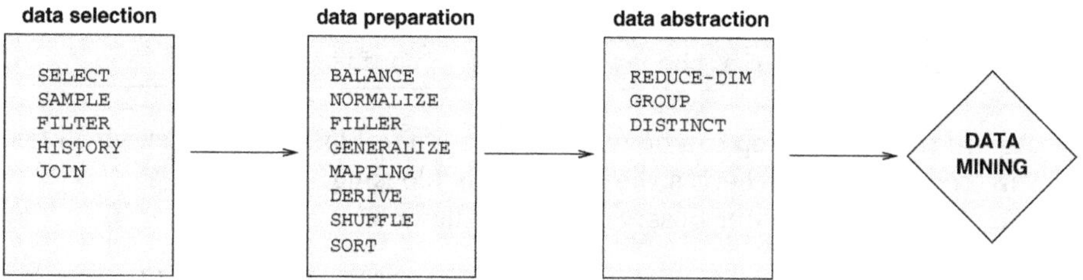

Fig. 1. KDD operators in our Algebra

To identify the blocking and resource intensive steps, we have to look deeper into the implementations of those operators.

One possible implementation for the `GROUP` operator is the following `collect-Group` which aggregates the tuples in R according to the attribute list T_U while applying the aggregation function f:

```
collectGroup(R, T_U, f)
   1. [collect] Collect tuples with the same value in T_U (e.g. in a
      hash table)
   2. [aggregate] Calculate the aggregated attributes for each
      collected group of tuples
```

This implementation consists of two steps, a collect step and an aggregation step. Note that the second step runs with little resource demand in linear time and is not blocking.

The `sortMergeJoin` implementation for the `JOIN` operator works in a similar fashion. It sorts the tuples of both relations according to the join attributes and merges them like a zipper:

```
sortMergeJoin(R,S, join attributes T_U,T_V)
   1a. [sort] Sort the first relation R by T_U
   1b. [sort] Sort the second relation S by T_V
   2. [merge] Merge the relations
```

Here, the sorting steps have been separated from the rest. Then, the following merge step can be executed, again with little resource demand, and is not blocking. Sorting, however, is blocking.

If we consider the `countPickSample` implementation for the `SAMPLE` or the `rangeMinmaxNormalize` implementation for the `NORMALIZE` operator we observe a similar result:

```
countPickSample(Relation R, sample size g)
   1. [count] Determine the number M of tuples
   2. [pick] Let 1 of n = M/g tuples pass
```

```
rangeMinmaxNormalize(R, A_Ri, new interval [a,b])
   1. [range] Determine the minimum and maximum of the values of
A_Ri
   2. [minmax] Rescale the value of A_Ri to fit into [a,b]
```

Again, the second step can be performed in a non-blocking manner with little resource consumption whereas the first step is blocking.

We may summarize our observations as follows.

Dividing the Operator Implementations is Often Possible. Many of the complex operators seem to possess algorithms that process the data stream in two steps. The first step serves to prepare the data for the following steps, which will then process it in a fast, non-blocking manner with little resource consumption.

Data Preparation Steps are Limited. There seem to be only a few standard operators for data preparation steps. In the examples presented there were *or-*

dering according to selected attributes (achieved by [**sort**]), *grouping* by selected attributes (achieved by [**collect**]), *counting tuples* and determining the *range of values* of an attribute. In principle, it is possible to treat those preprocessing steps as separate operators. What is important is that the preparation levels attained by executing a single pre-processing step may be useful in more than one subsequent processing step, provided it is not destroyed by the following step. This opens the chance to avoid the preparation step in some cases.

Avoiding Data Preparation is the Clue to Interaction. Blocking operators can slow down the response time and are therefore not suitable for interactive data processing. Hence, our goal must be to find a way to substitute non-blocking data preparation steps for the blocking ones, or to avoid the latter altogether. The idea is to keep track of the already achieved preparation levels in the pipeline. By cleverly combining and reusing these preparation levels we hope to avoid many costly preparation steps and go right on to process the tuples in the non-blocking calculation step. The next section introduces the necessary concepts.

4 Data Stream Quality

The previous section served to provide an intuitive understanding of the problem. What we need now is a systematic approach to describe the state of data preparation. We introduce data stream quality as the central concept. The previous section already indicated some useful data qualities: sortedness, continuity, and the knowledge of cardinality or extremal values. Below we formalize these qualities and present theorems that capture useful relationships between the different qualities.

4.1 Definitions

To describe continuity and sortedness, we assume equality (denoted by $d_1 =_{T_U} d_2$) and (lexicographic) ordering on a pair of tuples (denoted by $d_1 <_{T_U} d_2$) with regard to a list of attributes T_U, both with the standard meaning. In consequence, we also use $\leq_{T_U}$ in the usual manner.

The most important data stream quality is the lexicographic order of the tuples in a data stream with regard to a list of attributes, for example resulting from the `SORT` operator:

Definition 1. *A data stream $D = [d_1, \ldots, d_M]$ has data stream quality* **sorted ascendingly** *regarding a list T_U of attributes, in characters $\mathbf{S}^+([T_U])$ iff $i < j \Longrightarrow d_i \leq_{T_U} d_j$ for all $i, j \in \{1, \ldots, M\}$.*

Decreasing order ($\mathbf{S}^-$) can be defined analogously.

The `COLLECT` operator produces a data stream where all tuples having the same T_U values occur in a consecutive block, a property useful, e.g., for aggregation:

Definition 2. *A data stream $D = [d_1, \ldots, d_M]$ has data stream quality* **continuous** *regarding the list $T_U = [A_{U_1}, \ldots, A_{U_n}]$ of attributes, written as* $\mathbf{C}([T_U])$, *iff for all $i, j \in \{1, \ldots, M\}$, $d_i = d_j \Longrightarrow d_i =_{T_U} d_k$ for all k with $i \leq k \leq j$.*

As a special case of continuity, we have uniqueness. We call a data stream distinct, if no two tuples are equal regarding to T_U.

Definition 3. *A data stream $D = [d_1, \ldots, d_M]$ has data stream quality* **distinct** *regarding the list $T_U = [A_{U_1}, \ldots, A_{U_n}]$ of attributes, written as* $\mathbf{D}([T_U])$, *iff $d_i =_{T_U} d_j \Longrightarrow i = j$ for all $i, j \in \{1, \ldots, M\}$.*

In contrast to the above data stream qualities, the following ones are not related to the sequence of the tuples in the stream, but describe the data stream as a whole.

The data stream quality achieved by the `COUNT`-operator is the knowledge of the tuple count. It is independent from any list T_U of attributes.

Definition 4. *A data stream $D = [d_1, \ldots, d_M]$ has data stream quality* **known tuple count**, *in characters* **num**, *iff M is known with the arrival of d_1.*

Similar to the knowledge of the number of tuples, the a-priori knowledge of extreme values increases the quality of a data stream. In contrast to the first two definitions, it is defined on individual attributes.

Definition 5. *A data stream $D = [d_1, \ldots, d_M]$ has data stream quality* **known attribute range** *regarding an attribute A_{R_i} in characters* **min** *(A_{R_i}) or* **max** *(A_{R_i}), iff $min_{A_{R_i}}(D)$ or $max_{A_{R_i}}(D)$ is known with the arrival of d_1.*

4.2 Theorems

The following theorems cover dependencies between the different data stream qualities. The proofs are straightforward. T_U denotes a list of attributes.

Theorem 1. *If a data stream is sorted regarding T_U, it is also sorted with regard to every prefix of T_U.*

Theorem 2. *If a data stream is sorted regarding T_U, it is also continuous with regard to T_U.*

Theorem 3. *If a data stream is distinct regarding T_U, it is also distinct with regard to every permutation of T_U.*

Theorem 4. *If a data stream is continuous regarding T_U, it is also continuous with regard to every permutation of T_U.*

Theorem 5. *If a data stream is distinct regarding T_U, it is also distinct in an extension of this list by further attributes.*

Theorem 6. *If a data stream is distinct regarding T_U, it is also continuous in T_U.*

Theorem 7. *If a data stream is sorted in increasing(decreasing) order, its minimum (maximum) is also known with the arrival of the first tuple.*

4.3 Data Stream Quality Requirements and Transformations

Given the data stream qualities, we now examine how different operator implementations can exploit them. We concentrate – as an example – on the GROUP operator as a complex operator that is particularly important to KDD. For each implementation we present the input data stream quality that is at least necessary for the correct working of the implementation, and the change of data quality as a result of the transformation by the implementation.

The GROUP operator has three implementations which differ in their minimally required input data stream quality, their memory requirements and their blocking capabilities. The traditional implementation is hashGroup, which uses a hash table to calculate the aggregation results for each combination of T_U.

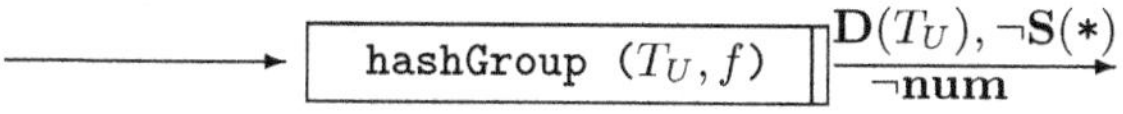

$$\longrightarrow \boxed{\texttt{hashGroup}\ (T_U, f)} \xrightarrow[\neg\mathbf{num}]{\mathbf{D}(T_U),\ \neg\mathbf{S}(*)}$$

For this implementation no data stream quality requirements are needed on input. Because each combination of T_U values produces exactly one result tuple, the output stream is distinct with regard to T_U. However, the number of tuples is not known beforehand. Furthermore, the hash function usually destroys all existing orders. The disadvantages of this implementation are the large amount of memory necessary to simultaneously store all results, and the blocking output (denoted by the bar near the end of the box), which restricts the use in a pipeline.

If the data stream quality at the input is at least continuous with respect to the list of the grouping attributes, the blockGroup implementation can be used.

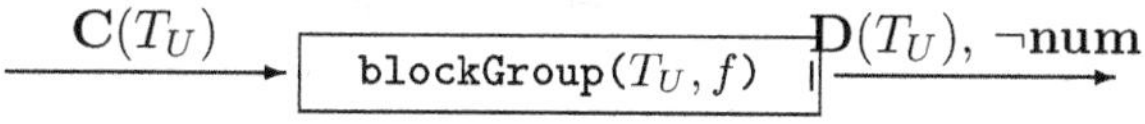

$$\xrightarrow{\mathbf{C}(T_U)} \boxed{\texttt{blockGroup}(T_U, f)} \xrightarrow{\mathbf{D}(T_U),\ \neg\mathbf{num}}$$

It applies the aggregation function block by block and therefore does not need to store more than one block of continuous tuples at the same time. Therefore, the amount of required memory is usually far less than with the hashGroup implementation. Moreover, if the input is sorted according to a list of attributes, that property also holds for the output. The knowledge about minimum and maximum values for attributes which are not included in T_U are lost along with these attributes. This implementation is delaying rather than blocking (denoted by the short bar at the end of the box) because the first result tuple cannot be output until the first block has been processed.

The noopGroup is used if the data stream is already distinct in T_U. Here, the aggregation function can be applied tuple by tuple without blocking or delaying.

As a result, the different implementations show that with rising input data stream quality memory requirements decrease and pipelining capabilities increase. Analysis of further operators show that this principle hold for most of the complex operators of the KDD process.

5 Providing Data Stream Qualities

Data stream quality is a prerequisite for implementations that are resource-economical and non-blocking. However, the prior steps to produce the qualities seem to be blocking and need many resources. In this section we look for an access method that can produce data streams with desired data stream qualities right off the disk. We will see that there is a trade-off between the efficiency of the index and the level of delivered data stream quality. Therefore, we introduce the notion of pseudo-quality that allows us to control the desired degree of data stream quality.

5.1 The Index Structure

Traditional clustering indices or combinations of several one-dimensional indices support only one or very few attributes and in general allow sequential reading in just one dimension. Our needs are different, though. We have to deal with mass data and do not know in advance which attributes will be the subject of interest during query evaluation. Therefore, in order to achieve equal treatment of multiple attributes and allow for sequential disk access, we choose a multidimensional index structure based on the Peano (Z) space filling curve [13] for physical clustering of the multidimensional data. The Peano curve was chosen over other mappings that promise somewhat better clustering properties (such as e.g. the Hilbert curve [6,8]) because of the efficient translation between operations in the one-dimensional Z-address space and operations in the multidimensional coordinate space by algorithms operating on the bit representation of the coordinates. The basic ideas are similar to the UB-tree [1] with its range query algorithm and TETRIS algorithm [12]: The space-filling curve is used to map the multidimensional data tuples (represented as points in multidimensional space) to one-dimensional addresses reflecting the point's position on this curve while preserving locality as far as possible. These addresses are then used to store the tuples using a conventional one-dimensional access method (such as in our case, a clustering B*-tree). The range-query algorithm for this data structure works as illustrated in Figure 2a: The segments of the space filling curves lying in the query box (shaded) are calculated and the data belonging to these segments can then be read sequentially from disk using the one-dimensional addresses of the entry/exit points of the curve with the query box. The small numbers in the figures denote the order in which these curve segments are processed. For this range query, we see that four segments are processed, which typically requires four random accesses.

In our work, we extended the UB-tree algorithms by some optimizations for sequential reading in a clustering index and the support for various data stream qualities in arbitrary dimensions. To deliver the data satisfying a particular range query according to a specified data stream quality, e.g. sorted, our algorithm works as follows. To provide the sort order, the query box is segmented into regions which are processed region by region in the desired sorting order. For each region, the standard range query algorithm is used. Figure 2b shows how the range query results of Figure 2a can be presented in sorted order along dimension y (data stream quality $\mathbf{S}^{+}([y])$). In order to deliver the data in this order, the index splits the query box into flat regions of height 1 and processes them one after another. As we can see by counting the corresponding line seg-

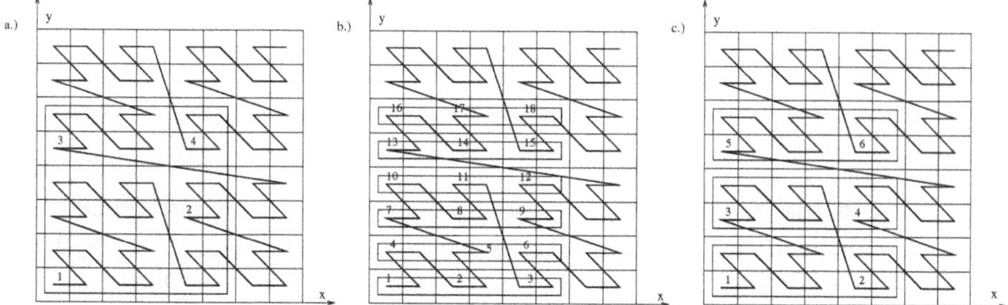

Fig. 2. Range query, sorted, pseudo-sorted

ments, this approach is not very efficient: In our simplified example, 18 segments are processed (and correspondingly 18 index accesses are performed).

The solution to this problem is shown in Figure 2c: By a controlled increase in the block size of the segmentation, we are able to improve the index performance at the expense of a controlled degradation of our data stream quality criteria. The resulting data stream can be seen as a concatenation of blocks. Data stream quality criteria (in our case sort order) are only valid between these blocks, but within these blocks, the values appear in arbitrary order. We call such degraded data stream qualities *pseudo-qualities*. These qualities will be defined in the following section. Note that the number of different values per block is limited by the block size. In our example, we can see that by allowing the index to deliver several (in our example 2) values of the sorting attribute per block) the number of index accesses can be reduced to 6.

This approach can easily be generalized to higher dimensions. Instead of two-dimensional ranges, we create higher dimensional boxes which are processed one after another.

In the case of continuity we take a similar approach. The difference is that the continuity property does not impose a special sort order. It simply requests that no value combination in the grouping attributes appears in more than one block, so there are more degrees of freedom to the shape of those blocks in multidimensional space. This is exploited by our index to reduce the number of entry/exit points with the space filling curve, rendering the index more efficient, especially for higher dimensions.

5.2 Pseudo Qualities

In the last section we have seen what data stream qualities look like if we extend our UB-index to tolerate slight degradations from our strict quality criteria for the sake of efficiency. We include these in our formal framework by adding the following definitions.

Definition 6. [Pseudo-Sorting]

A data stream D_R has the data stream quality **pseudo-sorted$_k$ ascendingly** *with respect to T_U and k, in short* $\mathbf{PS}_k^+([A_{U_1}, \ldots, A_{U_n}])$, *iff there* *exists a partitioning of* D_R *into consecutive streams* $D_R = D_1 \circ \ldots \circ D_b$, *with*

1. *for all* $g \in \{1, \ldots, b\} : |\{\Pi_{T_U}(d_i) \mid d_i \in D_g\}| \leq k$
2. *for all* $g, h \in \{1, \ldots, b\}$ *with* $g < h$ *and for all* $d \in D_g$ *and* $e \in D_h$: $d <_{T_U} e$

In other words, a data stream is pseudo-sorted if it can be partitioned into consecutive blocks such that each block has at most k distinct values of the sort key (in arbitrary order) and the key ranges of successive blocks form a non-overlapping ascending sequence of ranges.

Similar to the above definition, we define pseudo-continuity. Here it suffices to require that each value combination appears in exactly one block:

Definition 7. [Pseudo-Continuity]

A data stream D_R *has the data stream quality* **pseudo-continuous**$_k$ *with respect to attribute list* T_U *and k, in short* $\mathbf{PC}_k([A_{U_1}, \ldots, A_{U_n}])$, *iff there exists a partitioning of* D_R *into consecutive streams* $D_R = D_1 \circ \ldots \circ D_b$, *with*

1. *for all* $g \in \{1, \ldots, b\}$: $|\{\Pi_{T_U}(d_i) \mid d_i \in D_g\}| \leq k$
2. *for all* $g, h \in \{1, \ldots, b\}$ *with* $g \neq h$ *and for all* $d \in D_g$ *and* $e \in D_h$: $d \neq_{T_U} e$

With these definitions we can now formulate the following theorems. T_U denotes a list of attributes. Theorems 8 to 10 show that sorting is a special case of pseudo-sorting ($k = 1$).

Theorem 8. *If a data stream is sorted with respect to* T_U, *then it is also pseudo-sorted*$_k$ *for every k.*

Theorem 9. *A data stream is sorted with respect to* T_U, *iff it is pseudo-sorted*$_1$ *with respect to* T_U.

Theorem 10. *If a data stream is pseudo-continuous*$_k$ *(pseudo-sorted*$_k$*) with respect to* T_U, *then it also pseudo-continuous (pseudo-sorted) for every multiple of k.*

Theorem 11. *If a data stream is pseudo-sorted*$_k$ *with respect to* T_U *then it also is pseudo-continuous*$_k$.

Theorem 12. *If a data stream is continuous with respect to* T_U, *then it is pseudo-continuous*$_k$ *for every k.*

Theorem 13. *A data stream is continuous with respect to* T_U, *iff it is pseudo-continuous*$_1$ *with respect to* T_U.

Theorem 14. *If a data stream is pseudo-continuous*$_k$ *with respect to* T_U, *then it is pseudo-continuous*$_k$ *with respect to every permutation of* T_U.

5.3 Operators Dealing with Pseudo-Qualities

Since most of the implementations for the calculation steps cannot make direct use of the pseudo-qualities described in the previous section, we still need preparation operators in order to transform pseudo data qualities into strong data qualities. These operators work on each block in the data stream (the index inserts markers into the data stream to delimit individual blocks) and release the result with the desired quality to the next operator. Our goal is to replace the blocking operators SORT and COLLECT by non-blocking operators that transform pseudo-sorted and pseudo-continuous data streams to sorted and continuous data streams, respectively.

For the first case, we introduce the `k-Sort` operator. This operator is a specialization of SORT for pseudo-sorted inputs, and respects the block boundaries: After sorting an entire block with a maximum of k different values for the sort key, the operator knows that all tuples in the following blocks are greater with respect to the sort attributes, and therefore it can release the result and clear its internal data structures. In our notation, the `k-Sort` operator can be characterized as shown below. Note that in general, all data stream qualities are lost with the exception of those that refer to a prefix or an extension of sort attribute list T_U.

$$\xrightarrow{\mathbf{PS}_k(T_U)} \boxed{\texttt{k-Sort } (T_U)} \xrightarrow[\neg\mathbf{S}(\mathrm{X}),\ \neg\mathbf{C}(\mathrm{Y})]{\mathbf{S}(T_U)}$$

X refers to all attribute sequences that are neither a prefix of T_U nor an extension of T_U. Y denotes all attribute lists that are neither a prefix nor a superset of T_U.

Similarly, `k-Collect` transforms pseudo-continuous data streams into continuous streams. After collecting the values of an entire block (e.g., in a hash table) and reading a block boundary, it can be certain that none of the value combinations in T_U seen in this block will ever reappear in subsequent blocks. Hence, it can release the continuous result and clear its internal data structures. The operator is shown below in our notation. Here in general, all data stream qualities involving attribute lists different from T_U are lost.

$$\xrightarrow{\mathbf{PC}_k(T_U)} \boxed{\texttt{k-Collect } (T_U)} \xrightarrow[\neg\mathbf{S}(*),\ \neg\mathbf{C}(\mathrm{X})]{\mathbf{C}(T_U)}$$

X refers to all attribute lists that contain only a subset of the attributes from T_U.

With our specialized index and the operators described above, it is now possible to provide all desired data stream qualities in a non-blocking and resource-efficient way.

6 Parallel and Interactive Preprocessing

6.1 Basic Example

We have to demonstrate the positive effects of our approach in query execution. Consider the following simple query that computes from a table of account movements the average amount each customer (denoted by FirstName and LastName) was credited in each transaction.

```
SELECT FirstName, LastName, AVG(sum) as meanCash
FROM Accounting
GROUP BY FirstName, LastName
ORDER BY LastName, meanCash
```

The resulting operator tree is displayed in Figure 3. To be able to use an efficient `blockGroup` implementation for the GROUP BY that processes blocks of continuous values, the data stream should be continuous with respect to [**F**irstName, **L**astName] (in short, **C**([F,L])). For the `Sort` operator, the data stream should either be sorted by LastName (then we would use the `blockSort` implementation that sorts blocks with the same value in its first attribute), or pseudo-sorted (then we would use the new `k-Sort` implementation). In our case, it turns out that **PS**([L,F]) is the optimum configuration for the index access: Because of Theorem 11 we know that this also means **PC**([L,F]), which can be used by the `k-collect` operator to provide continuity to the `blockGroup` implementation. The output of this operator is reduced in cardinality, but remains pseudo-sorted in ([L,F]. It can therefore be sorted by the `k-Sort` operator to provide strict sorting in LastName to the `blockSort` operator, which now produces the desired order.

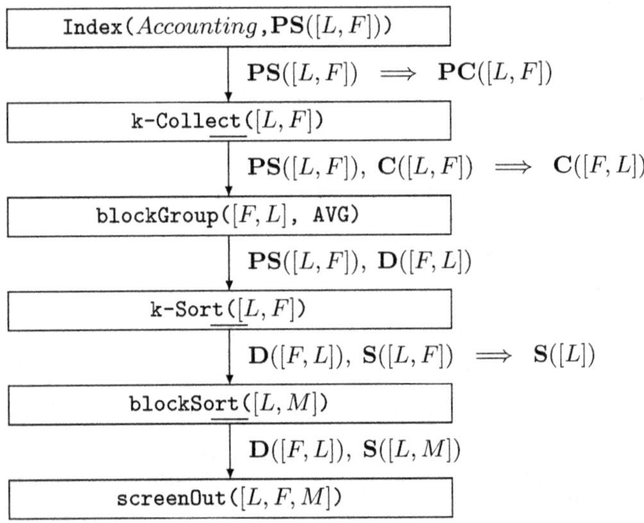

Fig. 3. Pipeline for Example Query

This simple example shows the power of our framework: The index structure can efficiently provide initial data stream qualities, the k-implementations of data preparation operators can be non-blocking, because the block markers in the pseudo-ordered data stream allow them to release their actual results and proceed with the next block, resource consumption can be effectively controlled by varying the block size k, data preparation operators can be inserted where they are really needed (e.g., be placed after initial selections), and the delivered data stream qualities (that formerly existed only within individual implementations of complex operators) are made explicit and can be shared by several operators.

6.2 Experiments

We have implemented our extended algebra with data stream qualities in JavaParty [14] on our parallel platform. In our first experiments with complex queries on the 1 GB-TPC-D data set, a pipelined implementation of the algebra could indeed reduce response times by a factor of up to 20. Against pure in-memory execution without data stream qualities, the improvements in response entail a penalty of less than 30% in overall execution time. For datasets that do not fit into main memory (which is the more realistic assumption, especially in multiuser environments) conventional execution has to swap intermediate results to disk. In this case there is barely any penalty, instead, we even get better overall execution times because of the possibility of effective control over resource consumption. These results demonstrate that we achieve interactive query processing and at the same time enhance scalability and efficiency.

7 Parallel and Interactive Data Mining

After having gained some insight into preprocessing, we now look at how this approach can be combined with data mining functionality to achieve parallel and interactive data mining. As an example we look at the problem of finding distance-based outliers.

7.1 Distance-Based Outliers

We define distance-based outliers in accordance with [10]. A data tuple is called a (p,D)-outlier if at least fraction p of the tuples in the dataset are beyond a distance of D from the observed tuple. A simple operator to calculate these outliers would be the nested-loop operator: For every tuple, the distances from this tuple to all the others are calculated and those lying within distance D are counted. If we get over the threshold defined by p, the tuple is no outlier and the next tuple is analyzed. Clearly, this algorithm is not efficient and only usable if the whole data fits in memory because index accesses to find neighboring tuples would make things only worse. In [10] the problem is alleviated by several block- and cell based approaches aiming to reduce disk accesses and complexity.

Our data stream qualities allow us to take a different approach: If the data stream has the quality `sorted` according to some attribute, we are able to exploit this quality to increase interactivity and save resources. The corresponding `windowOutlier` operator works as shown in figure 4. On the data stream ordered by attribute A_{U1}, we define a window of size D, which is shifted over this stream

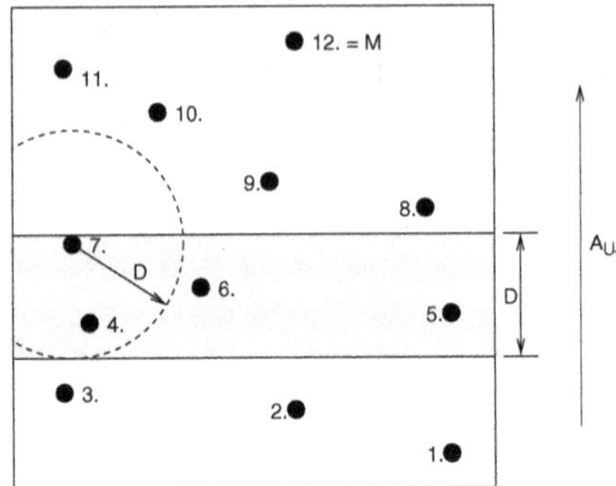

Fig. 4. Pipelining with the windowOutlier operator

as data flows through the operator. As a tuple enters the window (e.g., tuple 7) it is compared to all others within the window. Each time the distance between two tuples is lower than D, each tuple's counter is increased. As a tuple leaves this window (as in this case, tuple 3), it is marked as an outlier if its counter is below the threshold.

We see that this implementation is not blocking and resource efficient because only the tuples within D need to be stored in memory and compared at a time. Therefore we can add the following implementations for finding outliers with respect to a set T_U of attributes to our set of algorithms:

num, $\mathbf{S}(A_U)$ → `standardOutlier` (T_U) → $\mathbf{S}(A_U)$

num, $\mathbf{S}(A_U)$ → `windowOutlier` (T_U) → $\mathbf{S}(A_U)$

The `standardOutlier` implementation needs no specific data stream quality but is blocking. The `windowOutlier` implementation needs the data stream quality `sorted` on some Attribute A_U with $A_U \in T_U$ and the knowledge of the number of incoming tuples in order to deliver outliers in a non-blocking way. Note that this implementation preserves the quality `sorted` in the resulting data stream.

7.2 Evaluation

We implemented both variants of the operator in our platform and generated a dataset of 10000 tuples to measure the effects. The data consists of seven attributes of the following types, cardinalities and distributions:

```
key (Integer)
int1 0-99, evenly distributed, 2 Tuples at 200, 300, 400
int2 0-9, evenly distributed, 3 tuples at 20, 30, 40
float1 0,08 - 9999,75, evenly distributed
float2 -36,58 - 44,52, Gaussian distribution
String1 1-5 Characters, evenly distributed
String2 3 Characters, Names of the German federal states
```

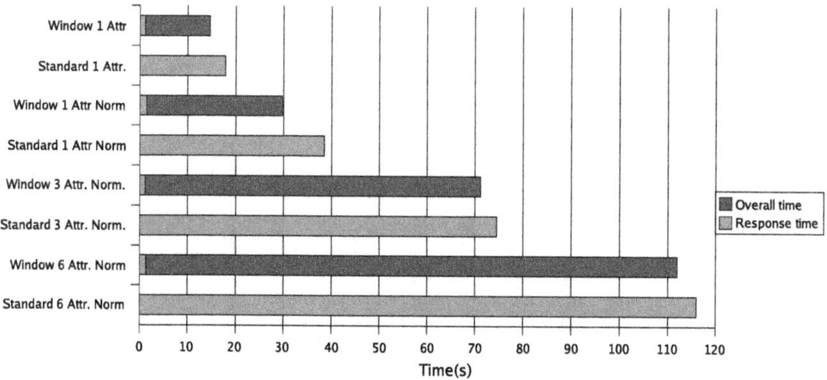

Fig. 5. Execution Times for standard and windowed Outlier operator

Figure 5 displays the results comparing the `standardOutlier` implementation and the `windowOutlier` implementation. For both implementations we tested the following attribute combinations:

- the `int1` attribute (1 Attr),
- the `int1` attribute normalized (1 Attr Norm)
- the `int1, int2, Float1` attribute set (3 Attr. Norm) and
- the `int1, int2, Float1, Float2, String1, String2` attribute set (6 Attr. Norm)

with a percentage p of 0.9 and a distance D of 70. The bars show the response times (i.e., the time until the first result is produced) and the remaining overall execution time (which is equal to the response time for the `standardOutlier` implementation, of course). We used `int1` as the sorting attribute (e.g. the dimension in which the window is moved) in the `windowOutlier` implementation. We see that the `windowOutlier` implementation shows substantial improvements in response time. In most cases, the first results are presented almost immediately. The total execution time is slightly better as well for the `windowOutlier` implementation. That means that the additional work for sorting the input by means of our index structure and an additional sort operator is more than compensated for by the reduced number of comparisons of the `windowOutlier` operator.

In order to further investigate the effect of the window size on overall execution time for the `windowOutlier` implementation, we measured the execution times for different window sizes D. Figure 6 shows the execution times for the normalized attribute sets used above. We see that the implementation can use the smaller window sizes to significantly reduce execution time. For a window size of 10, execution time can be reduced by up to 80% compared to the maximum execution times occurring for window sizes of 100 and greater[1]. For the

[1] Because we again used `int1` as our sorting attribute, there is no effect of window sizes greater than 100 because nearly all tuples fit into one window.

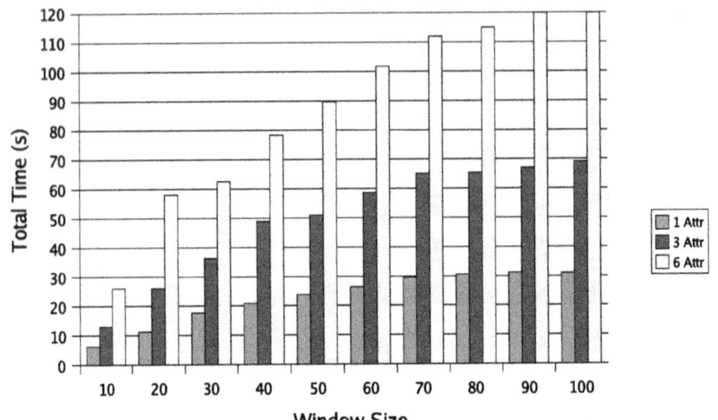

Fig. 6. Different window sizes for the `windowOutlier` operator

`standardOutlier` implementation (not displayed here), execution times stay nearly constant for all window sizes. The average time is about 10% higher than the displayed maximum execution time of the `windowOutlier` variant. The response times in this experiment stayed under 6% of the overall execution time throughout all window sizes.

Because our multidimensional index structure allows the use of every index attribute for sorting, the third experiment shows the influence of the sorting attribute on execution times. We compare the response time and overall execution time for the detection of six-attribute-outliers with each of the six attributes chosen as sort attribute. The results (see figure 7) are quite interesting: Whereas the results for the two integer attributes are nearly identical (because of their nearly identical normalized cardinality distribution), the `float1` attribute performs much better, whereas the `float2` attribute lies between the `float1` result and the integer results. The reason is that in contrast to the `int1` attribute, the `float1` attribute is equally distributed throughout the normalized attribute range, so there are fewer tuples in each window. The `float2` attribute has similar properties, but its Gaussian distribution leads to a degradation in execution time. The explanation for the high response time for the String attributes is that the relatively high window size (again 70) combined with the chosen mapping of Strings into the normalized domain place nearly all the tuples within one window, so the `windowOutlier` degenerates to a `standardOutlier`. This result shows the utmost importance of the choice of the right sort attribute especially for high distance values D. It is the task of the query optimizer to choose an attribute with sufficiently high cardinality and well-balanced distribution within the attribute range in order to render the implementation interactive and efficient. The necessary information can be obtained from optimizer statistics.

These experiments demonstrate that the concept of using data stream qualities to improve interactivity and efficiency holds promise for a seamless extension from preprocessing to data mining tasks. The multidimensional index structure

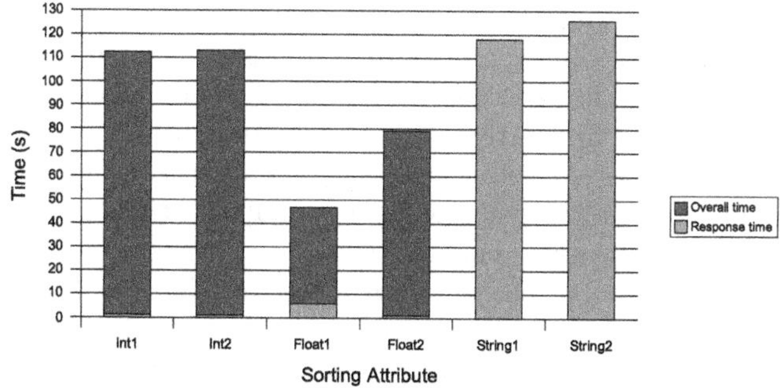

Fig. 7. The influence of the sorting attribute

allows to deliver several sort orders efficiently. The optimizer can then choose the optimum data stream qualities for the operators. Interactive and resource-efficient processing of whole streams of preprocessing and data mining operators thus became an achievable reality.

7.3 Enhancing Scalability

The experiments demonstrated that the Outlier Operator now is interactive, but still rather expensive. In order to further enhance scalability, the approach described above can be improved by cascading several `windowOutlier` implementations into a pipeline: By using the fact that

$$O(p, D') \supseteq O(p, D) \text{ for } D' < D$$

indicating that an outlier with respect to a given distance D is always an outlier for each smaller distance D' as well, the pipeline starts with small window sizes (and therefore fewer comparisons) marking all potential outliers for this window size. Only these potential outliers have to be tested in the subsequent steps on whether their outlier property still holds for larger window sizes as well. In effect, the early stages in this pipeline work with higher throughput because of the small window sizes, and the later stages are more efficient as well because they have fewer potential outliers to observe. In this way we are able to distribute the work across several implementations and, by balancing the load within this pipeline, increase overall pipeline throughput.

8 Conclusion

We set out to improve efficiency, scalability and interactivity in KDD. The first idea was to isolate the blocking, resource-intensive operators. We identified the data preparation operators as the ones that tend to interrupt the continuity of a data stream. By introducing the notion of data stream quality and by employing advanced index structures that support an algebra of data-stream-quality-aware operators we were able to demonstrate substantial improvements in sequential and parallel query execution. The results gained with the Outlier operator demonstrate that this not only holds for preprocessing but can be extended to parallel and interactive data mining as well.

These results together with the possibility to control resource consumption at optimization time seem to open a wide area of applications for our techniques beyond KDD. Such applications may range from the integration of analysis functionality into mainstream commercial databases all the way across embedded databases, up to pipelined distributed query processing in Grid Computing. Still, we feel we just scratched the surface. We plan to extend our work towards an extensive investigation of the combination of our index structure with various data stream quality levels. In addition we plan to further enhance scalability making use of replicated input data by means of our index structure. Only then do we feel confident enough to tackle the issue of interleaving more complex and multi-stage data mining algorithms with our improved algebra.

References

1. R. Bayer. The universal B-tree for multidimensional indexing. Technical Report TUM-I9637, TU München, November 1996.
2. D.J. DeWitt and J. Gray. Parallel database systems: The future of high performance database systems. *Communications of the ACM*, 35(6):85–98, June 1992.
3. Jens-Peter Dittrich, Bernhard Seeger, David Scot Taylor, and Peter Widmayer. Progressive merge join: A generic and non-blocking sort-based join algorithm. In *Proceedings of the 28th VLDB Conferende*, 2002.
4. P.J. Haas and J.M. Hellerstein. Ripple joins for online aggregation. In A. Delis, C. Faloutsos, and S. Ghandeharizadeh, editors, *SIGMOD 1999, Proceedings ACM SIGMOD International Conference on Management of Data, June 1–3, 1999, Philadephia, Pennsylvania, USA*, pages 287–298. ACM Press, 1999.
5. J. Han, Y. Fu, W. Wang, K. Koperski, and O. Zaiane. Dmql: A data mining query language for relational databases. In *Proceddings of the SIGMOD'96 Workshop on Research Issues on Data Mining and Knowledge Discovery*, pages 27–34, Montreal, Kanada, June 1996.
6. D. Hilbert. Über die stetige Abbildung eine Linie auf ein Flächenstück. *Mathematische Annalen*, 1891.
7. Z. Ives, D. Florescu, M. Friedmann, A. Levy, and D.S. Weld. An adaptive query execution system for data integration. In *Proceddings of the ACM SIGMOD Conference*, 1999.
8. H. V. Jagadish. Linear clustering of objects with multiple atributes. In Hector Garcia-Molina and H. V. Jagadish, editors, *Proceedings of the 1990 ACM SIGMOD International Conference on Management of Data, Atlantic City, NJ, May 23–25, 1990*, pages 332–342. ACM Press, 1990.
9. T. Johnson, L.V.S. Lakshmanan, and R.T. Ng. The 3w model and algebra for unified data mining. In *Proceedings of the 26th VLDB Conference*, Kairo, Egypt, 2000.

10. E.M. Knorr and R.T. Ng. Algorithms for mining distance-based outliers in large datasets. In *Proceedings of the 24th VLDB Conference*, New York, USA, 1998.
11. S. Manegold, F. Waas, and M.L. Kersten. On optimal pipeline processing in parallel query execution. Technical report, CWI, Amsterdam, February 1998. http://www.cwi.nl/ftp/CWIreports/INS/INS-R9805.ps.Z.
12. V. Markl, M. Zirkel, and R. Bayer. Processing operations with restrictions in rdbms without external sorting: The tetris algorithm. In *Proceedings of the 15th International Conference on Data Engineering, 23–26 March 1999, Sydney, Austrialia*, pages 562–571. IEEE Computer Society, 1999.
13. J.A. Orenstein and T.H. Merrett. A class of data structures for associative searching. In *Proceedings of the Third ACM SIGACT-SIGMOD Symposium on Principles of Database Systems, April 2–4, 1984, Waterloo, Ontario, Canada*, pages 181–190. ACM, 1984.
14. M. Philippsen and M. Zenger. Javaparty - transparent remote objects in java. *Concurrency: Practice and Experience*, 1997.
15. V. Raman, B. Raman, and J.M. Hellerstein. Online dynamic reordering for interactive data processing. In *Proceedings of the 25th VLDB Conference, Edinburgh, Scotland*, 1999.
16. Patricia G. Selinger, Morton M. Astrahan, Donald D. Chamberlin, Raymond A. Lorie, and Thomas G. Price. Access path selection in a relational database management system. In *Proc. of the SIGMOD Conference*, 1979.
17. M. Spiliopoulou, M. Hatzopoulos, and C. Vassilakis. A cost model for the estimation of query execution time in a parallel environment supporting pipeline. *Computers and Artificial Intelligence*, 1996.
18. T. Urhan and M.J. Franklin. Xjoin: A reactively-scheduled pipelining join operator. *IEEE Data Engineering Bulletin*, 2000.
19. T. Urhan and M.J. Franklin. Dynamic pipeline scheduling for improving interactive performance of online queries. In *Proceedings of the 27th Intl. Conference on Very Large Data Bases*, 2001.
20. A.N. Wilschut and P.M.G. Apers. Dataflow query execution in a parallel main-memory environment. In *Proceedings of the First International Conference on Parallel and Distributed Information Systems*, pages 68–77, Miami Beach, December 1991.
21. A.N. Wilschut and S.A. van Gils. A model for pipelined query execution. In *Proceedings of the MASCOTS93 Syposium*, 1993.

Frequent Itemset Discovery with SQL Using Universal Quantification

Ralf Rantzau

Department of Computer Science, Electrical Engineering
and Information Technology, University of Stuttgart
Universitätsstraße 38, 70569 Stuttgart, Germany
rrantzau@acm.org

Abstract. Algorithms for finding frequent itemsets fall into two broad categories: algorithms that are based on non-trivial SQL statements to query and update a database, and algorithms that employ sophisticated in-memory data structures, where the data is stored in flat files. Most performance experiments have shown that SQL-based approaches are inferior to main-memory algorithms. However, the current trend of database vendors to integrate analysis functionalities into their query execution and optimization components, i.e., "closer to the data," suggests to revisit these results and to search for new, potentially better solutions.

We investigate approaches based on SQL-92 and present a new approach called *Quiver* that employs universal and existential quantifications. In the table schema for itemsets of our approach, a group of tuples represents a single itemset. Such a "vertical" layout is similar to the popular layout used for the transaction table, which is the input of frequent itemset discovery. We show that current DBMS do not provide efficient query processing strategies for dealing with quantified queries, mostly due to the lack of an adequate SQL syntax for set containment tests. Performance tests using a query processor prototype and a novel query operator, called set containment division, promise an improved performance for quantified queries like those used for Quiver.

1 Introduction

The discovery of frequent itemsets is a computationally expensive preprocessing step for association rule discovery [1], which finds rules in large transactional data sets. Frequent itemsets are combinations of items that appear frequently together in a given set of transactions. Association rules characterize, for example, the purchase pattern of retail customers or the click pattern of web site visitors. Such information can be used to improve marketing campaigns, retailer store layouts, or the design of a web site's contents and hyperlink structure.

Most commercial data mining systems and research prototypes employ algorithms that run on data stored in flat files. However, database vendors begin to act against the general lack of mining functionality to support business intelli-

R. Meo et al. (Eds.): Database Support for Data Mining Applications, LNAI 2682, pp. 194–213, 2004.

gence and integrate new "primitives" into their systems [2]. Companies employing data mining tools for their business [3] realize the need for integrating data mining algorithms with DBMS.

Some authors in research as well as in industry have suggested special data mining languages like *MSQL* [4], *DMQL* [5] or a variant of DMQL [6], and *ATLaS* [7]. Others have enriched query languages with mining functionality, like the *MINE RULE* operator [8], *OLE DB for DM* [9]. However, the query processing power offered by modern database systems for mining purposes has been widely neglected in the past. Some research results show that algorithms for frequent itemset discovery based on SQL are less efficient than those based on sophisticated in-memory data structures [10]. Others claim that "even SQL *as is* is adequate even for complex data mining queries and algorithms" [11]. Nevertheless, it becomes ever more important for database system vendors to offer novel analytic functionalities to support business intelligence applications.

In this paper, we analyze several approaches to compute frequent itemsets using SQL. We also propose a new SQL-based approach and compare it to the other approaches.

1.1 The Problem of Frequent Itemset Discovery

We briefly introduce the widely established terminology relevant for frequent itemset discovery. An *item* is an object of analytic interest, like a product of a shop or a URL of a document on a web site. An *itemset* is a set of items and a *k-itemset* contains k items. A *transaction* is an itemset representing a fact, like a purchase of products or a collection of documents requested by a user during a web site visit.

Given a set of transactions, the *frequent itemset discovery problem* is to find itemsets within the transactions that appear at least as frequently as a given threshold, called *minimum support*. For example, a user can define that an itemset is frequent if it appears in at least 2% of all transactions.

Almost all itemset discovery algorithms consist of a sequence of steps that proceed in a bottom-up manner. The result of the kth step is the set of frequent k-itemsets, denoted as F_k. The first step computes the set of frequent items (1-itemsets). Each following step consists of two phases: First, the *candidate generation phase* computes a set of potential frequent k-itemsets from F_{k-1}. The new set is called C_k, the set of candidate k-itemsets. It is a superset of F_k. Second, the *support counting phase* filters out those itemsets from C_k that appear more frequently in the given set of transactions than the minimum support and stores them in F_k.

All known SQL-based algorithms follow this "classical" two-phase approach. There are other, non-SQL-based approaches, like *frequent-pattern growth*, which do not require a candidate generation phase [12]. The frequent-pattern growth algorithm, however, employs a (relatively complex) main-memory data structure, called frequent-pattern tree, which disqualifies it for a straightforward comparison with SQL-based algorithms.

1.2 Motivation for a New SQL-Based Approach

The key problem in frequent itemset discovery is: "How many transactions contain a certain given itemset?" This question can be answered in relational algebra using the division operator (÷), introduced in [13]. Suppose that we have a relation *Transaction*(*transaction*, *item*) containing a set of transactions and a relation *Itemset* (*item*) containing a single itemset, i.e., each tuple contains one item. We want to collect those *transaction* values in a relation *Contains* (*transaction*), where *for all* tuples in *Itemset* there is a corresponding tuple in *Transaction* that has a matching item value together with that *transaction*. In relational algebra, this problem can be stated as *Transaction*(*transaction*, *item*) ÷ *Itemset*(*item*) = *Contains*(*transaction*).

The example in Figure 1 illustrates the division operation. The *Transaction* table consists of three transactions and two of them contain all items of *Itemset*. We simply have to count the number of tuples in *Contains* to decide if the itemset is frequent. Using division terminology, *Transaction* plays the role of the dividend, *Itemset* represents the divisor, and *Contains* is the quotient.

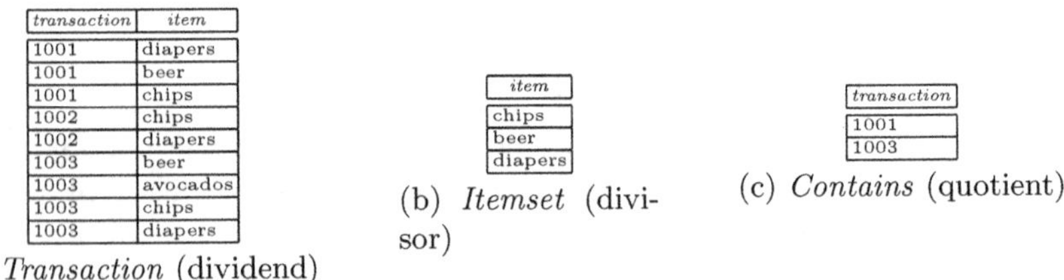

transaction	*item*
1001	diapers
1001	beer
1001	chips
1002	chips
1002	diapers
1003	beer
1003	avocados
1003	chips
1003	diapers

(a) *Transaction* (dividend)

item
chips
beer
diapers

(b) *Itemset* (divisor)

transaction
1001
1003

(c) *Contains* (quotient)

Fig. 1. Example showing the relationship between frequent itemset discovery and relational division: *Transaction* ÷ *Itemset* = *Contains*

Unfortunately, frequent itemset discovery poses the additional problem that we have to check *many* (candidate) itemsets for sufficient frequency, i.e., unlike in Figure 1, we usually do not have a constant divisor relation but we need many divisor relations.

We are not aware of any commercial database system that has implemented a division operator. One reason for this is that division is not a basic operator, i.e., it can be expressed by other operators of the relational algebra. Another reason is that it is difficult for a system to detect a division problem inside a SQL query because the language offers no dedicated syntax for division. Division problems within queries are often simulated either by indirect formulations using counting or by using a "NOT EXISTS" clause. Nevertheless, in theory there are several efficient algorithms for division [14,15]. For example, suppose that the divisor *Itemset* is sorted on *item*, and the dividend *Transaction* is grouped on *transaction* (as in Figure 1) and for each group, it is sorted on *item* in the same order as the dividend. Then we can employ an algorithm similar to sort-merge join that computes the quotient in one scan over the dividend and divisor tables.

The input data of the division may have such a data property for one of two reasons. First, the data was materialized in this way in tables on disk (clustered index). Second, the data is an intermediate result of another operator within a larger query. For example, a sort-merge join could have produced the dividend from two other input tables.

Based on the idea of using division to specify the itemset containment problem, we devised a complete algorithm in SQL using a vertical table layout and universal quantifications. We use the term "universal quantification" instead of "division" because we will first specify the queries of our approach using tuple relational calculus, which includes the mathematical universal quantifier ($\forall$). In addition, we will show SQL queries that are equivalent to the calculus expressions.

Although there are two main approaches to express universal quantification in SQL, as mentioned before, one based on counting, the other based on value comparisons, we focused on the latter approach. The reason is that the counting approach puts extra restrictions to the quantification problem and it is less intuitive. To see why, suppose we want to test if an itemset i is contained in a transaction t. The counting approach compares the number of items in i with those in t. This comparison makes sense only if we require that t contains only items that are contained in i. In other words, we have to remove those items from t in a preprocessing step that are not contained in i. After that, we count the number of items in i and t and find that $i \subseteq t$ if the numbers are equal. The SQL statements based on value comparisons that will be explained later in this paper are a more intuitive formulation of the quantification problem.

Note that the division operator is closely related to *set containment joins*, which can be implemented very efficiently, as shown for example in [16]. We currently investigate the relationship between these two classes of algorithms in detail as ongoing work [17]. A set containment join ($\bowtie_{\subseteq}$) is a join between set-valued attributes of two relations R and S: $R \bowtie_{\subseteq} S = \{(s,t) | R \bowtie S | s \subseteq t\}$. Hence, the items of each transaction or itemset would be represented by a set. The focus of this paper lies on algorithms based on SQL-92, which does not cover set-valued attributes. We have recently proposed a generalization of division, called *set containment division*, where the divisor is defined by a group of tuples. For example, the divisor relation *Itemset* (*itemset*, *item*) defines a divisor for each value of *itemset*. The result is the union of division operations between the dividend relation and a single group of the divisor relation [14]. This operator is equivalent to set containment join but processes relations in the first normal form.

The rest of this paper is organized as follows. In Section 2, we discuss alternative ways to store and process data in tables of a relational database system. Section 3 highlights important known approaches using SQL-92 before we introduce a new approach that makes use of a vertical table layout in Section 4. We compare the performance of several algorithms discussed before and assess a novel query processing technique, set containment division, for the new algorithm in Section 5. Section 6 concludes the paper.

2 Table Layout

Before data can be mined with SQL, it has to be made available as relational tables. Typically, the data is stored in tables within that database system or, if the data has to be kept outside of the database system, it has to be made accessible to the database system by using wrappers that provide a relational view on the data.

2.1 Layout Types

Two types of data objects are relevant for frequent itemset discovery: transactions and itemsets. For each type, there are basically two main layouts (schemas) for representing these objects in a table. In particular, the items of an object can be stored either in a single row, which we call a *horizontal* layout, or in several rows, which we call *vertical* layout, as illustrated in Table 1. Note that the horizontal layout for itemsets has a position attribute *pos* associated with each item. This is necessary because most algorithms assume a lexicographic order of items within an itemset and they need to access an item at a specific position.

Table 1. Table layout alternatives for storing the items of transactions and itemsets

<table>
<tr><th>Layout</th><th>Transactions</th><th>Itemsets</th></tr>
<tr><td>horizontal
(single-row/multi-column)</td><td><table><tr><th>transaction</th><th>$item_1$</th><th>$item_2$</th><th>$item_3$</th></tr><tr><td>1001</td><td>diapers</td><td>beer</td><td>chips</td></tr><tr><td>1002</td><td>chips</td><td>diapers</td><td>NULL</td></tr></table></td><td><table><tr><th>itemset</th><th>$item_1$</th><th>$item_2$</th></tr><tr><td>101</td><td>beer</td><td>chips</td></tr><tr><td>102</td><td>beer</td><td>diapers</td></tr></table></td></tr>
<tr><td>vertical
(multi-row/single-column)</td><td><table><tr><th>transaction</th><th>item</th></tr><tr><td>1001</td><td>diapers</td></tr><tr><td>1001</td><td>beer</td></tr><tr><td>1001</td><td>chips</td></tr><tr><td>1002</td><td>chips</td></tr><tr><td>1002</td><td>diapers</td></tr></table></td><td><table><tr><th>itemset</th><th>pos</th><th>item</th></tr><tr><td>101</td><td>1</td><td>beer</td></tr><tr><td>101</td><td>2</td><td>chips</td></tr><tr><td>102</td><td>1</td><td>beer</td></tr><tr><td>102</td><td>2</td><td>diapers</td></tr></table></td></tr>
</table>

Almost all known SQL-based approaches assume that transactions are stored in a vertical layout. To the best of our knowledge, only the approach proposed by Rajamani et al. [18], assumes a horizontal layout. In that approach, the horizontal/vertical layout for transactions is called multi-column/single-column data model, respectively. No layout alternatives for itemsets are discussed in that paper because the focus is on *input* data (i.e., only the transactions) for association rule discovery algorithms, not on *intermediate* or *result* data (itemsets) representations, as in this paper.

Analogous to transactions, there are two different table layouts for itemsets. All known approaches for frequent itemset discovery based on SQL-92 assume a horizontal itemset table layout.

A third, hybrid way of storing items is possible, combining the vertical and horizontal approaches. It may happen that the size of itemsets or, more likely, the size of some transactions is larger than the number of item attributes that have been defined for the tables of a horizontal layout. For example, if 99% of all transactions to be stored in a database have up to ten items but only 1% has ten items or more then a database designer may decide that a horizontal layout with ten item attributes is reasonable. For the few long transactions, however, the remaining items can be stored in additional rows. For example, if the transaction table layout has ten item attributes and we want to store an itemset of size 33, then at least four rows are required to store the entire transaction. Of course, all rows belonging to the same transaction/itemset must have a common transaction/itemset ID, as with the vertical approach. We will not further discuss this approach in this paper.

Further, rather exotic approaches have been proposed to represent transaction data. For example, the decomposed data structure described in [19] is a non-relational data layout, called *decomposed storage structure*, used in the database system *Monet* where the transactions are stored as follows. Let I denote the number of distinct items in the transactions. The data are stored in I columns(!) where each column contains the set of transaction IDs that contain the item. Hence, the frequent 1-itemsets are simply the columns that contain a sufficient number of transaction IDs. Other layouts are used by SQL-based algorithms with object-relational features like user-defined functions, as defined in SQL:1999. These layouts have a column with a container data type (like BLOB or VARCHAR) to store lists of objects. One example approach, called *Vertical* [10], uses a transaction table layout *Transaction(transaction, item_list)*, where *item_list* contains the items of a transaction. A similar approach, called *Horizontal* [10], uses the layout *Transaction(item, transaction_list)*, similar to the decomposed structure described above. For each distinct item, there is a list of all transaction IDs that contain the item. However, the Horizontal approach uses a row instead of a column for each item, as in the decomposed layout. In this paper, we restrict our discussion to approaches based on SQL-92, i.e., we focus on the vertical and horizontal layout.

2.2 Vertical vs. Horizontal Layout

In the following, we will use the term *object* to denote itemsets and transactions alike. The vertical approach differs from the horizontal approach in several ways, like the maximum object size and the number of tables and indexes used for the objects.

Object Size. The size of an object does not need to be specified in the vertical layout. If we want to store extremely large objects using a horizontal layout, it could happen that the maximum number of table columns allowed in the database system in use is lower than the desired object size.[1] Not only the storage of objects may be restricted in a horizontal layout but also the processing of queries may cause problems. The number of attributes allowed in a SELECT clause of SQL may also be lower than required by an SQL-based algorithm.

[1] For example, Microsoft SQL Server 2000 allows a maximum of 1,024 columns per table. IBM DB2 Universal Database 7.2 allows up to 1,012 columns per table.

Therefore, we have to take care of the fact that we should also avoid very long attribute lists in projections, i.e., the SQL-based approach should not produce an intermediate result that has a horizontal layout inside the queries even if the outcome of a query is in a vertical layout.

Number of Tables. Objects of different size can be stored in the same table if a vertical layout is used. We could even store all objects in a single table, i.e., the transaction table, provided that the data types of itemset IDs and transaction IDs are compatible, that the IDs of transactions and itemsets are unique, and that the position attribute values are set according to the lexicographical ordering of items for transactions. In a vertical approach, the support counters of frequent itemsets can be kept in a separate table *Support*(*itemset*, *count*), where the *itemset* value corresponds to that of the respective itemset table, for example by means of a foreign key *Support.itemset* referencing table *Itemset*. In a horizontal layout, the support counter can also be added as an additional attribute to the *Itemset* table: $(itemset, item_1, \ldots, item_n, count)$. It is reported that the horizontal layout for transactions seems to allow faster algorithms [18]. However, the vertical layout is much more common for market basket analysis, the most popular field of application for association rule discovery.

Number of Indexes. Fewer indexes come into consideration and are required to improve performance in the vertical layout. An itemset table in a vertical layout has three attributes. Hence, only 15 column combinations for indexes are possible[2]. The larger number of potential indexes for the horizontal layout requires a more thorough analysis on which subset of indexes could actually be exploited by the queries given the current characteristics of data to be mined.

3 SQL-Based Algorithms

There is a multitude of algorithms for frequent itemset discovery. Most approaches do not consider the query functionality of a database system but merely its storage capability, if at all. Implementations of these approaches, including commercial mining systems, use a database system like a file system for retrieving input transaction data and in rare cases also for storing intermediate and result itemsets. The focus of most research on new algorithms lies on main-memory data structures that allow an efficient candidate generation and support counting phase. Such algorithms have to provide scalability in addition to the core functionality itself. In contrast, SQL-based approaches can rely on the query execution engine to handle a scalable processing of the queries. However, they often lack the efficiency of main-memory based approaches.

[2] There are $n! \sum_{k=0}^{n-1} \frac{1}{k!}$ combinations for n attributes. This is at least exponential, since $2^n - 1 \leq n! \sum_{k=0}^{n-1} \frac{1}{k!} < en!$. Here, we do *not* take into account the types of indexes, like clustered, secondary, bitmap, etc., but we focus on the attribute combinations only.

Even the subclass of algorithms that use SQL queries is large. A couple of approaches employ queries containing user-defined functions, which are processed by an object-relational database system. Furthermore, several approaches do not employ any user-defined procedural code at all. Such algorithms use only queries that conform to the SQL-92 standard. In this paper, we focus on approaches based on SQL-92.

3.1 Overview of Algorithms Based on SQL-92

The *SETM* algorithm is the first SQL-based approach [20] described in the literature. Several researchers have suggested improvements of SETM. For example, in [21] the use of views is suggested instead of some of the tables employed, as well as a reformulation using subqueries. The performance of SETM on a parallel DBMS has been studied further [22]. The results have shown that SETM does not perform well on large data sets and new approaches have been devised, like for example *K-Way-Join*, *Three-Way-Join*, *Subquery*, and *Two-Group-Bys* [10]. These new algorithms differ only in the statements used for support counting. They use the same SQL statement for generating C_k, as illustrated in Figure 2 for the example value $k = 4$. The statement creates a new candidate k-itemset by exploiting the fact that all of its k subsets of size $k - 1$ have to be frequent. This condition is called *Apriori property*. It was originally introduced in the *Apriori* algorithm [23,24]. Two frequent subsets are picked to construct a new candidate. These itemsets must have the same items from position 1 up to $k - 1$. The new candidate is further constructed by adding the kth items of both itemsets in a lexicographically ascending order. In addition, the statement checks if the $k - 2$ remaining subsets of the new candidates are frequent as well.

The algorithms presented in [10] perform differently for different data characteristics. Subquery is reported to be the best algorithm overall compared to the other approaches based on SQL-92. The reason is that it exploits common prefixes between candidate k-itemsets when counting the support.

Another approach presented in [10] called *K-Way-Join* uses k instances of the transaction table and joins it k times with itself and with a single instance of C_k. An example SQL statement of this approach is shown on the left of Figure 5, where it is contrasted with an equivalent approach using a vertical layout on the right.

```
INSERT INTO C4 (itemset, item1, item2, item3, item4)
SELECT newid(), item1, item2, item3, item4
FROM (
  SELECT a1.item1, a1.item2, a1.item3, a2.item3
  FROM   F3 AS a1, F3 AS a2, F3 AS a3, F3 AS a4
  WHERE  a1.item1 = a2.item1 AND
         a1.item2 = a2.item2 AND
         a1.item3 < a2.item3 AND
         -- Apriori property.
         -- Skip item1.
         a3.item1 = a1.item2 AND
         a3.item2 = a1.item3 AND
         a3.item3 = a2.item3 AND
         -- Skip item2.
         a4.item1 = a1.item1 AND
         a4.item2 = a1.item3 AND
         a4.item3 = a2.item3) AS temporary;
```

Fig. 2. Generation of candidate 4-itemsets in SQL-92. Such a statement is used by all known algorithms that have a horizontal table layout

More recently, an approach called *Set-oriented Apriori* has been proposed [25]. The authors argue that too much redundant computation is involved in each support counting phase. They claim that it is beneficial to save the information about which item combinations are contained in which transaction, i.e., Set-oriented Apriori generates an additional table $T_k(transaction, item_1, \ldots, item_k)$ in the kth step of the algorithm. The algorithm derives the frequent itemsets by grouping on the k items of T_k and it generates T_{k+1} using T_k. Their performance results have shown that Set-oriented Apriori performs better than Subquery, especially for high values of k.

4 A New Approach Using Universal Quantification

In this section, we suggest a new approach called *Quiver (QUantified Itemset discovery using a VERtical table layout)* for computing frequent itemsets using SQL. It requires a vertical table layout for computing candidate and frequent itemsets, as defined in Section 2. In addition, it employs universal and existential quantifications of tuple variables. In the following, we will discuss the two phases of frequent itemset discovery according to Quiver, candidate generation and support counting.

4.1 Candidate Generation Phase

Before we show how to accomplish the entire candidate generation in SQL, we explain the key ideas of the Quiver approach using the tuple relational calculus notation used in [26]. We do this because the calculus is more concise than SQL and the universal quantification used in Quiver becomes apparent.

Calculus Expressions. The generation of candidate k-itemsets can be expressed in a single calculus expression *Candidate*(k). However, we have decomposed this expression into several subexpressions for a clearer presentation.

In the following, we assume that we have computed F_{k-1}, the set of $(k-1)$-itemsets for some $k \geq 2$ during the previous support counting phase. All expressions use the tuple variables a_1, a_2, a_3, b_1, b_2, and c referring to the same relation F_{k-1}. All k-candidates have to fulfill the calculus query $C_k = \{c \in F_{k-1} | \mathit{Candidate}(k)\}$. Note that the expressions shown below are actually *templates* of expressions because they are parameterized. For example, the template $A(k, p)$, explained below, has the input parameter k, the size of the candidate itemsets to be created, as well as the parameter p, the item position within an itemset. However, in the rest of the paper we will use term "expression" instead of "expression templates" for simplicity.

The *Candidate* expression relies on two subexpressions, *unique* and *Prefix-Pair*. The *unique* expression can be regarded as a function that creates a new itemset ID that must be distinct from all existing IDs and that is guaranteed to be different from any ID that is returned for a different input value pair. When we use SQL for generating such a unique ID, we could simply use the current timestamp.

$$
\begin{aligned}
C_k &= \{c \in F_{k-1} | \mathit{Candidate}(k)\} \\
\mathit{Candidate}(k) &= \exists a_1 \in F_{k-1} \exists a_2 \in F_{k-1} (\\
&\quad [c.\mathit{itemset} = \mathit{unique}()] \wedge \\
&\quad [((c.\mathit{pos} = a_1.\mathit{pos}) \wedge (1 \leq a_1.\mathit{pos} \leq k-1) \wedge (c.\mathit{item} = a_1.\mathit{item})) \vee \\
&\quad ((c.\mathit{pos} = k) \wedge (a_2.\mathit{pos} = k-1) \wedge (c.\mathit{item} = a_2.\mathit{item}))] \wedge \\
&\quad \mathit{PrefixPair}(k)) \\
\mathit{PrefixPair}(k) &= \forall b_1 \in F_{k-1} \forall b_2 \in F_{k-1} (\\
&\quad [(b_1.\mathit{itemset} = a_1.\mathit{itemset}) \wedge (b_2.\mathit{itemset} = a_2.\mathit{itemset}) \wedge \\
&\quad (b_1.\mathit{pos} < k-1) \wedge (b_1.\mathit{pos} = b_2.\mathit{pos})] \rightarrow (b_1.\mathit{item} = b_2.\mathit{item})) \wedge \\
&\quad \exists b_1 \in F_{k-1} \exists b_2 \in F_{k-1} (\\
&\quad [(b_1.\mathit{itemset} = a_1.\mathit{itemset}) \wedge (b_2.\mathit{itemset} = a_2.\mathit{itemset}) \wedge \\
&\quad (b_1.\mathit{pos} = k-1) \wedge (b_1.\mathit{pos} = b_2.\mathit{pos})] \rightarrow (b_1.\mathit{item} < b_2.\mathit{item})) \wedge \\
&\quad \mathit{Apriori}(k) \\
\mathit{Apriori}(k) &= \bigwedge_{p=1}^{k-2} A(k,p) \\
A(k,p) &= \exists a_3 \in F_{k-1} \forall b_1 \in F_{k-1} \forall b_2 \in F_{k-1} (\\
&\quad [((b_1.\mathit{itemset} = a_1.\mathit{itemset}) \wedge (b_2.\mathit{itemset} = a_3.\mathit{itemset}) \wedge \\
&\quad (b_2.\mathit{pos} < p) \wedge (b_1.\mathit{pos} = b_2.\mathit{pos})) \rightarrow (b_1.\mathit{item} = b_2.\mathit{item})] \wedge \\
&\quad [((b_1.\mathit{itemset} = a_1.\mathit{itemset}) \wedge (b_2.\mathit{itemset} = a_3.\mathit{itemset}) \wedge \\
&\quad (p = b_2.\mathit{pos} < k-1) \wedge (b_1.\mathit{pos} = b_2.\mathit{pos}+1)) \rightarrow (b_1.\mathit{item} = b_2.\mathit{item})] \wedge \\
&\quad [((b_1.\mathit{itemset} = a_2.\mathit{itemset}) \wedge (b_2.\mathit{itemset} = a_3.\mathit{itemset}) \wedge \\
&\quad (b_2.\mathit{pos} = k-1) \wedge (b_1.\mathit{pos} = b_2.\mathit{pos})) \rightarrow (b_1.\mathit{item} = b_2.\mathit{item})])
\end{aligned}
$$

The second subexpression of *Candidate*, called *PrefixPair*, finds itemset ID pairs of frequent $(k-1)$-itemsets (a_1, a_2) that have a common prefix of size $k-2$. Such an itemset pair has the same *item* value at each position from 1 to $k-2$, and the *item* value of the first itemset at position $k-1$ is lexicographically ordered before that of the second itemset. For example, we will create a new itemset $ABCD$ for C_4 if we find the itemsets ABC and ABD in F_3, which have the common prefix AB.

The *PrefixPair* calculus expression contains universal quantifications and logical implications. An implication of the form $f \rightarrow g$ expresses the fact that if f holds then must g hold, too. For example, we can phrase the for-all expression in *PrefixPair* as follows: "For all item combinations $(b_1.\mathit{item}, b_2.\mathit{item})$ of itemsets a_1 and a_2, if we look at the same position of any but the last item of both itemsets, then they must have the same *item* value at this position." The existential expression that follows the universal expression can be phrased as: "At position $k-1$, the item value of the first itemset that we aim to find has to be lexicographically less than that of the second itemset."

The Quiver approach tries to reduce the number of candidates by ignoring all itemsets that do not fulfill the Apriori property, like in the K-Way-Join algorithm, described in Section 3.1. Thus, *PrefixPair* contains another expression called *Apriori*(k), which has to hold as well. This expression checks if, apart from a_1 and a_2, which we know are frequent, all other subsets of size $k-1$ are frequent, too. For example, for the potential candidate $ABCD$, *Apriori*(4) checks if the subsets ACD and BCD are contained in F_3. We already know that ABC and ABD have to be frequent because we used them for the construction of the potential candidate. Hence, we can ignore these checks in *Apriori*(4). In general, *Apriori*(k) tests the existence in F_{k-1} of those $(k-1)$-itemsets that we get when skipping a single item at the positions 1 to $k-2$ from a potential candidate k-itemset. For candidate itemset $ABCD$, for example, we skip position 1 to get BCD and position 2 to get ACD. This Apriori check is represented by a conjunction of k similar expressions $A(k,p)$, each having a different value of the position parameter p, where $1 \le p \le k-2$, i.e., $\mathit{Apriori}(k) = A(k,1) \wedge A(k,2) \wedge \ldots \wedge A(k,k-2)$. Each expression $A(k,p)$ checks if such a $(k-1)$-itemset is frequent that we get when we skip the item at position p of the potential candidate k-itemset.

From Calculus to SQL. We can easily derive SQL statements from the tuple relational calculus expressions specified in the previous section. For example, an implication can be replaced by a disjunction, i.e., we transform $f \rightarrow g \equiv \neg f \vee g$ into "NOT f OR g." Unfortunately, there is no universal quantifier available in SQL. Therefore, we translate $(\forall x \in T : f(x)) \equiv (\neg\exists x \in T : \neg f(x))$ into "NOT EXISTS (SELECT * FROM T AS x WHERE NOT $f(x)$)." In addition, we can use De Morgan's rule for pushing a negation into a conjunction or a disjunction, for example $\neg(f \wedge g) = \neg f \vee \neg g$.

Figure 3 shows the SQL-92 statements of the candidate generation phase in Quiver that are equivalent to the tuple relational calculus given above. While the calculus expressions to compute C_k are generic, the SQL statements are shown for the example value $k = 3$.

The first SQL query shown in Figure 3 populates the pefix-pair table P (*itemset_new*, *itemset_old*$_1$, *itemset_old*$_2$), where *itemset_new* is a newly created unique ID and the other attributes belong to the ID pair of frequent $(k-1)$-itemsets with a common $(k-2)$-prefix. In this query, we have merged the ID generation with the computation of prefix pairs, i.e., the calculus expression *Unique* is translated into the *newid()* function returning a unique identifier (offered by Microsoft SQL Server 2000, for example).

The second statement derives the candidate k-itemsets. For each record in the table P, we copy each *item* and its corresponding *pos* value belonging to *itemset_old*$_1$ into the target table $C_k(\mathit{itemset}, \mathit{pos}, \mathit{item})$, together with the newly created itemset ID value *itemset_new*. In addition, we add another record $(\mathit{ID}, k, \mathit{item})$ to C_k, where the item value of *itemset_old*$_2$ is taken from position $k-1$. This procedure is illustrated with example data in Figure 4, where a single candidate 4-itemset is generated.

4.2 Support Counting Phase

We have seen how universal quantification is used to generate C_k in the Quiver approach. These candidates now have to be checked if they appear frequently

```
SET @k = 3;
INSERT INTO P (itemset_new, itemset_old1, itemset_old2)
SELECT newid(), itemset1, itemset2
FROM   (
SELECT DISTINCT a1.itemset AS itemset1, a2.itemset AS itemset2
FROM   F2 AS a1, F2 AS a2
WHERE  NOT EXISTS (
          SELECT * FROM F2 AS b1, F2 AS b2
          WHERE  ((b1.itemset = a1.itemset) AND (b2.itemset = a2.itemset) AND
                  (b1.pos < @k-1) AND (b1.pos = b2.pos)) AND NOT (b2.item = b1.item)
        ) AND
        EXISTS (
          SELECT b1.item, b2.item
          FROM   F2 AS b1, F2 AS b2
          WHERE  (b1.itemset = a1.itemset) AND (b2.itemset = a2.itemset) AND
                 (b1.pos = @k-1) AND (b1.pos = b2.pos) AND (b1.item < b2.item)
        )
        AND
        -- In the following, we skip the item at position p of itemset a1.
        -- The following EXIST clause has to be added for each value of p,
        -- where 1 <= p <= k-2.

        -- Skip item at position p = 1.
        EXISTS (
          SELECT a3.itemset
          FROM   F3 AS a3
          WHERE  NOT EXISTS (
                    SELECT b1.item, b2.item
                    FROM   F3 AS b1, F3 AS b2
                    WHERE  NOT (
                              -- Condition 1: 1 <= i < p
                              ( NOT (
                                  (b1.itemset = a1.itemset) AND (b2.itemset = a3.itemset) AND
                                  (1 <= b2.pos) AND (b2.pos < 1) AND (b1.pos = b2.pos)
                                ) OR
                                (b1.item = b2.item)
                              ) AND
                              -- Condition 2: p <= i < k-1
                              ( NOT (
                                  (b1.itemset = a1.itemset) AND (b2.itemset = a3.itemset) AND
                                  (1 <= b2.pos) AND (b2.pos < @k-1) AND (b1.pos = b2.pos + 1)
                                ) OR
                                (b1.item = b2.item)
                              ) AND
                              -- Condition 3: i = k-1
                              ( NOT (
                                  (b1.itemset = a2.itemset) AND (b2.itemset = a3.itemset) AND
                                  (b2.pos = @k-1) AND (b1.pos = b2.pos)
                                ) OR
                                (b1.item = b2.item)
                              )
                            )
                 )
        )
) AS temporary;

INSERT INTO C3 (itemset, pos, item)
SELECT p.itemset_new, f.pos, f.item
FROM   F2 AS f, P AS p
WHERE  f.itemset = p.itemset_old1
UNION
SELECT p.itemset_new, @k, f.item
FROM   F2 AS f, P AS p
WHERE  f.itemset = p.itemset_old2 AND
       f.pos     = @k-1;
```

Fig. 3. Candidate generation with SQL according to the Quiver approach

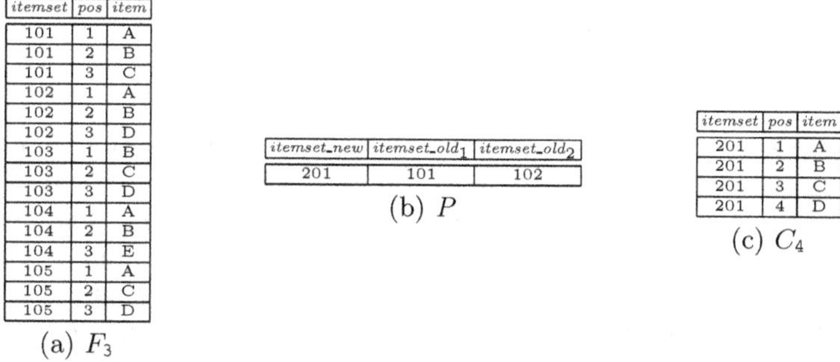

itemset	*pos*	*item*
101	1	A
101	2	B
101	3	C
102	1	A
102	2	B
102	3	D
103	1	B
103	2	C
103	3	D
104	1	A
104	2	B
104	3	E
105	1	A
105	2	C
105	3	D

(a) F_3

itemset_new	*itemset_old*$_1$	*itemset_old*$_2$
201	101	102

(b) P

itemset	*pos*	*item*
201	1	A
201	2	B
201	3	C
201	4	D

(c) C_4

Fig. 4. Example data for candidate generation using the Quiver approach

enough in the transactions to qualify for F_k. We propose a new approach for support counting that uses universal quantification as well.

Before we discuss the new approach, we show a vertical approach for support counting that is equivalent to the horizontal approach K-Way-Join, described in

```
INSERT INTO S3 (itemset, support)
SELECT   a1.itemset, COUNT(*)
FROM     C3 AS c, T AS t1, T AS t2, T AS t3
WHERE    c.item1           = t1.item AND
         c.item2           = t2.item AND
         c.item3           = t3.item AND
         t1.transaction  = t2.transaction AND
         t1.transaction  = t3.transaction
GROUP BY c.itemset
HAVING   COUNT(*) >= @minimum_support;

INSERT INTO F3 (itemset, item1, item2, item3)
SELECT c.itemset, c.item1, c.item2, c.item3
FROM   C3 AS c, S3 AS s
WHERE  c.itemset = s.itemset;
```

(a) Original, horizontal version of K-Way-Join

```
INSERT INTO S3 (itemset, support)
SELECT   a1.itemset, COUNT(*)
FROM     C3 AS c1, C3 AS c2, C3 AS c3,
         T  AS t1, T  AS t2, T  AS t3
WHERE    c1.itemset = c2.itemset             AND
         c1.itemset = c3.itemset             AND
         t1.transaction = t2.transaction AND
         t1.transaction = t3.transaction AND
         c1.item         = t1.item           AND
         c2.item         = t2.item           AND
         c3.item         = t3.item           AND
         c1.pos          = 1                 AND
         c2.pos          = 2                 AND
         c3.pos          = 3
GROUP BY c1.itemset
HAVING   COUNT(*) >= @minimum_support;

INSERT INTO F3 (itemset, pos, item)
SELECT c.itemset, c.pos, c.item
FROM   C3 AS c, S3 AS s
WHERE  c.itemset = s.itemset;
```

(b) Vertical version of K-Way-Join

Fig. 5. The support counting phase in the SQL algorithms *K-Way-Join* and *Vertical K-Way-Join*. This example derives F_3, the set of frequent 3-itemsets

Section 3.1. The direct translation into the vertical layout is given in Figure 5. We show this mapping because the Quiver approach for support counting is similar to the vertical version of K-Way-Join: We replace the explicit check of an item value at each of the k positions (*pos* values 1, 2, and 3 in Figure 5) with a universal quantification that checks *all* positions available in the candidate k-itemset. The figure also shows the intermediate result table S_k containing only the support count information for each itemset. The final result F_k is derived by joining C_k with S_k. If we are not interested in storing the support counters into a separate table then the statement for deriving S_k can be merged with the second query, which computes F_k.

Calculus Expressions. The new approach for support counting is defined in tuple relational calculus as follows:

$$
\begin{aligned}
\mathit{Query} &= \{c_1 \in C, t_1 \in T \mid \mathit{Contains}\} \\
\mathit{Contains} &= \forall c_2 \in C \exists t_2 \in T(\\
&\qquad (c_2.\mathit{itemset} = c_1.\mathit{itemset}) \rightarrow \\
&\qquad ((t_2.\mathit{transaction} = t_1.\mathit{transaction}) \wedge (t_2.\mathit{item} = t_1.\mathit{item})))
\end{aligned}
$$

The expression *Contains* derives combinations of transactions and candidates. It has two free tuple variables c_1 and t_1, where c_1 represents a candidate itemset and t_1 is a transaction that contains the itemset. The quantified (bound) tuple variables c_2 and t_2 represent the items corresponding to c_1 and t_1, respectively. The universal quantification lies in the condition that *for each* item $c_2.\mathit{item}$ belonging to itemset $c_1.\mathit{itemset}$, there must be an item $t_2.\mathit{item}$ belonging to transaction $t_1.\mathit{transaction}$ that matches with c_2.

A combination (c_1, t_1) fulfilling the calculus query $\{c_1 \in C, t_1 \in T \mid \mathit{Contains}\}$ indicates that the itemset $c_1.\mathit{itemset}$ is contained in the transaction $t_1.\mathit{transaction}$. We can find the support of each candidate by counting the number of distinct values $t_1.\mathit{transaction}$ that appear in a combination $c_1.\mathit{itemset}$. We do not show the actual counting because the basic tuple relational calculus does not include aggregate functions.

From Calculus to SQL. The calculus query can be translated into SQL in the same way as explained in Section 4.1. It is important to note that the resulting query in Figure 6 applies the aggregation on the set of *unique* transaction IDs because duplicates can occur as a result of the query processing.

```
INSERT INTO S (itemset, support)
SELECT   itemset, COUNT(DISTINCT transaction) AS support
FROM     (
  SELECT c1.itemset, t1.transaction
  FROM   C AS c1, T AS t1
  WHERE  NOT EXISTS (
    SELECT *
    FROM   C AS c2
    WHERE  NOT EXISTS (
      SELECT *
      FROM   T AS t2
      WHERE  NOT (c1.itemset = c2.itemset) OR
                 (t2.transaction = t1.transaction AND
                  t2.item        = c2.item)))
         ) AS Contains
GROUP BY itemset
HAVING   support >= @minimum_support;

INSERT INTO F (itemset, pos, item)
SELECT c.itemset, c.pos, c.item
FROM   C AS c, S AS s
WHERE  c.itemset = s.itemset;
```

Fig. 6. Support counting phase in Quiver as a SQL query. The parameter k for the candidate table C_k is omitted, i.e., this query is the same for every iteration of the algorithm

5 Performance Experiments

We first present the results of several experiments using SQL-based algorithms on a commercial database system. It is no surprise that they revealed a performance of the Quiver approach that is sometimes several orders of magnitude worse than the fastest known SQL-based approaches. This discouraging result is due to the absence of operators realizing efficient set containment tests of Quiver as well as an adequate syntax in SQL to express these tests.

We then discuss experiments using our own implementation of query execution plans (QEPs) using a Java class library for building query processors that demonstrate that a QEP using a set containment division operator can result in higher performance than equivalent plans derived from a query optimizer of a commercial DBMS, which do not offer such an operator.

5.1 Experiments with a Commercial DBMS

We compared the Quiver approach to K-Way-Join, Subquery, and Set-oriented Apriori, discussed in Section 3.1, as well as to a vertical version of K-Way-Join, proposed in Section 4.2. Remember that Quiver and Vertical K-Way-Join use a vertical table layout for both itemsets and transactions, while the other approaches use a horizontal table layout for itemsets and a vertical layout for transactions.

The commercial DBMS used for the experiments was Microsoft SQL Server 2000 Standard Edition running on a 4-CPU Intel Pentium-III Xeon PC with 900 MHz, 4 GB main memory, and Windows 2000 Server.

The synthetic data sets used for the experiments have been produced using a Java implementation of the well-known IBM data generation tool [23]. Similar data sets have been used in numerous publications for association rule discovery algorithms. By convention, our data sets are called T5.I5.10k and T5.I5.D100k, respectively. The data set T5.I5.D10k/T5.I5.D100k consists of 10,000/100,000 transactions (D), each containing 5.896/5.445 items (T) on the average, resulting in 58,964/544,590 rows. The largest size of a transaction is 17/12 items.

Several indexes have been created on the intermediate and result tables of the algorithm, summarized in Table 2. Although it is possible to experiment with all attribute combinations for indexing tables of the vertical layout like P (the equivalent of the calculus expression *PrefixPair* in Section 4.1), it requires much effort to do so for large values of k for horizontal tables like C_k. Hence, all we can say is that our index choice is one out of many. However, we have made a careful analysis on what indexes to offer to the optimizer. Note that the performance results reported in [10] on algorithms based on SQL-92, mentioned in Section 3.1, are very detailed but they do not cover the entire number of indexes neither. Hence, we leave as future work a comprehensive study on selecting the most promising indexes for the algorithms.

For each algorithm, we give the tables used by the algorithm: K-Way-Join (T^v, C_k^h, S_k^h, F_k^h), Subquery (T^v, T_f^v, Q_k^h, C_k^h, S_k, F_k^h), Set-oriented Apriori (T^v, T_f^v, T_k^h, Q_k^h, C_k^h, S_k, F_k^h), Vertical K-Way-Join (T^v, P, C_k^v, S_k, F_k^v), and Quiver (T^v, P, C_k^v, S_k, F_k^v). The attributes of each table are also listed in Table 2. Tables storing itemsets in a horizontal/vertical layout are marked by superscript h/v, respectively.

Figure 7 shows the results of our experiments that we briefly summarize in the following.

Candidate Itemset Generation. Experiments with both data sets have shown that the candidate generation phase based on a vertical table layout (Quiver and

Table 2. Overview of indexes created on tables used for SQL-based algorithms

Layout	Table	PI: Primary/clustered index SI: Secondary indexes
vertical	T^v (*transaction, item*)	PI: (*transaction, item*) SI: (*transaction*), (*item*), (*item, transaction*)
vertical	T_f^v (*transaction, item*)	PI: (*transaction, item*) SI: (*transaction*), (*item*), (*item, transaction*)
horizontal	T_k^h (*transaction*, $item_1, \ldots, item_k$)	PI: (*transaction*, $item_1, \ldots, item_k$) SI: (*transaction*), ($item_1, \ldots, item_k$)
horizontal	Q_k^h (*itemset*, $item_1, \ldots, item_k$)	PI: (*itemset*, $item_1, \ldots, item_k$) SI: (*itemset*), ($item_1$), ..., ($item_k$)
horizontal	C_k^h (*itemset*, $item_1, \ldots, item_k$)	PI: (*itemset*) SI: ($item_1$), ..., ($item_k$)
horizontal	F_k^h (*itemset*, $item_1, \ldots, item_k$)	PI: (*itemset*) SI: ($item_1$), ..., ($item_k$)
vertical	C_k^v (*itemset, pos, item*)	PI: (*itemset*) SI: (*item*), (*pos*)
vertical	F_k^v (*itemset, pos, item*)	PI: (*itemset, item*) SI: (*itemset*), (*item*), (*pos*)
–	S_k (*itemset, support*)	PI: (*itemset*) SI: (*support*), (*itemset, support*), (*support, itemset*)
–	P (*itemset_new*, *itemset_old*$_1$, *itemset_old*$_2$)	PI: (*itemset_new*) SI: (*itemset_old*$_2$), (*itemset_old*$_2$)

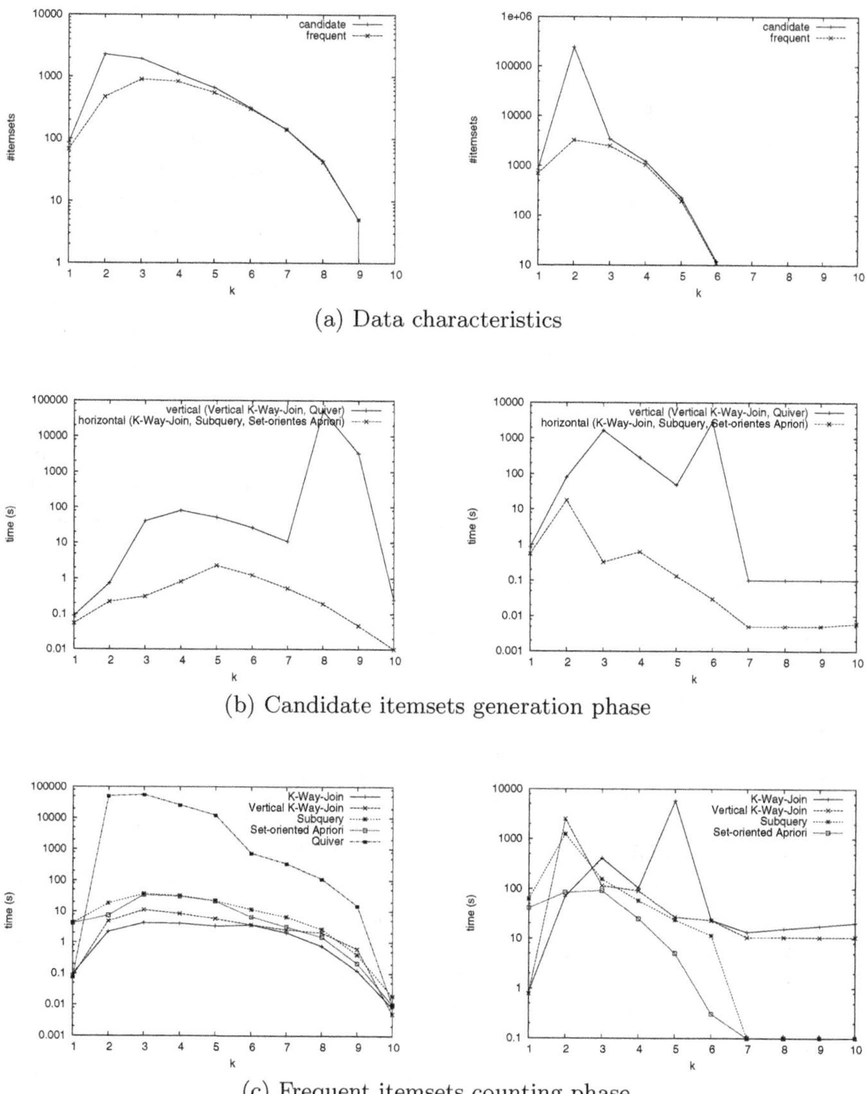

(a) Data characteristics

(b) Candidate itemsets generation phase

(c) Frequent itemsets counting phase

Fig. 7. Results of performance experiments with a commercial DBMS. The data sets are T5.I5.D10k, minimum support of 1% (100 transactions) on the left and T5.I5.D100k, minimum support of 0.1% (100 transactions) on the right. Note the logarithmic scale of the y-axes

Vertical K-Way-Join) was slower than the horizontal approach. For some values of k, the response time differed by more than two orders of magnitude. This difference is mainly caused by an expensive processing of the numerous correlated (NOT) EXISTS subqueries that compute the table P, shown in Figure 3. This query has a growing complexity for increasing itemset cardinalities. Due to space restrictions, we cannot elaborate here on the different characteristics of the QEPs produced by the optimizer of Microsoft SQL Server.

Frequent Itemset Counting. The results show that for the smaller data set (T5.I5.D10k), K-Way-Join is superior to all approaches in all but the last iteration of the frequent itemset counting phase. All algorithms but Quiver had a performance that stayed within a corridor of at most an order of magnitude from each other. Quiver was far from an acceptable response time.

The experiment with the larger data set (T5.I5.D10k) has confirmed the claim in [25] that Set-oriented Apriori performs better than Subquery for late passes of the algorithm, i.e., for high values of k. We do not give the response time for Quiver on the right of Figure 7(c). The experiment was running an unacceptably long time and we had to stop its execution.

These experiments have demonstrated that universal quantification is not well supported by modern commercial database systems. We have performed similar experiments with another commercial DBMS but we cannot report on the details here. However, our observation has been confirmed by the other experiments.

5.2 Experiments with XXL

The aim of our research is to investigate if Quiver queries can be executed more efficiently in a database system that provides implementations of a set containment division or set containment join operator. We have implemented algorithms for these operators in a testbed based on the open source Java class library *XXL* [27], designed for building query processors.

We have reimplemented the QEP for a certain Quiver query chosen by SQL Server in our prototype and recreated the data set and indexes used by the DBMS. The QEP realized the query deriving S_4 (Figure 6) from the data set T5.I5.D10k with a minimum support of 1%. In addition, we have reimplemented the original, horizontal version of K-Way-Join for the same problem and data setting. We compared the execution time for these QEPs to an alternative QEP that employs a set containment division operator. Due to lack of space we cannot show the QEPs here but refer to [17] for a more thorough discussion of the implementations.

Figure 8 shows that for a small number of candidates, the QEP employing set containment division performed close to or better than the other plans.

The experiments with XXL have been conducted on a 4-CPU Sun Ultra 80 server with Sun UltraSPARC-II 450 MHz CPUs, 4 GB main memory, Sun Solaris, and JRE 1.4.1.

6 Conclusion

In this paper, we have investigated the problem of frequent itemset discovery and compared several solutions based on SQL-92. The discovery of frequent itemsets is generally composed of two phases: candidate generation and itemset counting.

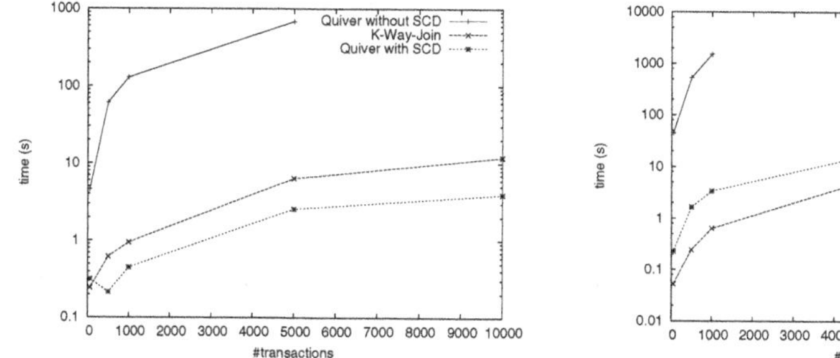

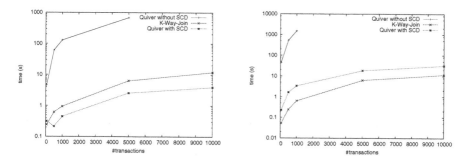

Fig. 8. Results of performance experiments with query execution plans implemented using XXL. The number of candidate itemsets is 4 on the left and 32 on the right. Note the logarithmic scale of the y-axes

Itemset counting can be regarded as a relational division problem. Several algorithms are available that allow a database system to efficiently process division within a query.

Today, commercial database systems have no implementation of the division operator, which is partially because it is hard to detect a division inside a query and there is no keyword available in any SQL implementation that would facilitate the specification of a division. Furthermore, division is not considered a frequently occurring problem in query processing [26] and it can be circumvented by indirect query formulations.

There are two main approaches to represent transactions and itemsets in SQL-92: a horizontal table layout, where all items of an object are stored in a single row, and a vertical table layout, where an object spans as many rows as its number of items. We have presented a new approach called *Quiver* that employs a vertical table layout for *both* transactions and itemsets. All known algorithms based on SQL-92 use a horizontal table layout for itemsets. Because of the vertical table layout, the queries for both phases of frequent itemset discovery can employ universal quantification, as shown in our Quiver approach. The reason why we investigated such an approach using for-all quantifiers is because it allows a natural formulation of the frequent itemset discovery problem—counting the number of transactions that contain *all* elements of a given itemset.

Ongoing and future work is devoted to realizing efficient algorithms for set containment division and set containment join operators. The latter is based on set-valued attributes, i.e., a nested, non-1NF representation of data. As mentioned in Section 1.2, we currently believe that a nested representation cannot add substantial performance gains over an unnested representation. The crucial containment test problem has to be solved either way and some effort has to go into initially transforming data into a nested representation, if it is not grouped or even sorted already.

Even if Quiver does not prove to be an appropriate solution for real data mining applications, we believe that similar approaches based on a natural representation of the mining problem in SQL will narrow the gap between data mining algorithms and database systems.

References

1. Agrawal, R., Imielinski, T., Swami, A.: Mining association rules between sets of items in large databases. In: Proceedings SIGMOD, Washington DC, USA. (1993) 207–216
2. Clear, J., Dunn, D., Harvey, B., Heytens, M., Lohman, P., Mehta, A., Melton, M., Rohrberg, L., Savasere, A., Wehrmeister, R., Xu, M.: Nonstop SQL/MX primitives for knowledge discovery. In: Proceedings KDD, San Diego, California, USA. (1999) 425–429
3. Hipp, J., Günzer, U., Grimmer, U.: Integrating association rule mining algorithms with relational database systems. In: Proceedings ICEIS, Setubal, Portugal. (2001) 130–137
4. Imielinski, T., Virmani, A.: MSQL: A query language for database mining. DMKD **3** (1999) 373–408
5. Han, J., Fu, Y., Koperski, K., Zaiane, O.: DMQL: A data mining query language for relational databases. In: Proceedings DMKD Workshop, Montreal, Canada. (1996)
6. Hyong, T., Indriyati, A., Lup, L.: Towards ad hoc mining of association rules with database management systems. Research report, School of Computing, National University of Singapore (2000)
7. Wang, H., Zaniolo, C.: Atlas: A native extension of sql for data mining and stream computations. Technical report, Computer Science Department, UCLA, USA (2002)
8. Meo, R., Psaila, G., Ceri, S.: A new SQL-like operator for mining association rules. In: Proceedings VLDB, Bombay, India. (1996) 122–133
9. Netz, A., Chaudhuri, S., Fayyad, U., Bernhardt, J.: Integrating data mining with SQL databases: OLE DB for data mining. In: Proceedings ICDE, Heidelberg, Germany. (2001) 379–387
10. Sarawagi, S., Thomas, S., Agrawal, R.: Integrating association rule mining with relational database systems: Alternatives and implications. Research report rj 10107 (91923), IBM Almaden Research Center, San Jose, California, USA (1998)
11. Zaniolo, C.: Extending SQL for decision support applications. In: Presentation slides of keynote address at DMDW, Toronto, Ontario, Canada. (2002)
12. Han, J., Kamber, M.: Data Mining: Concepts and Techniques. Morgan Kaufmann Publishers (2001)
13. Codd, E.: Relational completeness of database sub-languages. In Rustin, R., ed.: Courant Computer Science Symposium 6: Database Systems. Prentice-Hall (1972) 65–98
14. Rantzau, R., Shapiro, L., Mitschang, B., Wang, Q.: Algorithms and applications for universal quantification in relational databases. Information Systems Journal, Elsevier **28** (2003) 3–32
15. Graefe, G., Cole, R.: Fast algorithms for universal quantification in large databases. TODS **20** (1995) 187–236
16. Melnik, S., Garcia-Molina, H.: Adaptive algorithms for set containment joins. TODS **28** (2003)

17. Rantzau, R.: Processing frequent itemset discovery queries by division and set containment join operators. In: Proceedings ACM Workshop DMKD, San Diego, California, USA. (2003)
18. Rajamani, K., Cox, A., Iyer, B., Chadha, A.: Efficient mining for association rules with relational database systems. In: Proceedings IDEAS, Montreal, Canada. (1999) 148–155
19. Holsheimer, M., Kersten, M., Mannila, H., Toivonen, H.: A perspective on databases and data mining. In: Proceedings KDD, Montreal, Quebec, Canada. (1995) 150–155
20. Houtsma, M., Swami, A.: Set-oriented data mining in relational databases. DKE **17** (1995) 245–262
21. Yoshizawa, T., Pramudiono, I., Kitsuregawa, M.: SQL based association rule mining using commercial RDBMS (IBM DB2 UDB EEE). In: Proceedings DaWaK, London, UK. (2000) 301–306
22. Pramudiono, I., Shintani, T., Tamura, T., Kitsuregawa, M.: Parallel SQL based association rule mining on large scale PC cluster: Performance comparison with directly coded C implementation. In: Proceedings PAKDD, Beijing, China. (1999) 94–98
23. Agrawal, R., Srikant, R.: Fast algorithms for mining association rules. In: Proceedings VLDB, Santiago, Chile. (1994) 487–499
24. Mannila, H., Toivonen, H., Verkamo, A.I.: Efficient algorithms for discovering association rules. In: AAAI Workshop on Knowledge and Discovery in Databases, Seattle, Washington, USA. (1994) 181–192
25. Thomas, S., Chakravarthy, S.: Performance evaluation and optimization of join queries for association rule mining. In: Proceedings DaWaK, Florence, Italy. (1999) 241–250
26. Ramakrishnan, R., Gehrke, J.: Database Management Systems. Second edn. McGraw-Hill (2000)
27. Bercken, J.V.d., Blohsfeld, B., Dittrich, J.P., Krämer, J., Schäfer, T., Schneider, M., Seeger, M.: XXL – A library approach to supporting efficient implementations of advanced database queries. In: Proceedings VLDB, Rome, Italy. (2001) 39–48

Deducing Bounds on the Support of Itemsets

Toon Calders*

University of Antwerp
Universiteitsplein 1, B-2610 Wilrijk, Belgium
toon.calders@ua.ac.be

Abstract. Mining *Frequent Itemsets* is the core operation of many data mining algorithms. This operation however, is very data intensive and sometimes produces a prohibitively large output. In this paper we give a complete set of rules for deducing tight bounds on the support of an itemset if the supports of all its subsets are known. Based on the derived bounds $[l, u]$ on the support of a candidate itemset I, we can decide not to access the database to count the support of I if l is larger than the support threshold (I will certainly be frequent), or if u is below the threshold (I will certainly fail the frequency test). We can also use the deduction rules to reduce the size of an adequate representation of the collection of frequent sets; all itemsets I with bounds $[l, u]$, where $l = u$, do not need to be stored explicitly. To assess the usability in practice, we implemented the deduction rules and we present experiments on real-life data sets.

1 Introduction

Mining frequent itemsets is a core operation in many data mining problems. Since their introduction [1], many algorithms have been proposed to find frequent itemsets, especially in the context of association rule mining [1,2,12].

The *frequent itemset problem* is stated as follows. Assume we have a finite set of items $\mathcal{I}$. A *transaction* is a subset of $\mathcal{I}$, together with a unique identifier. A *transaction database* $\mathcal{D}$ is a finite set of transactions. A subset of $\mathcal{I}$ is called an *itemset*. We say that an *itemset I is s-frequent in a transaction database* $\mathcal{D}$ if the number of transactions in $\mathcal{D}$ that contain all items of I is at least s. The number of transactions that contain all items of I is called the *absolute support of I*. The frequent itemset problem is, given a support threshold s and a transaction database $\mathcal{D}$, find all s-frequent itemsets. In the remainder of the paper, we will always assume that we are working over a transaction database $\mathcal{D}$ with items in $\mathcal{I}$.

All algorithms for mining frequent itemsets rely heavily on the following *monotonicity principle* [16] to prune the search space:

> Let $J \subseteq I$ be two itemsets. In every transaction database $\mathcal{D}$, the support of I will be at most as high as the support of J.

* Research Assistant of the Fund for Scientific Research - Flanders (FWO-Vlaanderen).

R. Meo et al. (Eds.): Database Support for Data Mining Applications, LNAI 2682, pp. 214–233, 2004.

Thus, based on the support of a set that is below the support threshold, we can *deduce*, using the monotonicity rule, that also the support of its supersets will be below the threshold. This simple rule of *deduction* has successfully been used in practice. Because of the success of this simple rule, much more attention went into efficient counting schemes than into finding additional ways to prune the search space. The standard example of an algorithm exploiting this monotonicity is the well-known Apriori-algorithm [2]. Apriori traverses the itemset-lattice level by level; in the ith loop, itemsets of cardinality i are counted in the database. Because of the monotonicity principle, all itemsets in loop i that have at least one subset that failed the support-test can be *pruned*; we know *a priori* that they will be infrequent. In this way we will never count itemsets that could be pruned using the monotonicity rule.

In this paper we present deduction rules, additional to the monotonicity rule, that calculate lower and upper bounds on the support of a candidate. As such, we continue work initiated in [9]. Based on the supports of all subsets of an itemset I, the deduction rules we present, will compute bounds $[l, u]$ on the support of I. We show that the rules calculate the best possible such bounds; that is, both l and u are possible as supports of I, and thus, the interval cannot be made more tight. Based on these bounds we can limit the number of candidates we need to count. For example, if l is above the support threshold, then we know *without counting its support in the database* that I is frequent. If there is no need to know the support of I exact, we can thus, in this case, omit counting I. If u is below the threshold, then we know for sure that I is not frequent, and we can prune it.

Besides reducing the number of candidate itemsets, we can also use the deduction rules to make *concise representations* [15] of the frequent itemsets. We call an itemset *derivable* if its lower and upper bound are the same. Thus, an itemset is derivable if its support is uniquely determined by the supports of its subsets. Therefore, for the derivable itemsets, it is not necessary to count their supports. There is also no need to store them; we can later always find the missing supports with the deduction rules. Based on this observation, the NDI-representation is defined. We shortly discuss relations with other concise representations in the literature, including *free sets* [5], *closed sets* [18,4,19], and *disjunction-free sets* [6].

The organization of the paper is as follows. In Section 2 we give an example showing that the monotonicity rule is not complete for the deduction of supports. This example also gives a sketch of the general approach we follow to derive the deduction rules. In Section 3 we formally define important notions we will use throughout the paper. In Section 4, the deduction rules are given, and it is proven that they are complete. In Section 5 we present a concise representation based on the deduction rules. Section 6 gives the results of experiments with the deduction rules. In Section 7 we discuss related work and Section 8 concludes the paper.

2 Motivating Example

Apriori does not prune perfectly. Consider the following database.

$$\mathcal{D} \; = \; \begin{array}{|c|c|}\hline \texttt{TID} & \texttt{Items} \\ \hline 1 & A,B \\ 2 & A,C \\ 3 & B,C \\ \hline \end{array} \qquad (1)$$

Suppose we are running the Apriori-algorithm on this database $\mathcal{D}$ with minimal absolute support equal to 1. Apriori starts with counting the supports of the singleton-itemsets in $C_1 = \{\{A\},\{B\},\{C\}\}$. Since they are all frequent, in its second loop, Apriori will consider the candidates in $C_2 = \{\{A,B\},\{A,C\},\{B,C\}\}$. Again all candidates are frequent, and thus, Apriori counts $C_3 = \{\{A,B,C\}\}$ in its third loop. However, the following observation shows that from the supports counted so far, we can derive that $\{A,B,C\}$ must be infrequent.

Let for each itemset I, $\mathcal{F}_I(\mathcal{D})$ denote the set of transactions

$$\mathcal{F}_I(\mathcal{D}) \;=_{def}\; \{(tid, I') \in \mathcal{D} \mid I' = I\} \;,$$

and let f_I be the cardinality of $\mathcal{F}_I(\mathcal{D})$. Hence, in the database $\mathcal{D}$ given in (1), $\mathcal{F}_{AB}(\mathcal{D}) = \{(1, AB)\}$, $\mathcal{F}_{AC}(\mathcal{D}) = \{(2, AC)\}$, and $\mathcal{F}_{BC}(\mathcal{D}) = \{(3, BC)\}$. For all other itemsets I, $\mathcal{F}_I(\mathcal{D})$ is empty. Notice that $\{\mathcal{F}_I(\mathcal{D}) \mid I \subseteq \mathcal{I}\}$ forms a partition of $\mathcal{D}$. This partition is illustrated in Fig. 1. The dots in this figure represent the transactions of $\mathcal{D}$. With every item a set is associated. The set associated with item A consists of all transactions that contain A. The partition defined by the sets $\mathcal{F}_I(\mathcal{D})$ is indicated in the figure.

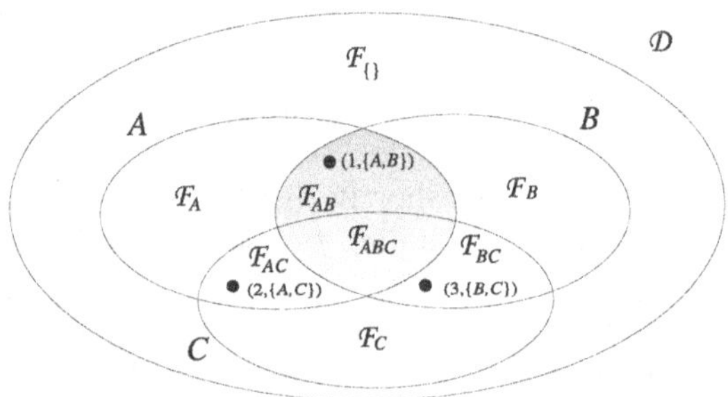

Fig. 1. Partition of $\mathcal{D}$

The next lemma expresses the supports of the itemsets in function of the numbers f_I, $I \subseteq \mathcal{I}$.

Lemma 1. *For each itemset I,*

$$support(I, \mathcal{D}) \;=\; \sum_{I \subseteq I' \subseteq \mathcal{I}} f_{I'}(\mathcal{D}) \;.$$

Proof.

$$\begin{aligned} support(I,\mathcal{D}) &= |\{(tid,I') \in \mathcal{D} \mid I \subseteq I' \subseteq \mathcal{I}\} \\ &= \sum_{I \subseteq I' \subseteq \mathcal{I}} |\{(tid,I'') \in \mathcal{D} \mid I'' = I'\}| \\ &= \sum_{I \subseteq I' \subseteq \mathcal{I}} |\mathcal{F}_{I'}(\mathcal{D})| = \sum_{I \subseteq I' \subseteq \mathcal{I}} f_{I'}(\mathcal{D}) \ . \end{aligned}$$

□

In Fig. 1, the grey region indicates the set of all transactions that contain the itemset AB. The transactions that contain AB are exactly those of the form $(tid, \{A,B\})$ and $(tid, \{A,B,C\})$. Hence, the set of transactions in $\mathcal{D}$ containing AB is $\mathcal{F}_{AB}(\mathcal{D}) \cup \mathcal{F}_{ABC}(\mathcal{D})$.

In the running example, after the second loop, we have the following information:

$$\begin{array}{lll} support(\{\},\mathcal{D}) = 3 & support(A,\mathcal{D}) = 2 & support(B,\mathcal{D}) = 2 \\ support(C,\mathcal{D}) = 2 & support(AB,\mathcal{D}) = 1 & support(AC,\mathcal{D}) = 1 \\ support(BC,\mathcal{D}) = 1 & & \end{array} \quad (2)$$

Therefore, the following equalities hold[1]:

$$\left\{ \begin{array}{ll} f_{\{\}} + f_A + f_B + f_C + f_{AB} + f_{AC} + f_{BC} + f_{ABC} = 3 & (\{\}) \\ f_A + f_{AB} + f_{AC} + f_{ABC} = 2 & (A) \\ f_B + f_{AB} + f_{BC} + f_{ABC} = 2 & (B) \\ f_C + f_{AC} + f_{BC} + f_{ABC} = 2 & (C) \\ f_{AB} + f_{ABC} = 1 & (AB) \\ f_{AC} + f_{ABC} = 1 & (AC) \\ f_{BC} + f_{ABC} = 1 & (BC) \end{array} \right. \quad (3)$$

For example, the equation $f_A + f_{AB} + f_{AC} + f_{ABC} = 2$ expresses that the support of A equals 2. (3) expresses the same information as (2).

Furthermore, since $f_I = |\mathcal{F}_I|$, it is also true that

$$f_{\{\}}, f_A, f_B, f_C, f_{AC}, f_{BC}, f_{AB}, f_{ABC} \geq 0 \quad (4)$$

We now show how we can derive from (3) and (4) that f_{ABC} must be 0, and hence that the support of ABC is 0. Rewriting (3) gives:

$$\left\{ \begin{array}{ll} f_{\{\}} = -f_{ABC} & f_{AB} = 1 - f_{ABC} \\ f_A = f_{ABC} & f_{AC} = 1 - f_{ABC} \\ f_B = f_{ABC} & f_{BC} = 1 - f_{ABC} \\ f_C = f_{ABC} & \end{array} \right. \quad (5)$$

Since both $f_{\{\}}$ and f_A are greater than or equal to 0 (Cfr. (4)), the first two lines of (5) imply respectively $f_{ABC} \leq 0$ and $f_{ABC} \geq 0$. Thus, from the information in (2), it can be derived that $support(ABC,\mathcal{D})$ must be 0, and hence we know *a priori* that ABC cannot be frequent. Nevertheless, Apriori does not prune ABC. This example shows that pruning can be improved beyond monotonicity.

[1] A similar representation is also used in [9,7,8].

3 Definitions

In this section we formalize the notions used in the example of last section. We introduce *support expressions* to model information about the supports of the itemsets. The notions of *implication* and *tight implication* express what can be derived from a set of support expressions.

Definition 1. *A* support expression over $\mathcal{I}$ *is an equality*

$$support(I) = s \ ,$$

with I an itemset over $\mathcal{I}$ and s an integer greater than or equal to 0.

A transaction database $\mathcal{D}$ over $\mathcal{I}$ is said to satisfy a support expression *$support(I) = s$ if and only if $support(I, \mathcal{D}) = s$.*

A transaction database is said to satisfy a set of support expressions *$\mathcal{S}$ if and only if it satisfies every expression in $\mathcal{S}$.* □

During the execution of the Apriori-algorithm, the support of ever larger itemsets is counted. In the theory we develop, the knowledge of the supports of the itemsets accumulated in the previous counting steps is modelled as a set of support expressions. In the example in Section 2, the knowledge given in (2) is expressed by the following set of support expressions:

$$\mathcal{S} = \left\{ \begin{array}{llll} support(\{\}) &= 3, & & \\ support(A) &= 2, & support(B) \;\; = 2, & support(C) \;\; = 2, \\ support(AB) &= 1, & support(AC) = 1, & support(BC) = 1 \end{array} \right\} .$$

In the candidate generation and pruning phases, it is decided which sets to count in the next iteration. The decision of which sets will be counted is based solely on the supports of the itemsets counted so far. For example, a set ABC is a candidate in the next loop, only if all three sets AB, AC, and BC were found frequent. If, for example, AB is infrequent, then it can be derived that the support of ABC is below the support threshold as well. Indeed, from support expression $support(AB) = s$ it follows that the support of ABC must be in the interval $[0, s]$. Such deductions are formalized as *logical implication* in the next definition.

Definition 2. *Let I be an itemset over $\mathcal{I}$, and let $l, u \geq 0$ be integers.*

A set of support expressions $\mathcal{S}$ is said to imply bounds $[l, u]$ on the support of I*, denoted $\mathcal{S} \models support(I) \in [l, u]$, if in every transaction database $\mathcal{D}$ that satisfies $\mathcal{S}$, $l \leq support(I, \mathcal{D}) \leq u$ holds.*

The bounds $[l, u]$ are said to be tight*, denoted $\mathcal{S} \models_{\text{tight}} support(I) \in [l, u]$, if there does not exist a smaller interval $[l', u'] \subset [l, u]$ such that $\mathcal{S} \models support(I) \in [l', u']$.* □

Implication denotes what we can derive from a set of support expressions. Given a set of support expressions $\mathcal{S}$, the deduction of $support(I) \in [l, u]$ is correct —or *sound*— if and only if it is true in *every* database that satisfies $\mathcal{S}$.

Tight implication denotes that the bounds cannot be improved; $\mathcal{S} \models_{\text{tight}} \mathit{support}(I) \in [l,u]$ indicates that, given $\mathcal{S}$, both $\mathit{support}(I) = l$ as $\mathit{support}(I) = u$ are possible. Hence, based on $\mathcal{S}$, we cannot improve the interval $[l,u]$. Therefore, a deduction mechanism is *complete* if, given $\mathcal{S}$ and a target set I, it *always* produces the *tight* interval for I.

Example 1.

$$\mathcal{S} = \left\{ \begin{array}{llll} \mathit{support}(\{\}) &= 3, & & \\ \mathit{support}(A) &= 2, & \mathit{support}(B) = 2, & \mathit{support}(C) = 2, \\ \mathit{support}(AB) &= 1, & \mathit{support}(AC) = 1, & \mathit{support}(BC) = 1 \end{array} \right\}.$$

From the monotonicity rule we know that $\mathcal{S} \models \mathit{support}(ABC) \in [0,1]$. The interval $[0,1]$ however, is not tight for the support of ABC. From the reasoning in Section 2, we know that in every database that satisfies $\mathcal{S}$, the support of ABC must be 0. Hence, $\mathcal{S} \models_{\text{tight}} \mathit{support}(ABC) \in [0,0]$. □

The next lemma makes a similar connection between support expressions and systems of linear inequalities as in the example in Section 2.

Lemma 2. *Let $\mathcal{S}$ be a collection of support expressions over $\mathcal{I}$. There exists a transaction database $\mathcal{D}$ over $\mathcal{I}$ that satisfies $\mathcal{S}$, if and only if the following system of inequalities has an integer solution in the variables x_I, $I \subseteq \mathcal{I}$:*

$$\mathit{Sys}(\mathcal{S}) \;=_{def}\; \left\{ \begin{array}{ll} x_I \geq 0 & \forall I \subseteq \mathcal{I} \\ \sum\limits_{I \subseteq I' \subseteq \mathcal{I}} x_{I'} = s_I & \forall (\mathit{support}(I) = s_I) \in \mathcal{S} \end{array} \right.$$

Proof. If: Consider an integer solution of $\mathit{Sys}(\mathcal{S})$. In such a solution all x_I's are integers greater than or equal to 0. Let now $\mathcal{D}$ be the transaction database that for all $I \subseteq \mathcal{I}$ contains x_I transactions of the form (tid, I). Hence, for all $I \subseteq \mathcal{I}$, $f_I(\mathcal{D}) = x_I$. Using Lemma 1, we obtain:

$$\forall I \subseteq \mathcal{I} : \mathit{support}(I,\mathcal{D}) \;=\; \sum_{I \subseteq I' \subseteq \mathcal{I}} f_{I'}(\mathcal{D}) \;=\; \sum_{I \subseteq I' \subseteq \mathcal{I}} x_{I'} \tag{6}$$

For all support expressions $\mathit{support}(I) = s_I$ in $\mathcal{S}$, $\mathit{Sys}(\mathcal{S})$ contains the equality $\sum_{I \subseteq I' \subseteq \mathcal{I}} x_{I'} = s_I$, and hence, via (6), $\mathit{support}(I,\mathcal{D}) = s_I$. Thus, $\mathcal{D}$ satisfies $\mathcal{S}$.
Only if: Let $\mathcal{D}$ be a transaction database that satisfies $\mathcal{S}$. Then $x_I = f_I(\mathcal{D})$, for all $I \subseteq \mathcal{I}$ is an integer solution of the system $\mathit{Sys}(\mathcal{S})$. Indeed, for all I, $f_I(\mathcal{D})$ is greater than or equal to 0. Furthermore, since $\mathcal{D}$ satisfies $\mathcal{S}$, for all support expressions $\mathit{support}(I) = s_I$ in $\mathcal{S}$, $\mathit{support}(I,\mathcal{D}) = s_I$. Because of Lemma 1, $\mathit{support}(I,\mathcal{D}) = \sum_{I \subseteq I' \subseteq \mathcal{I}} f_{I'}(\mathcal{D})$, and hence $\sum_{I \subseteq I' \subseteq \mathcal{I}} f_{I'}(\mathcal{D}) = s_I$. □

Example 2. There exists a transaction database $\mathcal{D}$ with $\mathit{support}(\{\},\mathcal{D}) = 3$, $\mathit{support}(A,\mathcal{D}) = 2$, $\mathit{support}(B,\mathcal{D}) = 2$, and $\mathit{support}(AB,\mathcal{D}) = 0$ if and only if the following system of inequalities has a solution:

$$\left\{ \begin{array}{ll} x_{\{\}}, x_A, x_B, x_{AB} \geq 0 & x_B + x_{AB} = 2 \\ x_{\{\}} + x_A + x_B + x_{AB} = 3 & x_{AB} = 0 \\ x_A + x_{AB} = 2 & \end{array} \right.$$

From the last three equalities we derive that $x_A = x_B = 2$, and $x_{AB} = 0$. This however conflicts with $x_{\{\}} + x_A + x_B + x_{AB} = 3$, since all variables must be greater than or equal to 0. Hence, we conclude that there does not exist a transaction database satisfying the given support expressions. □

Problem Statement. In the remainder we will concentrate on implication problems for a set I, based on a set $\mathcal{S}$ of support expressions that contains exactly one expression for each strict subset of I. We do not consider cases in which $\mathcal{S}$ contains support expressions for supersets of I, or in which subsets are missing. Hence, given an integer s_J for all $J \subset I$, tight implication of the following type is studied:

$$\{support(J) = s_J \mid J \subset I\} \models_{\text{tight}} support(I) \in [l, u] \ .$$

Notice that the information $\{support(J) = s_J \mid J \subset I\}$ is available for every candidate itemset I in the Apriori-algorithm.

4 Deduction Rules

In this section we describe sound and complete rules for deducing tight bounds on the support of a set I if the supports of all its subsets are given. Because we do not consider itemsets that are not subsets of I, we can assume that all items in the database are elements of I. Since "projecting away" the other items in a transaction database does not change the supports of subsets of I, we can assume without loss of generality that $\mathcal{I} = I$. The correctness of this observation follows from the next lemma.

Definition 3. *Let $I \subseteq \mathcal{I}$ be an itemset.*

- *The* projection *of a transaction (tid, I') over $\mathcal{I}$ on I, denoted $\pi_I(tid, I')$, is the transaction $(tid, I' \cap I)$.*
- *The* projection *of a transaction database $\mathcal{D}$ over $\mathcal{I}$ on I, denoted $\pi_I\mathcal{D}$, is defined as $\pi_I\mathcal{D} =_{def} \{\pi_I T \mid T \in \mathcal{D}\}$.*

Lemma 3. *Let $\mathcal{I}$ be a set of items, and let $J \subseteq I$ be itemsets. For every transaction database $\mathcal{D}$ over $\mathcal{I}$ it holds that*

$$support(J, \mathcal{D}) = support(J, \pi_I\mathcal{D}) \ .$$

Proof. For $J \subseteq I$,

$$\begin{aligned} support(J, \mathcal{D}) &= |\{(tid, I') \in \mathcal{D} \mid J \subseteq I'\}| \\ &= |\{(tid, I') \in \mathcal{D} \mid J \subseteq (I' \cap I)\}| && (J \subseteq I) \\ &= |\{(tid, I'') \in \pi_I\mathcal{D} \mid J \subseteq I''\}| \\ &= support(J, \pi_I\mathcal{D}) \end{aligned}$$

□

This lemma allows for an important reduction of the system $Sys(\mathcal{S})$ associated with a set of support expressions $\mathcal{S}$ that contains an expression for every strict subset of I. Instead of having a variable x_J for every itemset $J \subseteq \mathcal{I}$, with Lemma 3 we can restrict the variables to only those x_J such that $J \subseteq I$.

Corollary 1. *Given an itemset $I \subseteq \mathcal{I}$, and integer $s_J \geq 0$, for every $J \subseteq I$. There exists a transaction database $\mathcal{D}$ satisfying $\forall J \subseteq I : support(J, \mathcal{D}) = s_J$ if and only if the following system of inequalities has a solution:*

$$\begin{cases} x_J \geq 0 & \forall J \subseteq I \\ \displaystyle\sum_{J \subseteq I' \subseteq I} x_{I'} = s_J & \forall J \subseteq I \end{cases}$$

Proof. Because of Lemma 3, the existence of a database $\mathcal{D}$ over $\mathcal{I}$ satisfying the given expressions implies the existence of such a database over I, namely $\pi_I \mathcal{D}$. The corollary now follows from Lemma 2. □

Let $I \subseteq \mathcal{I}$ be an itemset. We assume that all supports of the strict subsets J of I are known, let s_J denote $support(J, \mathcal{D})$. We will now derive optimal bounds on the support of I. These bounds can be determined as follows: the best possible lower bound is the smallest integer l such that the system of support expressions

$$\{support(J) = s_J \mid J \subset I\} \cup \{support(I) = l\}$$

is satisfiable. The best upper bound is the largest integer u such that

$$\{support(J) = s_J \mid J \subset I\} \cup \{support(I) = u\}$$

is satisfiable. Let now s_I be an arbitrary integer. From Corollary 1, we know that the system of support constraints

$$\{support(J) = s_J \mid J \subset I\} \cup \{support(I) = s_I\}$$

is satisfiable if and only if the following system of inequalities has an integer solution:

$$\begin{cases} x_J \geq 0 & \forall J \subseteq I \\ \displaystyle\sum_{J \subseteq I' \subseteq I} x_{I'} = s_J & \forall J \subseteq I \end{cases}$$

This system can be solved for the x_J's as follows:

$$\begin{cases} s_I & = x_I \\ s_{I-A} & = x_I + x_{I-A} \\ s_{I-B} & = x_I + x_{I-B} \\ s_{I-AB} & = x_I + x_{I-A} \\ & \quad + x_{I-B} + x_{I-AB} \\ \ldots \end{cases} \rightarrow \begin{cases} x_I & = s_I \\ s_{I-A} & = x_I + x_{I-A} \\ s_{I-B} & = x_I + x_{I-B} \\ s_{I-AB} & = x_I + x_{I-A} \\ & \quad + x_{I-B} + x_{I-AB} \\ \ldots \end{cases}$$

$$\rightarrow \begin{cases} x_I & = s_I \\ x_{I-A} & = s_{I-A} - s_I \\ x_{I-B} & = s_{I-B} - s_I \\ s_{I-AB} & = x_I + x_{I-A} \\ & \quad + x_{I-B} + x_{I-AB} \\ \ldots \end{cases} \rightarrow \begin{cases} x_I & = s_I \\ x_{I-A} & = s_{I-A} - s_I \\ x_{I-B} & = s_{I-B} - s_I \\ x_{I-AB} & = s_{I-AB} - s_{I-A} \\ & \quad - s_{I-B} + s_{I-AB} \\ \ldots \end{cases}$$

In general, $x_J = \sum_{J\subseteq J'\subseteq I}(-1)^{|J'-J|}s_{J'}$, as the following lemma shows.

Lemma 4. *Let I be an itemset, and for all $J \subseteq I$, s_J, x_J be integers. The following are equivalent*

(1) $\forall J \subseteq I : s_J = \sum_{J\subseteq I'\subseteq I} x_{I'}$
(2) $\forall J \subseteq I : x_J = \sum_{J\subseteq J'\subseteq I}(-1)^{|J'-J|}s_{J'}$

(The proof of this lemma can be found in Appendix A.)

Therefore, the system of support constraints

$$\{support(J) = s_J \mid J \subset I\} \cup \{support(I) = s_I\}$$

is satisfiable if and only if the following system of inequalities has an integer solution:

$$\begin{cases} x_J \geq 0 & \forall J \subseteq I \\ x_J = \sum\limits_{J\subseteq J'\subseteq I}(-1)^{|J'-J|}s_{J'} & \forall J \subseteq I \end{cases}$$

Hence, if

$$\sum_{J\subseteq J'\subseteq I}(-1)^{|J'-J|}s_{J'} \geq 0 \qquad \forall J \subseteq I$$

or, equivalent,

$$\begin{cases} s_I \leq \sum\limits_{J\subseteq J'\subset I}(-1)^{|I-J'|+1}s_{J'} & \forall J \subseteq I, |I-J| \text{ odd} \\ s_I \geq \sum\limits_{J\subseteq J'\subset I}(-1)^{|I-J'|+1}s_{J'} & \forall J \subseteq I, |I-J| \text{ even} \end{cases}$$

Let $\sigma_I(J,\mathcal{D})$ denote the sum

$$\sigma_I(J,\mathcal{D}) =_{def} \sum_{J\subseteq J'\subset I}(-1)^{|I-J'|+1}support(J',\mathcal{D})$$

and let $\mathcal{R}_I(J,\mathcal{D})$ denote the rule $support(I) \leq \sigma_I(J,\mathcal{D})$ if $|I-J|$ is odd, and $support(I) \geq \sigma_I(J,\mathcal{D})$ if $|I-J|$ is even. We obtain the following theorem that states that the bounds for itemset I found by the rules $\mathcal{R}_I(J)$, for all $J \subseteq I$, are the best bounds possible; that is, the interval found is tight.

Theorem 1. *Let $\mathcal{D}$ be a transaction database, and let I be an itemset. s_J denotes $support(J,\mathcal{D})$.*

$$\{support(J) = s_J \mid J \subset I\} \models_{\text{tight}} support(I) \in [l,u]$$

with

$$\begin{aligned} l &= \max\{\sigma_I(J,\mathcal{D}) \mid J \subset I, J \text{ even}\} \\ u &= \min\{\sigma_I(J,\mathcal{D}) \mid J \subset I, J \text{ odd}\} \end{aligned}$$

Hence, the rules $\mathcal{R}_I(J), J \subseteq I$ are sound and complete for implication of the support of I, based on the supports of the strict subsets of I.
(The proof of this theorem can be found in Appendix A.)

The rules $\mathcal{R}_{ABCD}(J)$ have been given in Figure 2.

$$\begin{cases} support(ABCD) \geq s_{ABC} + s_{ABD} + s_{ACD} + s_{BCD} \\ \qquad\qquad -s_{AB} - s_{AC} - s_{AD} - s_{BC} - s_{BD} - s_{CD} \\ \qquad\qquad +s_A + s_B + s_C + s_D - 1 \\ support(ABCD) \leq s_A - s_{AB} - s_{AC} - s_{AD} + s_{ABC} + s_{ABD} + s_{ACD} \\ support(ABCD) \leq s_B - s_{AB} - s_{BC} - s_{BD} + s_{ABC} + s_{ABD} + s_{BCD} \\ support(ABCD) \leq s_C - s_{AC} - s_{BC} - s_{CD} + s_{ABC} + s_{ACD} + s_{BCD} \\ support(ABCD) \leq s_D - s_{AD} - s_{BD} - s_{CD} + s_{ABD} + s_{ACD} + s_{BCD} \\ support(ABCD) \geq s_{ABC} + s_{ABD} - s_{AB} \\ support(ABCD) \geq s_{ABC} + s_{ACD} - s_{AC} \\ support(ABCD) \geq s_{ABD} + s_{ACD} - s_{AD} \\ support(ABCD) \geq s_{ABC} + s_{BCD} - s_{BC} \\ support(ABCD) \geq s_{ABD} + s_{BCD} - s_{BD} \\ support(ABCD) \geq s_{ACD} + s_{BCD} - s_{CD} \\ support(ABCD) \leq s_{ABC} \\ support(ABCD) \leq s_{ABD} \\ support(ABCD) \leq s_{ACD} \\ support(ABCD) \leq s_{BCD} \\ support(ABCD) \geq 0 \end{cases}$$

Fig. 2. Tight bounds on *support*$(ABCD)$

Example 3. Consider the following transaction database.

$\mathcal{D} =$

TID	items
1	A, B
2	A, C, D
3	A, B, D
4	C, D
5	B, C, D
6	A, D
7	B, D
8	B, C, D
9	B, C, D
10	A, B, C, D

$$\begin{array}{lll} s_{\{\}} = 10, & s_A = 5, & s_B = 7, \\ s_C = 6, & s_D = 9, & s_{AB} = 3, \\ s_{AC} = 2, & s_{AD} = 4, & s_{BC} = 4, \\ s_{BD} = 6, & s_{CD} = 6, & s_{ABC} = 1, \\ s_{ABD} = 2, & s_{ACD} = 2, & s_{BCD} = 4. \end{array}$$

Figure 2 gives the rules to determine tight bounds on the support of $ABCD$. Based on these deduction rules we derive the following bounds on the support of $ABCD$ *without counting in the database* $\mathcal{D}$.

$support(ABCD, \mathcal{D}) \geq 1$ (Rule $support(ABCD) \geq s_{ABC} + s_{ACD} - s_{AC}$)
$support(ABCD, \mathcal{D}) \leq 1$ (Rule $support(ABCD) \leq s_{ABC}$)

Therefore, we can conclude, without having to count, that the support of $ABCD$ in $\mathcal{D}$ must be 1. In the experiments we will see that this exactness is not very unusual; even in real-life data, and for small itemsets, we will be able to derive very narrow intervals. □

5 Concise Representation

5.1 Derivable Itemsets

Definition 4. *Let I be an itemset, and $\mathcal{D}$ a transaction database. Let for all itemsets $J \subseteq I$, s_J denote $support(J, \mathcal{D})$. We say that I is a derivable itemset w.r.t. $\mathcal{D}$, if*

$$\{support(J) = s_J \mid J \subset I\} \models_{\text{tight}} support(I) \in [s_I, s_I]$$

□

Notice that I derivable means that we do not have to count I in the database to know its support. Based on the supports of the subsets of I we can derive the support of I exactly.

5.2 NDI-Representation

Based on the notion of derivable itemsets we propose a *concise representation.* A concise representation [15] is a subset of the set of frequent itemsets, extended with supports, that allows for deriving the whole set of frequent sets and their supports. Such a concise representation is typically much smaller than the whole set of frequent itemsets, even though it contains the same amount of information. Therefore, in situations where the number of frequent itemsets is very large, it is often better to only mine a concise representation. Let $[l_I, u_I]$ be the bounds we can derive for itemset I, based on the supports of its subsets. We now define the NDI-representation as follows:

$$\text{NDI}(\mathcal{D}, s) \;=_{def}\; \{(I, support(I, \mathcal{D})) \mid (l_I \neq u_I), support(I, \mathcal{D}) \geq s\}.$$

That is, NDI only contains those sets that are both frequent in $\mathcal{D}$ and not derivable w.r.t. $\mathcal{D}$.

Theorem 2. *Let $\mathcal{D}$ be a transaction database, and let s be a support threshold.* NDI$(\mathcal{D}, s)$ *is a concise representation for the s-frequent itemsets in $\mathcal{D}$.*

5.3 Algorithm

In [8], the following theorem has been proven:

Theorem 3 (Anti-monotonicity of Derivability). *Let $I \subseteq J$ be itemsets over $\mathcal{I}$, and $\mathcal{D}$ be a transaction database over $\mathcal{I}$. If I is a derivable itemset, then J must be a derivable itemset as well.*

Based on this theorem we come up with the following algorithm to find all frequent non-derivable itemsets.

```
NDI(D,s)
    i := 1; NDI := {}; C_1 := {{i} | i ∈ I};
    for all I in C_1 do I.l := 0; I.u := |D|;
    while C_i not empty do
        Count the supports of all candidates in C_i in one pass over D;
        F_i := {I ∈ C_i | support(I, D) ≥ s};
        NDI := NDI ∪ F_i;
        PreC_{i+1} := AprioriGenerate(F_i);
        C_{i+1} := {};
        for all J ∈ PreC_{i+1} do
            Compute bounds [l, u] on support of J;
            if l ≠ u then J.l := l; J.u := u; C_{i+1} := C_{i+1} ∪ {J};
        i := i + 1
    end while
    return NDI
```

For a more elaborated description of the algorithm we refer the interested reader to [8].

6 Experiments

The experiments were performed on a 1.5GHz Pentium IV PC with 256MB of main memory. To empirically evaluate the proposed NDI-algorithm and deduction rules, we performed several tests on the datasets summarized in the following table. For each dataset the table shows the number of transactions, the number of items, and the average transaction length.

Dataset	# trans.	# items	Avg. length
BMS-POS	515 597	1 656	6.53
T40I10D100K	100 000	1 000	39.6
Connect-4	67 557	125	42
BMS-Webview-1	59 602	497	2.51
Pumsb	49 046	2 112	74
Mushroom	8 124	120	23

These datasets are all well-known benchmarks for frequent itemset mining. The *BMS-Webview* and *BMS-POS* datasets are click-stream data from a small dot-com company that no longer exists. These two datasets were donated to the research community by Blue Martini Software. The *Pumsb*-dataset is based on census data, the *Mushroom* dataset contains characteristics from different species of mushrooms. The *Connect-4* dataset contains different game positions. The Pumsb dataset is available in the UCI KDD Repository [13], and the Mushroom and Connect-4 datasets can be found in the UCI Machine Learning Repository [3]. The *T40I10D100K* dataset was generated using the IBM synthetic data generator.

The NDI-algorithm differs slightly from the algorithm presented in Section 5. In order to avoid the generation of pairs of items in the second loop, the candidates are only generated while iterating over the dataset. In this way the generation of pairs that do not appear in the database is avoided.

6.1 Overhead of Rule Evaluation

We first study the influence of limiting the depth of the rules we evaluate. In Fig. 3, the number of sets that are derivable when we evaluate rules up to depth k, and the time needed to find them are indicated for different k. As can be seen, the number of NDIs drops quickly from depth 1 to depth 2. In the Mushroom experiment, the test with $k = 1$ was even not feasible. From depth 3 on, higher depths result in only a slight decrease of the number of NDIs. This is not that remarkable since the number of NDIs of these sizes is small. The running times in Fig. 3 show that for these limited depths, the cost of evaluating all rules is rather small.

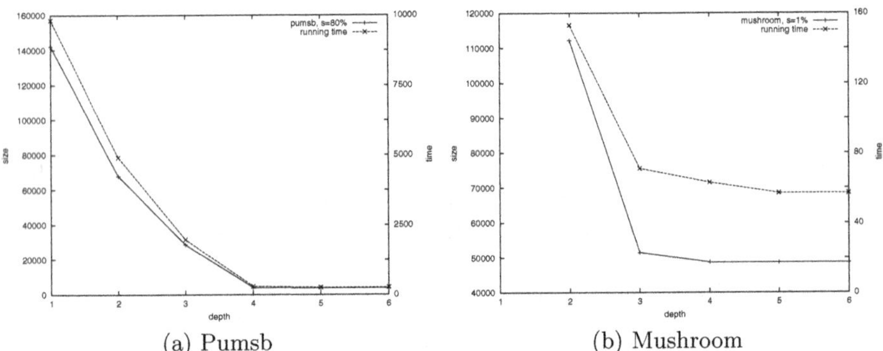

(a) Pumsb (b) Mushroom

Fig. 3. Size of the representations when k is limited

In Tab. 1, four experiments with the BMS-POS dataset are reported in detail. For each iteration of the algorithm, the number of candidates and the number these candidates that turn out to be frequent NDIs are reported, together with the computation time of respectively candidate generation with rule evaluation and counting. In Tab. 1 (a) and (b), all rules are evaluated, while in Tab. 1 (c) and (d) no rules are evaluated. Hence, the experiments in Tab. 1 (c) and (d) are in fact plain Apriori. The BMS-POS dataset is interesting, since it is the only dataset that contains almost no derivable itemsets. Therefore, these examples show very well the cost of evaluating the rules. The rule evaluation time is included in the generation time for the candidates. Tab. 1 shows that the evaluation times for the rules are very reasonable. This is especially so when the number of transactions becomes very high. For iteration 2, the number of candidates and the generation time of these candidates is not reported, because they are generated on the fly to avoid pairs of items with support 0.

Table 1. Example runs on the BMS-POS dataset

	$\|C_k\|$		$\|NDI\|$		Width
1	1656	6.437s	510	0s	515597
2			9553	35.828s	11610
3	187839	6.625s	39912	80.859s	29167
4	157481	14.329s	74768	120.828s	11831
5	117797	28.437s	77353	126.922s	3462
6	62233	41.922s	47499	93.266s	998
7	18369	38.656s	16276	46.297s	250
8	2230	18s	2141	16.547s	88
9	10	1.906s	9	8.579s	19
	268021	686.9			

(a) BMS-POS, support = 361, all rules

	$\|C_k\|$		$\|NDI\|$		Width
1	1656	6.609s	461	36.047s	515597
2			7554	75.437s	116102
3	126338	4.719s	27904	103.188s	29167
4	95701	8.922s	46115	98.25s	11831
5	63578	15.5s	42047	67.641s	3462
6	29226	20.578s	22300	29.937s	998
7	7075	14.86s	6315	12.031s	250
8	704	5.234s	685	7.985s	88
9	1	0.343s	1	8.579s	9
	153382	508.234s			

(b) BMS-POS, support = 465, all rules

	$\|C_k\|$		$\|NDI\|$		Width
1	1656	5.531s	510	0s	n/a
2			9553	35.719s	n/a
3	187839	5.031s	39912	80.672s	n/a
4	157481	7.375s	74768	126s	n/a
5	117929	8.984s	77361	128.484s	n/a
6	63981	7.406s	47741	95.422s	n/a
7	21335	3.812s	17293	49.203s	n/a
8	3765	1.109s	3283	19.797s	n/a
9	255	0.282s	228	9.625s	n/a
	270649	594.141s			

(c) BMS-POS, support = 361, no rules

	$\|C_k\|$		$\|NDI\|$		Width
1	1656	5.5s	461	0s	n/a
2			7554	34.86s	n/a
3	126338	3.344s	27904	73.547s	n/a
4	95701	4.359s	46115	102.469s	n/a
5	63641	4.656s	42048	95.297s	n/a
6	30114	3.421s	22341	63.532s	n/a
7	7967	1.422s	6480	29.953s	n/a
8	962	0.359s	849	13.079s	n/a
9	30	0.188s	29	8.312s	n/a
	153781	453.093s			

(d) BMS-POS, support = 465, no rules

6.2 Comparison with Mining All Frequent Sets

Since the overhead of calculating the rules is small, the running times of the Apriori-algorithm and the NDI-algorithm are almost linear in the size of their respective output. Therefore, the gain in speed of the NDI-algorithm over the extraction of all frequent itemsets with the Apriori-algorithm is more or less the ratio between the number of frequent sets and the number of frequent NDIs. This claim is supported by Fig. 4. In Fig. 4, the running time of the NDI-algorithm, Apriori, and FPGrowth is given, together with the number of the frequent and

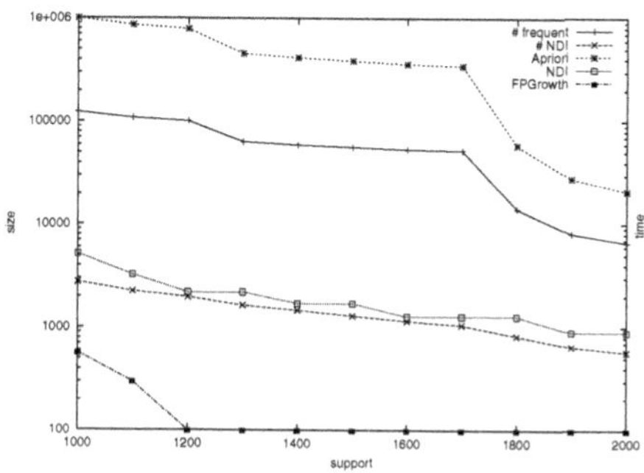

Fig. 4. Running time on the Mushroom dataset

the non-derivable sets, for different minimal supports. In this example, the execution time of FPGrowth is much lower than for the other algorithms. As long as the number of frequent sets is not too high, FPGrowth will be more efficient than mining the NDIs. As soon as the number of frequent sets becomes very high however, NDI will become more efficient.

Since in most of the experiments we present the number of frequent sets is very high, the execution of the Apriori-algorithm was not always possible. Instead we present a comparison with FPGrowth, and we report for some of the experiments the number of frequent sets and the number of NDIs. The results are presented in Fig. 5. For the Connect-4 and the Pumsb dataset it was not possible to perform the FPGrowth algorithm within reasonable time for the lowest supports.

In Fig. 5, it can clearly be seen that in most datasets once the support threshold becomes too low, and the number of frequent sets explodes, the NDI-algorithm becomes more efficient than mining all frequent itemsets. In Fig. 6, the number of frequent NDIs is compared with the total number of frequent sets. The only exceptional dataset in this perspective was the BMS-POS dataset, in which the performance of the NDI and FPGrowth algorithms stays more or less comparable. The explanation for this is in Tab. 1. In the BMS-POS dataset there are almost no derivable itemsets of low cardinality.

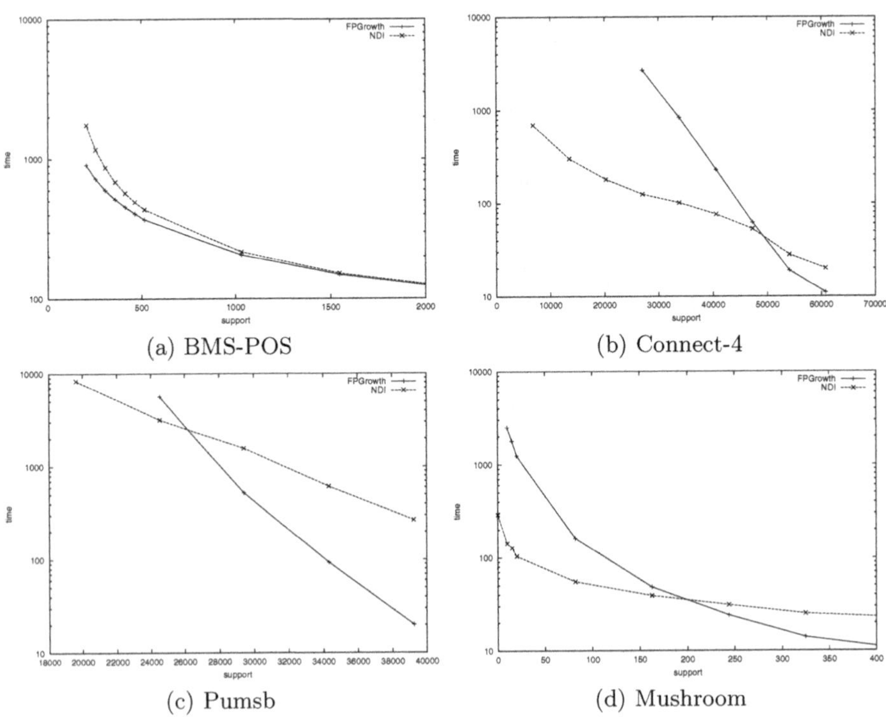

(a) BMS-POS (b) Connect-4

(c) Pumsb (d) Mushroom

Fig. 5. Comparison of running times of NDI and FPGrowth

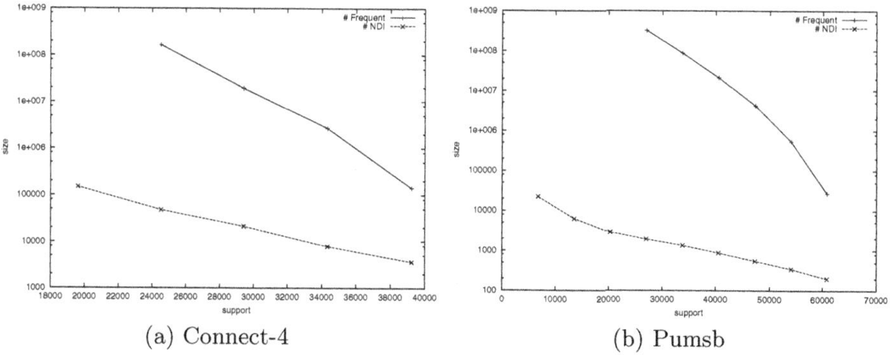

(a) Connect-4 (b) Pumsb

Fig. 6. Comparison of the number of frequent sets and NDIs

6.3 Comparison with Other Concise Representations

We compared the NDI-representation with the following other concise representations:

- *Free sets representation* FreeRep [5],
- *Disjunction-free sets representation* DFreeRep [6],
- *Disjunction-free generators representation* DFreeGenRep [14], and
- *Closed sets representation* Closed [18].

In Figure 7, the sizes of these representations and $|NDI|$ are reported on different datasets. The experiments show that on these datasets, the NDI-representation is often the smallest representation. Only in the BMS-Webview-1 dataset, the NDI-representation is slightly larger than the Closed sets representation.

7 Related Work

7.1 Concise Representations

Closed itemsets [18] received a lot of attention in the literature [4,19,20]. They can be introduced as follows: the *closure* of an itemset I is the largest superset of I such that its support equals the support of I. This superset is unique and is denoted by $cl(I)$. An itemset is called *closed* if it equals its closure. In [18], the authors show that the set of frequent closed sets is a concise representation for the frequent itemsets.

Free sets [5] *or Generators* [14] (Free sets [5] and generators [18,14] are the same.) An itemset I is called *free* if it has no subset with the same support. The free-set representation is based on the fact that is $support(A) = support(AB)$, also $support(AC) = support(ABC)$. This deduction can also be made with the following two deduction rules presented in this paper:

$$support(ABC) \leq support(AC), \text{ and}$$
$$support(ABC) \geq support(AB) + support(AC) - support(A) .$$

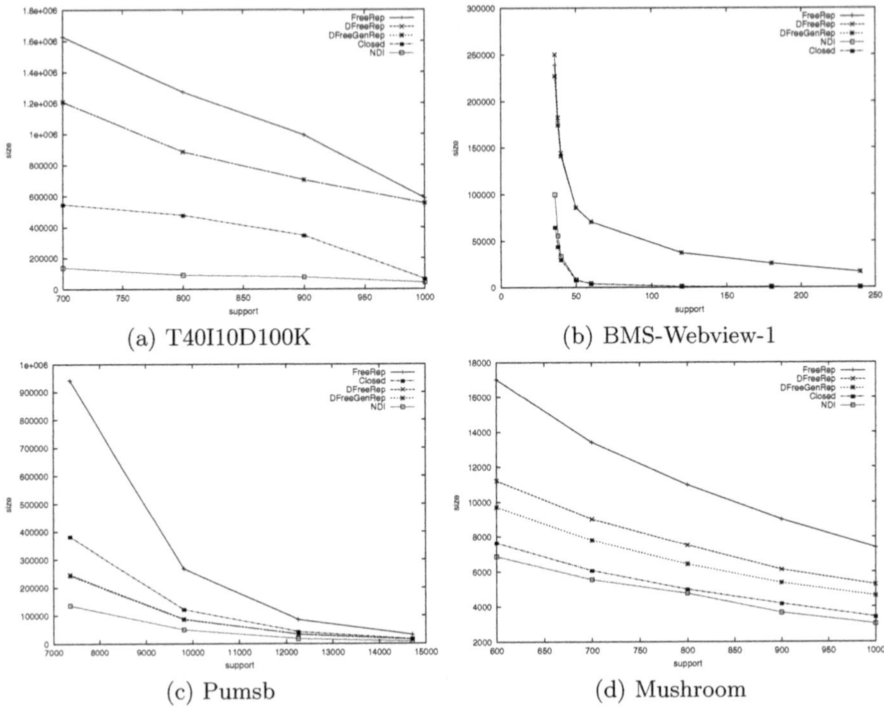

(a) T40I10D100K (b) BMS-Webview-1

(c) Pumsb (d) Mushroom

Fig. 7. Size of different concise representations

Disjunction-free sets [6] *or disjunction-free generators* [14] are an extension of free sets. A set I is called disjunction-free if there does not exist two items A, B in I such that

$$support(I) = support(I - A) + support(I - B) - support(I - AB) \ .$$

Free sets are a special case of disjunction-free sets, namely when $A = B$. The representation is based on the fact that when

$$support(ABC) = support(AB) + support(AC) - support(A) \ ,$$

it is also true that

$$support(ABCD) = support(ABD) + support(ACD) - support(AD) \ .$$

Again, this deduction follows from the rules presented in this paper:

$$\begin{array}{l} s_{ABCD} \geq s_{ABD} + s_{ACD} - s_{AD} \ , \ and \\ s_{ABCD} \leq s_{ABC} + s_{ABD} + s_{ACD} - s_{AB} - s_{AC} - s_{AD} + s_A \ . \end{array}$$

7.2 Deduction

Another application of deduction rules is developed in [11]. Based on the observation that highly frequent items tend to blow up the output of a data mining query by an exponential factor, the authors develop a technique to leave out these highly frequent items, and to reintroduce them after the mining phase by using a deduction rule, the *multiplicative* rule. The multiplicative rule can be stated as follows: let I, J be itemsets, then

$$support(I \cup J, \mathcal{D}) \geq support(I, \mathcal{D}) + support(J, \mathcal{D}) - support(\{\}, \mathcal{D}) \ .$$

This rule can be derived from the rules in our framework.

Also in the field of artificial intelligence, much work has been done around inferring knowledge. Interesting related work in artificial intelligence concentrates on logics for reasoning about probabilities, such as the probabilistic logic of *Nilsson* [17] and of *Fagin et al.* [10].

8 Conclusions and Further Work

We presented sound and complete rules for deducing bounds on the support of an itemset. These rules have many possible applications, such as improving the pruning in the Apriori-algorithm, making concise representations, and deducing the result of a data mining query based on previous query results. We evaluated the rules against different real-life data set. The experiments showed the usefulness of the deduction rules for mining concise representations of the frequent itemsets.

For the deduction rules presented in this paper, we need to know the supports of all subsets exactly. Interesting further work includes finding deduction rules for situations in which some of the subsets are missing, and when we only have partial knowledge of the supports.

Acknowledgement

We thank *Bart Goethals* of the *Helsinki Institute for Information Technology* for the implementation of many of the algorithms used in the experiments section and for his help with the experiments. The number of closed sets is produced by the ChARM-algorithm of *Mohammed Zaki* [20].

References

1. R. Agrawal et al. Mining association rules between sets of items in large databases. In *Proc. ACM SIGMOD*, pages 207–216, 1993.
2. R. Agrawal and R. Srikant. Fast algorithms for mining association rules. In *Proc. VLDB*, pages 487–499, 1994.
3. C.L. Blake and C.J. Merz. *UCI Repository of machine learning databases [http://www.ics.uci.edu/mlearn/MLRepository.html]*. Irvine, CA: University of California, Dept. of Inf. and CS., 1998.
4. J.-F. Boulicaut and A. Bykowski. Frequent closures as a concise representation for binary data mining. In *Proc. PaKDD*, pages 62–73, 2000.
5. J.-F. Boulicautet al. Approximation of frequency queries by means of free-sets. In *Proc. PKDD*, pages 75–85, 2000.
6. A. Bykowski and C. Rigotti. A condensed representation to find frequent patterns. In *Proc. PODS*, 2001.

7. A. Bykowski et al. Model-independent bounding of the supports of boolean formulae in binary data. In *Proc. ECML-PKDD Workshop KDID*, pages 20–31, 2002.
8. T. Calders and B. Goethals. Mining all non-derivable frequent itemsets. In *Proc. PKDD*, pages 74–85. Springer, 2002.
9. T. Calders and J. Paredaens. Axiomatization of frequent sets. In *Proc. ICDT*, pages 204–218, 2001.
10. R. Fagin et al. A logic for reasoning about probabilities. *Information and Computation*, 87(1,2):78–128, 1990.
11. D. Groth and E. Robertson. Discovering frequent itemsets in the presence of highly frequent items. In *Proc. Workshop RBDM, in Conjunction with 14th Intl. Conf. On Applications of Prolog*, 2001.
12. J. Han, J. Pei, and Y. Yin. Mining frequent patterns without candidate generation. In *Proc. ACM SIGMOD*, pages 1–12, 2000.
13. S. Hettich and S. D. Bay. *The UCI KDD Archive. [http://kdd.ics.uci.edu]*. Irvine, CA: University of California, Dept. of Inf. and CS., 1999.
14. M. Kryszkiewicz. Concise representation of frequent patterns based on disjunction-free generators. In *Proc. ICDM*, pages 305–312, 2001.
15. H. Mannila and H. Toivonen. Multiple uses of frequent sets and condensed representations. In *Proc. KDD*, 1996.
16. H. Mannila and H. Toivonen. Levelwise search and borders of theories in knowledge discovery. *DMKD*, 1(3):241–258, 1997.
17. N. Nilsson. Probabilistic logic. *Artificial Intelligence*, 28:71–87, 1986.
18. N. Pasquier et al. Discovering frequent closed itemsets for association rules. In *Proc. ICDT*, pages 398–416, 1999.
19. J. Pei et al. Closet: An efficient algorithm for mining frequent closed itemsets. In *ACM SIGMOD Workshop DMKD*, Dallas, TX, 2000.
20. M.J. Zaki and C. Hsiao. ChARM: An efficient algorithm for closed association rule mining. In *Proc. ICDM*, 2002.

Appendix A : Proofs

Proof of Lemma 4

Proof. (1)⇒(2): Proof by induction on $|I-J|$ that $x_J = \sum_{J\subset I'\subseteq I}(-1)^{|I-J|}s_{I'}$.
Base case ($J=I$): for $J=I$, (2) reduces to $x_I = s_I$. $x_I = s_I$ is also in (1).
General case (J **arbitrary**): By induction hypothesis, for all I', such that $J \subset I' \subseteq I$, $x_{I'} = \sum_{I'\subseteq J'\subseteq I}(-1)^{|J'-I'|}s_{J'}$. From (1) we know that $\sum_{J\subseteq I'\subseteq I} x_{I'} = s_J$.

Hence,

$$
\begin{aligned}
x_J = s_J - \sum_{J\subset I'\subseteq I} x_{I'} &= s_J - \sum_{J\subset I'\subseteq I}\ \sum_{I'\subseteq J'\subseteq I}(-1)^{|J'-I'|}s_{J'} \\
&= s_J - \sum_{J\subset J'\subseteq I}\left(\sum_{J\subset I'\subseteq J'}(-1)^{|J'-I'|}\right)s_{J'} \\
&= s_J - \sum_{J\subset J'\subseteq I}\left(\sum_{J\subset I'\subseteq J'}(-1)^{|I'-J|}\right)(-1)^{|J'-J|}s_{J'}
\end{aligned}
$$

$$= s_J - \sum_{J \subset J' \subseteq I} \left(\sum_{i=1}^{|J'-J|} \binom{|J'-J|}{i} (-1)^i \right) (-1)^{|J'-J|} s_{J'}$$

$$= s_J - \sum_{J \subset J' \subseteq I} \left(-1 + \sum_{i=0}^{|J'-J|} \binom{|J'-J|}{i} (-1)^i \right) (-1)^{|J'-J|} s_{J'}$$

$$= s_J + \sum_{J \subset J' \subseteq I} (-1)^{|J'-J|} s_{J'} \;=\; \sum_{J \subseteq J' \subseteq I} (-1)^{|J'-J|} s_{J'}$$

(2)⇒(1): Proof by induction on $I - J$ that $s_J = \sum_{J \subseteq I' \subseteq I} x_{I'}$.
Base case ($J = I$): for $J = I$, (1) reduces to $s_I = x_I$. $x_I = s_I$ is also in (2).
General case (J **arbitrary**): By induction hypothesis, for all J', such that $J \subset J' \subseteq I$, $s_{J'} = \sum_{J' \subseteq I' \subseteq I} x_{I'}$. Since $x_J = \sum_{J \subseteq J' \subseteq I} (-1)^{|J'-J|} s_{J'}$, also $s_J = x_J - \sum_{J \subset J' \subseteq I} (-1)^{|J'-J|} s_{J'}$. Hence,

$$s_J = x_J - \sum_{J \subset J' \subseteq I} (-1)^{|J'-J|} \sum_{J' \subseteq I' \subseteq I} x_{I'} = x_J - \sum_{J \subset I' \subseteq I} \left(\sum_{J \subset J' \subseteq I'} (-1)^{|J'-J|} \right) x_{I'}$$

$$= x_J - \sum_{J \subset I' \subseteq I} (-1)\, x_{I'} \;=\; \sum_{J \subseteq I' \subseteq I} x_{I'}$$

Proof of Theorem 1

Proof. By definition, the integers l, u such that $[l, u]$ is the tight interval implied for *support*(I) by $\{support(J) = s_J \mid J \subset I\}$, are the minimal and maximal integer s_I such that the system

$$\mathcal{S} \;=\; \{support(J) = s_J \mid J \subset I\} \cup \{support(I) = s_I\}$$

is satisfiable. Using Corollary 1 and Lemma 4, we obtain that $\mathcal{S}$ has a solution if and only if

$$\begin{aligned} x_J &\geq 0 && \forall J \subseteq I \\ x_J &= \sum_{J \subseteq J' \subseteq I} (-1)^{|J'-J|} s_{J'} && \forall J \subseteq I \end{aligned}$$

has a solution. This system has a solution if and only if

$$\begin{aligned} s_I &\leq \sum_{J \subseteq J' \subset I} (-1)^{|I-J'|+1} s_{J'} && \forall J \subseteq I, |I-J| \text{ odd} \\ s_I &\geq \sum_{J \subseteq J' \subset I} (-1)^{|I-J'|+1} s_{J'} && \forall J \subseteq I, |I-J| \text{ even} \end{aligned}$$

Hence, the maximal solution is the minimum of the upper bounds as given by $\mathcal{R}_I(J)$, J odd, and the minimal solution is the maximum of the lower bounds as given by $\mathcal{R}_I(J)$, J even. □

Model-Independent Bounding of the Supports of Boolean Formulae in Binary Data

Artur Bykowski[1], Jouni K. Seppänen[2], and Jaakko Hollmén[2]

[1] LISI, INSA-Lyon, Bât. Blaise Pascal, 20, ave A. Einstein, F-69621 Villeurbanne Cedex, France
Artur.Bykowski@insa-lyon.fr

[2] Helsinki University of Technology, Laboratory of Computer and Information Science, P.O. Box 5400, 02015 HUT, Finland
{Jouni.Seppanen,Jaakko.Hollmen}@hut.fi

Abstract. Data mining algorithms such as the Apriori method for finding frequent sets in sparse binary data can be used for efficient computation of a large number of summaries from huge data sets. The collection of frequent sets gives a collection of marginal frequencies about the underlying data set. Sometimes, we would like to use a collection of such marginal frequencies instead of the entire data set (e.g. when the original data is inaccessible for confidentiality reasons) to compute other interesting summaries. Using combinatorial arguments, we may obtain tight upper and lower bounds on the values of inferred summaries. In this paper, we consider a class of summaries wider than frequent sets, namely that of frequencies of arbitrary Boolean formulae. Given frequencies of a number of any different Boolean formulae, we consider the problem of finding tight bounds on the frequency of another arbitrary formula. We give a general formulation of the problem of bounding formula frequencies given some background information, and show how the bounds can be obtained by solving a linear programming problem. We illustrate the accuracy of the bounds by giving empirical results on real data sets.

1 Introduction

Database management systems allow querying extensional data and intentional data. From the user's point of view, extensional data are the data explicitly input by the user into the system. Intentional data are not put in by the user—that information is derived from extensional data.

In *inductive* database management systems the intentional data are usually quite complex from various points of view and require a high computational effort to obtain. In case of typical patterns (frequent sets, association rules), the common problem is that the domain of patterns is prohibitively large and the inductive database management system cannot compute them all. The typical approach is to let the user guide the system to the interesting patterns interactively, e.g., through queries, limiting the search space to be considered.

Then, the question is if the result of one query could be reused at least partly for obtaining the next result as an alternative to re-computing the whole answer

R. Meo et al. (Eds.): Database Support for Data Mining Applications, LNAI 2682, pp. 234–249, 2004.

from extensional data. The following, more difficult question is whether there might be some summaries that the inductive database management system can gather off-line before the data mining process, to improve the on-line behavior of most mining processes. Typically, one would be interested in sufficient statistics, i.e., summaries that can be substituted for the whole extensional data to avoid repetitive probing of the selection predicate for all candidate patterns. On the other hand, we prefer gathering summaries that are known to be efficient to obtain.

In this paper we pinpoint how to take advantage in a particular context of a collection of summary queries that have been evaluated against the extensional data to bound the value of the evaluation functions of other queries. Providing bounds may be interesting when we have thresholds on the evaluation function, and a tight bound can enable us to make a correct decision about accepting or rejecting a pattern in the query answer. We focus on the context of reusing previous queries (without pre-selecting) and leave open the question of choosing beforehand which summaries should be used.

Several data mining algorithms can be used for efficient computation of a large number of summaries from data. Such methods include Apriori-type algorithms for finding frequent sets [AMS$^+$96] or episodes [MTV97] in binary or sequential data and methods for clustering large data sets [ZRL97]. The summary information given by such algorithms can then be used as an efficient condensed representation of the data set. When the available summaries are orders of magnitude smaller than the data set itself (typical in case of huge data sets in a data mining context), it could be worth using them instead of the entire data set to compute other interesting summaries. Typically, the information contained in a collection of summaries will not be sufficient to compute the precise value of all other summaries, but at least bounds could be inferred. If the accuracy of the estimated result is not enough, the partial quantitative information (bounds) can be used to better optimize the query execution plan (of a query to the original data set).

An interesting fundamental question is: how much information about the underlying data set does a collection of summaries give? In this paper we consider this question in the setting of frequent sets for binary data. Information of frequencies of different itemsets can have strong implications for the frequencies of other itemsets. For example, if we know that[1] $f(AB) = f(A)$, then we know that $f(XA) = f(XAB)$ for any set X, a result that has been shown to be surprisingly useful in the context of so-called closures [PBTL99] and free sets [BBR00,BR01]. Also, we know that the frequency $f(X)$ of an itemset X is bounded from above by the minimum support $f(Y)$ of a subset Y of X.

More generally, if we possess the information about the frequencies of some Boolean formulae (frequent itemsets being a particular case), the frequency of any other Boolean formula can be inferred, to some extent. The main question we pose in this paper is how we could efficiently construct upper and lower bounds

[1] Here we denote by $f(AB)$ the frequency of the itemset $\{A, B\}$.

for the frequencies of any Boolean formula, given the existing information. We show that this formula bounding problem can in fact be solved by transforming the question into a linear program and solving that problem. In the worst case the transformation leads to a program of exponential size, but we also give empirical results showing that the transformation in many cases is an efficient one.

The paper is organized as follows. In Section 2 we define the basic notions and the support bounding task. Section 3 gives the solution, and Section 4 describes the empirical results. Finally, in Section 5, we summarize the paper and discuss open problems.

2 Problem: Support Queries in Databases

The problem we want to solve involves Boolean queries on binary relational databases. In order to present the problem in an exact way, we first make some formal definitions.

Definition 1. *A* relation r *over the finite attribute set* X *is a finite multiset of* tuples, *subsets of* X. *The* degree *of* r *is the cardinality of* X, *and the* size *of* r *is the (multiset) cardinality* $|r|$ *of* r. *The set* X *is called the* schema *of* r.

In contrast to ordinary relational databases, we deal with binary data only. This allows a convenient notational shortcut: for example, instead of the tuple $(0, 1, 1, 0, 1)$ over the attributes A, B, C, D, E, we can talk about the tuple $\{B, C, E\}$. Also, a Boolean query such as "($A = 1$ and $B = 1$) or ($C = 1$ and $D = 0$)" can be written as "(A and B) or (C and not D)", or, in the algebraic notation, $AB + C\overline{D}$. The syntax and semantics of such queries are defined next.

Definition 2. *A* Boolean formula *over the attribute set* X *is one of:*

1. $\top$ *(the* true constant*),*
2. A *for some attribute* $A \in X$ *(an* atom*),*
3. $(\neg\phi)$ *for some Boolean formula* ϕ *over* X *(a* negation*),*
4. $(\phi\psi)$ *for some Boolean formulae* ϕ, ψ *over* X *(a* conjunction*),*
5. $(\phi + \psi)$ *for some Boolean formulae* ϕ, ψ *over* X *(a* disjunction*).*

We omit parentheses when there is no danger of ambiguity. Furthermore, the negation operator $\neg$ always binds to the shortest following subformula, and conjunction binds tighter than disjunction. In the case of negated atoms, we also write $\overline{A}$ for $\neg A$. Thus, $\overline{A}BC + A(B + \overline{C})$ means $((((\neg A)B)C) + (A(B + (\neg C))))$. These conventions leave it ambiguous in which direction conjunction and disjunction associate, but in fact all readings of an ambiguous formula have equivalent semantics by the following definition.

Definition 3. *Given a tuple* $t \in r$, *we define the truth value of all Boolean formulae over the schema of* r *as follows.*

1. $[\top]_t = 1$,
2. $[A]_t = 1$ *if* $A \in t$, $[A]_t = 0$ *if* $A \notin t$,
3. $[\neg\phi]_t = 1 - [\phi]_t$,
4. $[(\phi\psi)]_t = [\phi]_t\,[\psi]_t$,
5. $[(\phi + \psi)]_t = [\phi]_t + [\psi]_t - [\phi]_t\,[\psi]_t$.

Two formulae ϕ and ψ are *equivalent*, if they always have the same truth value on the same tuple. That all readings of an ambiguous formula such as $\phi\psi\theta$ are equivalent is a standard result in propositional logic. Different in our problem is that we extend the semantics to whole relations.

Definition 4. *Let r be a relation and ϕ a Boolean formula over a common schema. Then the* support *of ϕ in r is the proportion of tuples in r for which ϕ is true,*

$$[\phi]_r = |r|^{-1} \sum_{t \in r} [\phi]_t.$$

We write $[\phi]$ when the relation is clear from the context.

The data mining literature contains a wealth of material on *itemsets*, sets of attributes. After Boolean formulae have been defined, it is easy to give semantics to itemsets as simple conjunctions.

Definition 5. *Let X be a relation schema. A subset of X is an* itemset, *and it is identified with the conjunction of all its elements.*

The name "itemset" originated in association rule mining, whose traditional application is market-basket data: the attributes are items offered for sale at a supermarket, and the tuples are customer transactions. It turns out that to find association rules that are in a certain sense interesting, it suffices to compute all itemsets whose support exceeds a threshold. This is usually done by a breadth-first search algorithm called Apriori [AMS+96], but several variations have been proposed. For example, depth-first search can be performed using FP-trees [HPY00], and a sampling approach can avoid database scans [Toi96]. An active area of research is mining not all frequent itemsets but only an interesting subfamily; see e.g. [GZ01,CG02,PBTL99,BBR00,BR01].

As an example, Table 1 shows a small binary database. We have e.g. $[A]_t = [A]_u = [AB]_t = [AC]_u = 1$ and $[A]_v = [B]_u = [AB]_u = [AC]_v = 0$, and over the whole database $[A]_r = 2/3$, $[B]_r = 1/3$, $[C]_r = 1$, $[AC]_r = 2/3$, and $[ABC]_r = 1/3$. If the frequency threshold is $1/2$, the frequent itemsets are A, C, AC, and trivially the empty set, which corresponds to $\top$.

Table 1. An example database r

Tuple	A	B	C
t	1	1	1
u	1	0	1
v	0	0	1

We now come to the formula bounding problem. Given are a set Φ of Boolean formulae over a schema X, and their supports in an unknown relation r. The desired result is the support of another formula ψ. This support is sometimes completely determined by the givens, but this is rare; in general we want the set of all possible supports. As it turns out, the minimum and maximum support determine this set completely, and we can allow minima and maxima also as inputs. We denote by $\mathrm{Int}_{\mathbb{Q}}(0,1)$ the set of closed intervals $[a,b]$ of rational numbers where $0 \leq a \leq b \leq 1$.

Definition 6. *The* formula bounding task $\text{Bound}(X, \Phi, f, \psi)$ *has the following inputs: a relation schema X, a set Φ of Boolean formulae over X, a function f from Φ to $\mathrm{Int}_{\mathbb{Q}}(0,1)$, and a Boolean formula ψ over X. The solution of the task is the smallest set $I \subseteq [0,1]$ such that $[\psi]_r \in I$ for all relations r over X fulfilling the constraint $[\phi]_r \in f(\phi)$ for all $\phi \in \Phi$.*

In other words, we want a sound and complete inference procedure for the support bounds of Boolean formulae. We call a procedure *sound* if its result I rules out no possible solutions: for $q \notin I$, there should be no relation r fulfilling the constraints defined by f such that $[\psi]_r = q$. Conversely, the set I returned by a *complete* procedure is such that every solution $q \in I$ is realizable in some relation fulfilling the constraints. (A trivially complete but non-sound procedure returns $I = \emptyset$ for all inputs; the similar sound but non-complete procedure always returns $I = [0,1]$.) The problem is NP-hard, since it requires solving the satisfiability of ψ.

The following lemma shows that it suffices to find upper and lower bounds for the numbers in I. Thus, the task has a closure property: the output is in the same form as each element of the input.

Lemma 1. *If the solution I of* $\text{Bound}(X, \Phi, f, \psi)$ *is nonempty, then I is an interval of rational numbers.*

Proof. We must prove that given any three rationals $P, W, Q \in [0,1]$ with $P < W < Q$ and $P, Q \in I$, also $W \in I$. Since W lies between P and Q, there is a rational number $Z \in (0,1)$ such that $W = ZP + (1-Z)Q$. Since $P, Q \in I$, there exist relations p, q fulfilling the constraints of the bounding problem such that $[\psi]_p = P$ and $[\psi]_q = Q$. We will construct a relation w for which the support of all formulae θ over X is $[\theta]_w = Z\,[\theta]_p + (1-Z)\,[\theta]_q$. Since $[\theta]_w$ lies between the numbers $[\theta]_p$ and $[\theta]_q$, every inequality constraint $[\phi]_w \in f(\phi)$ will be satisfied. Further, $[\psi]_w = W$, as required.

To construct the relation w, we would like to take $Z/|p|$ copies of all tuples in p and $(1-Z)/|q|$ copies of all tuples in q. This is impossible in the general case, but if we multiply the numbers $Z/|p|$ and $(1-Z)/|q|$ by the least common multiple of their denominators, we can replace the numbers by integer multiples. It is then easy to check that $[\theta]_w = Z\,[\theta]_p + (1-Z)\,[\theta]_q$ for all formulae θ.

3 Solving the Bounding Task by Linear Programming

In this section, we describe a solution to the BOUND task of Definition 6. The solution is based on linear programming, and it is both sound and complete.

3.1 Change of Variables

Several kinds of equalities and inequalities hold in all relations. For example, $[A+B] = [A] + [B] - [AB]$ by the combinatorial inclusion-exclusion principle, and $0 \leq [AB] \leq [A] \leq 1$ by the anti-monotonicity of support. A procedure for the BOUND task has to incorporate all results of this type.

Let us analyze how these results could be proved. The middle inequality follows from the observation that $[A] = [AB] + [A\overline{B}]$ and that the support of $[A\overline{B}]$ lies in the interval $[0, 1]$. A similar idea gives a proof of the inclusion-exclusion formula:

$$\begin{aligned}[A+B] &= [AB] + [A\overline{B}] + [\overline{A}B] \\ &= \big([AB] + [A\overline{B}]\big) + \big([AB] + [\overline{A}B]\big) - [AB] = [A] + [B] - [AB].\end{aligned}$$

This suggests that a change of variables can make the needed results simpler to prove. To that end, we make the following definitions.

Definition 7. *Given an attribute $A \in X$, the* positive literal *based on A is the Boolean formula A, and the* negative literal *based on A is the Boolean formula $\overline{A}$. A* literal *based on A is either the positive literal or the negative literal based on A.*

Definition 8. *A* clause *over the attribute set X is a conjunction of zero or more literals based on different attributes.*

Our definition of a clause allows an attribute to appear at most once, in either a negative or a positive literal. For example, $B\overline{C}\overline{E}$ is a clause over the set $\{A, B, C, D, E\}$, but $B\overline{B}\overline{E}$ and $BB\overline{E}$ are not. The true constant $\top$ is a clause as the degenerate case of zero literals.

Definition 9. *A* full clause *over the attribute set X is a clause with exactly $|X|$ literals.*

In a full clause each attribute appears exactly once, either as a negative or a positive literal. For example, the conjunction $AB\overline{C}D\overline{E}$ is a full clause over the set $\{A, B, C, D, E\}$, whereas $B\overline{C}\overline{E}$ is not. In the language of propositional logic, a full clause fully describes a model over the given attribute set.

Full clauses are important for two reasons. First, there is a natural correspondence between relations and assignments of supports to full clauses. Given a relation r, any full clause θ over the schema of r is satisfied by some nonnegative integral number of identical tuples in r. Conversely, given an assignment of nonnegative rational supports for all full clauses summing up to 1, it is simple to construct a relation giving rise to these supports.

The second reason is that all formulae can be decomposed into full clauses (for formulae corresponding to typical queries it is easy). We record this in the following two propositions.

Proposition 1. *Every Boolean formula over an attribute set X can be equivalently written as a disjunction $C_1 + C_2 + \cdots + C_p$ of distinct full clauses C_i. We call this the* full disjunctive normal form.

Proposition 2. *The support of any Boolean formula ϕ over an attribute set X can be written as a sum of supports of distinct full clauses. That is, there is a set of full clauses $C_1, C_2, \ldots, C_p$ such that $[\phi]_r = [C_1]_r + [C_2]_r + \cdots + [C_p]_r$ for any relation r over X.*

These results enable us to untangle the complex interrelations of formulae. The supports of distinct full clauses are independent of each other[2], so any distribution of nonnegative supports for full clauses corresponds to a possible relation. Where the support of a Boolean formula appears in a constraint equality or inequality, we can invoke Proposition 2 to replace it by a sum of the supports of the corresponding full clauses. This amounts to a linear change of variables.

As an example, we consider an instance of $\text{BOUND}(X, \Phi, f, \psi)$ with $X = \{A, B\}$, $\Phi = \{\top, A, B, AB\}$, and $\psi = AB$. After the change of variables, we have the system depicted in Table 2 which we should solve for $[AB]$. We have the additional information that $0 \leq [\theta_i] \leq 1$ for all formulae θ_i, but we need not worry about the inclusion-exclusion principle or similar rules. We continue this example at the end of Section 3.2.

Table 2. Example bounding task with decomposition into full clauses

	AB	$A\overline{B}$	$\overline{A}B$	$\overline{AB}$
$[\top] = 1.0$	×	×	×	×
$[A] = 0.6$	×	×		
$[B] = 0.7$	×		×	
$[AB] \in [0, 0.5]$	×			

3.2 Linear Programming

We now turn to the classic optimization problem called linear programming. We only describe it briefly; see, e.g., Chapter 21 in [Kre93] for a good introduction to the subject, or the Linear Programming FAQ[3] for a comprehensive list of references. For the computational complexity of linear programming, see e.g. [MSW96]; briefly, common algorithms such as Simplex tend to be useful in practice although they have worst-case exponential complexity, but more sophisticated algorithms such as Karmarkar's algorithm [Kar84] achieve lower complexity.

---

[2] With the restriction that the supports of all full clauses sum up to 1; but this gives only a scaling factor.

[3] `http://www-unix.mcs.anl.gov/otc/Guide/faq/linear-programming-faq.html`.

Definition 10. *The linear programming problem* $\mathrm{LP}(\mathcal{A}, \mathcal{B}, \mathcal{C})$ *comprises an* $m \times n$ *matrix* $\mathcal{A}$*, an* m*-element column vector (*$m \times 1$ *matrix)* $\mathcal{B}$*, and an* n*-element row vector (*$1 \times n$ *matrix)* $\mathcal{C}$*. The solution of the problem is the vector* $\boldsymbol{x}$ *that minimizes the scalar value* $\mathcal{C}\boldsymbol{x}$ *subject to the restrictions* $\mathcal{A}\boldsymbol{x} \leq \mathcal{B}$, $\boldsymbol{x} \geq \mathbf{0}$. *We also denote by* $\mathrm{LP}'(\mathcal{A}, \mathcal{B}, \mathcal{C})$ *the problem that is otherwise similar but where the first restriction is replaced by* $\mathcal{A}\boldsymbol{x} = \mathcal{B}$.

The matrix $\mathcal{C}$ *is said to express the* objective function, *and* $\mathcal{A}$ *and* $\mathcal{B}$ *state the* constraints *of the problem.*

The problems $\mathrm{LP}(\mathcal{A}, \mathcal{B}, \mathcal{C})$ and $\mathrm{LP}'(\mathcal{A}, \mathcal{B}, \mathcal{C})$ are equivalent in expressive power and computational complexity. We use the first formulation in the fully general case of the formula bounding task BOUND (Definition 6). For the kinds of inputs we get from Apriori and similar procedures, we actually have equalities for all input formulae, so we can use $\mathrm{LP}'(\mathcal{A}, \mathcal{B}, \mathcal{C})$. Note that equalities $y = z$ can always be converted to the inequalities $y \leq z$ and $y \geq z$. We map the problem BOUND into an instance of a linear programming problem $\mathrm{LP}(\mathcal{A}, \mathcal{B}, \mathcal{C})$ or $\mathrm{LP}'(\mathcal{A}, \mathcal{B}, \mathcal{C})$ (depending on the kind of input). We talk about LP and inequalities in the following, but the case of LP$'$ and equalities is similar.

Assume now that I is the solution of an instance of $\mathrm{BOUND}(X, \Phi, f, \psi)$. By Lemma 1, we know that if the set I is nonempty, it is a subinterval of $[0, 1]$ (in rationals). Therefore, we proceed to compute its infimum; the case of the supremum is symmetric. Denote $n = |X|$, and denote the 2^n full clauses over X by $\theta_1, \theta_2, \ldots, \theta_{2^n}$.

As input to BOUND we have in effect a large set of inequalities that we will convert into one big matrix inequality $\mathcal{A}\boldsymbol{x} \leq \mathcal{B}$. The vector $\boldsymbol{x}$ will contain the unknowns: let $\boldsymbol{x} = ([\theta_1]\,[\theta_2]\,\ldots\,[\theta_{2^n}])^{\mathrm{T}}$. Then, Proposition 2 yields for every formula $\phi \in \Phi$ a binary vector $\boldsymbol{k} = (k_1\,k_2\,\ldots\,k_{2^n})$ such that the support of ϕ can be written as a matrix product, $[\phi] = \boldsymbol{k}\boldsymbol{x}$. Using this fact, we encode the constraint $[\phi] \in f(\phi)$ by adding into $\mathcal{A}$ two rows, $-\boldsymbol{k}$ and $\boldsymbol{k}$, and into $\mathcal{C}$ two numbers, $-a$ and b, where $[a, b] = f(\phi)$. Then any $\boldsymbol{x}$ satisfying $\mathcal{A}\boldsymbol{x} \leq \mathcal{B}$ must satisfy $a \leq \boldsymbol{k}\boldsymbol{x} \leq b$. Finally, as a necessary consistency constraint corresponding to the fact $[\top] = 1$, we add the rows $(-1\,-1\,\ldots\,-1)$ and $(1\,1\,\ldots\,1)$, and the numbers -1 and 1. All in all, the dimensions of $\mathcal{A}$ will be $2(|\Phi|+1) \times 2^n$, and the dimensions of $\boldsymbol{x}$ and $\mathcal{B}$ will be $2(|\Phi|+1) \times 1$. Ways to reduce these dimensions will be discussed after Theorem 1.

Having encoded all the constraints of the problem in $\mathcal{A}$ and $\mathcal{B}$, we now have to select $\mathcal{C}$ so that the solutions to the LP problem correspond to the supports of ψ. We once again invoke Proposition 2 to turn $[\psi]$ into a sum of supports of full clauses. Thus $\mathcal{C}$ will be a 0/1 vector with $\mathcal{C}\boldsymbol{x} = [\psi]$, and minimizing $\mathcal{C}\boldsymbol{x}$ subject to the constraints gives the required infimum. When the bounds for the supports of input formulae are rational numbers, linear programming yields a rational value for the infimum, since for example the Simplex algorithm [Kre93, §21.3] uses only sums, differences, products and ratios to solve LP. Thus, the infimum corresponds to an assignment of nonnegative rational values to the supports of the full clauses, summing to 1 and obeying all the constraints of the original problem. Multiplying all the supports by the least common multiple of their denominators gives integer counts, whence a relation can be constructed. Thus the infimum is actually a minimum.

We have now proved the following theorem.

Theorem 1. *The formula bounding task* $\textsc{Bound}(X, \Phi, f, \psi)$ *can be reduced to the linear programming task* $\mathrm{LP}(\mathcal{A}, \mathcal{B}, \mathcal{C})$. *The matrix* $\mathcal{A}$ *will have* $O(|\Phi|)$ *rows and* 2^n *columns, and the vectors* $\mathcal{B}$ *and* $\mathcal{C}$ *will respectively have* $O(|\Phi|)$ *and* 2^n *elements, where* $n = |X|$.

The output from our reduction has size $O(2^n|\Phi|)$, i.e., exponential in the number of attributes, where for the sake of simplicity we assume that all the numbers are represented using a fixed number of bits. Thus, a linear programming algorithm that requires polynomial time in the size of its input will take time that is polynomial in Φ but exponential in n. It would, therefore, be useful to diminish the exponential dependency on the number n of attributes.

First, if Φ consists of frequent itemsets, we can restrict X to only those attributes that appear in the query ψ. To see this, consider two full clauses θ and θ' whose only difference is that θ has A and θ' has $\overline{A}$, where A is an attribute that does not appear in ψ. The two coordinates in $\mathcal{C}$ corresponding to θ and θ' will be equal, and thus only the sum $[\theta] + [\theta']$ will be relevant to the objective function $\mathcal{C}\boldsymbol{x}$. If a frequent set $\phi \in \Phi$ has different coordinates at the positions corresponding to the two full clauses, it must include A; then there is a frequent set $\phi' \in \Phi$ that differs from ϕ only by excluding A. Thus in removing ϕ from Φ we lose no information relevant to $[\psi]$. Once all such frequent sets are gone, we can remove the attribute A from X.

Second, we discuss whether using the family of all 2^n full clauses is necessary. One of the reasons we used full clauses was that they can be used to answer any support queries of Boolean formulae. However, many other families of formulae have this property. For example, Proposition 1 of [MT96] implies that the family of all conjunctions of atoms can be used to determine the supports of all Boolean formulae. Let us define a *representation* Θ over X as a family of formulae such that the counts of all Boolean formulae over X can be determined from the counts of the formulae in Θ. In this context, we use integer counts $\mathrm{count}_r(\theta) = \sum_{t \in r} [\theta]_t$ instead of supports $[\theta]_r = \mathrm{count}_r(\theta)/\,\mathrm{count}_r(\top)$.

Any representation that works *for all* r must have 2^n formulae. Indeed, given the counts corresponding to a representation, we can use Proposition 2 to form a linear system of equations from which the counts of full clauses can be solved. If there are fewer than 2^n equations, the system is underdetermined, and since all its factors are integers, it will have infinitely many integral solutions. It is therefore relatively easy to construct two relations with the same counts of all formulae of the supposed representation but different counts of some full clauses.

However, this does not rule out smaller representations that work for specific relations. When storing the counts of the conjunctions-of-atoms representation, we can leave out some counts that can be derived from others. If, e.g., there are no tuples satisfying the conjunction AB, we can leave out the count of ABC, and if the counts of D and DE are equal, we need store only one of the counts of AD and ADE. Similar ideas have been studied in [ML98,BBR00,BR01].

In our problem, we use fractional supports, not counts, which removes one degree of freedom. Since the supports of full clauses must add up to 1, we can leave out one number from the full-clauses representation.

Third, in the case of LP′, where $\mathcal{A}$ is a 0/1 matrix, we can often reduce the problem. If some row $\boldsymbol{a}_i$ of the matrix $\mathcal{A}$ is less than or equal to another row $\boldsymbol{a}_j$, we can replace $\boldsymbol{a}_j$ by $\boldsymbol{a}_j - \boldsymbol{a}_i$, while doing the corresponding replacement in $\mathcal{B}$. Sometimes this will result in a zero in $\mathcal{B}$; we can then deduce that several unknowns are zero and remove them. Even if this doesn't occur, the matrix becomes sparser, which helps some algorithms that solve linear programming problems.

We now continue the example bounding task of Table 2. We reduce the system depicted in the table to LP$(\mathcal{A}, \mathcal{B}, \mathcal{C})$ with $\boldsymbol{x} = ([AB]\,[A\overline{B}]\,[\overline{A}B]\,[\overline{A}\overline{B}])^{\mathrm{T}}$. For example, the second equation is translated from $[AB]+[A\overline{B}] = 0.6$ to $(1\;1\;0\;0)\boldsymbol{x} \leq 0.6$ and $(-1\;-1\;0\;0)\boldsymbol{x} \leq -0.6$. These inequalities form the third and fourth lines of $\mathcal{A}$ and $\mathcal{B}$ (see below). In this case, the first equation already forms the consistency constraint $\sum[\theta] = 1$, so we need not add it now.

We obtain the values of $\boldsymbol{x}$, $\mathcal{A}$ and $\mathcal{B}$ listed in Table 3, and $\mathcal{C}$=(1 0 0 0) (resp. $\mathcal{C}$=(−1 0 0 0)) for finding the lower (resp. the upper) bound of $[AB]$. Solving these two LP problems gives the minimum 0.3 (with $\boldsymbol{x} = (0.3\;0.3\;0.4\;0.0)^{\mathrm{T}}$) and the maximum 0.5 (with $\boldsymbol{x} = (0.5\;0.1\;0.2\;0.2)^{\mathrm{T}}$). We can obtain actual relations by multiplying the values of $\boldsymbol{x}$ by 10.

Table 3. The example bounding task converted into a linear program

$$\boldsymbol{x} = \begin{pmatrix} [AB] \\ [A\overline{B}] \\ [\overline{A}B] \\ [\overline{A}\overline{B}] \end{pmatrix}, \quad \mathcal{A} = \begin{pmatrix} 1 & 1 & 1 & 1 \\ -1 & -1 & -1 & -1 \\ 1 & 1 & 0 & 0 \\ -1 & -1 & 0 & 0 \\ 1 & 0 & 1 & 0 \\ -1 & 0 & -1 & 0 \\ 1 & 0 & 0 & 0 \\ -1 & 0 & 0 & 0 \end{pmatrix}, \quad \mathcal{B} = \begin{pmatrix} 1 \\ -1 \\ 0.6 \\ -0.6 \\ 0.7 \\ -0.7 \\ 0.5 \\ 0 \end{pmatrix},$$

4 Experiments

We investigated the properties of the bounding procedure on two data sets. The first is *connect-4* containing some game-state descriptions, the second is *anpe*, a database about unemployed people, set up by the French unemployment agency. We describe the specific properties of the data sets along with our results in Sections 4.2 and 4.3.

We used as input to the bounding procedure different collections of frequent itemsets along with their supports [AMS+96,MT96]. As explained previously, an itemset is interpreted as the Boolean conjunction of items that it contains. Different collections of frequent itemsets correspond to different support thresholds, denoted by min_{supp}.

In the implementation of the experiments, we used a less voluminous, although totally equivalent, representation of frequent itemsets, first described in [BR01]. Since this representation is smaller than all frequent itemsets, the resulting Φ contains fewer queries. The equivalence of representations guarantees that the same information can be inferred from it as from all frequent itemsets and their supports. We verified the equivalence by repeating some of the experiments using the ordinary frequent itemsets, and got exactly the same results.

4.1 The Framework of the Experiments

We can compute the support of a Boolean formula over an itemset X exactly, if we know the supports of all subsets of X. The procedure for this computation in [MT96] is also applicable when we know the supports of frequent sets only, but then it will yield approximate bounds—it is sound but not complete. Thus, we test our new contribution using formulae over infrequent itemsets.

The protocol of the experiments can be simply put as following: we compare the average size of intervals inferred by Bound for 100 formulae, for which the combinatorial support-computing procedure of [MT96] is confronted with infrequent (thus missing) terms. The infrequent terms are due to the fact that the support threshold we use to mine frequent itemsets (considered further in the experiments with their corresponding supports as formulae with known supports) exceeds the support of some terms required by the procedure of [MT96].

The detailed protocol is the following. For each of the two data sets, we selected $k = 100$ random itemsets $X_1, ..., X_k$ that have 10 items each and whose supports do not exceed a predefined σ_{max} (10% for *connect-4* and 0.1% for *anpe*). To avoid selecting only itemsets with very low support, which typically account for the clobbering majority of all itemsets, we weighted the probability of selecting an itemset X proportionally to its support $[X]$. Even then, most of the selected itemsets have low support compared to σ_{max} (on average, 2.26% for *connect-4* and 0.010% for *anpe*).

Based on these itemsets, we randomly drew k Boolean formulae $\psi_1, ..., \psi_k$, one formula, ψ_i, over each X_i. To mimic formulae of interest in real life, for each X_i we first selected a subset $Y_i \subseteq X_i$ of items, each item of X_i with probability 0.7. Then we defined ψ_i as a disjunction of random full clauses over Y_i. We included each full clause θ in ψ_i with probability $0.5 - 0.04j$, where j is the number of negative literals in θ. Thus, we preferred clauses with more positive literals. For example, a clause with 10 negative literals had the probability of 0.1 to be included in ψ_i. Then we computed $\text{Bound}(X_i, \Phi_i, f_i, \psi_i)$ where Φ_i consists of the precomputed frequent sets among the subsets of X_i, and f_i assigns to each frequent set its known support. We report two scores, each an average over the 100 computations. Denoting the resulting lower and upper bounds by L_i and U_i for each computation, the first score is the average of $U_i - L_i$, the second the average of $(U_i - L_i)/U_i$, both averages over $i \in \{1, \ldots, 100\}$.

4.2 Experiments with *connect-4*

The *connect-4* data set is very dense. It contains relatively small number of items (129) and rows (67 557).

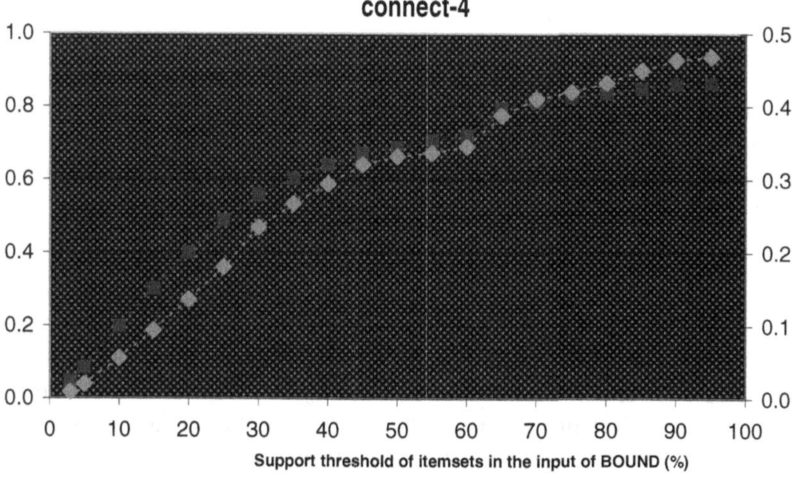

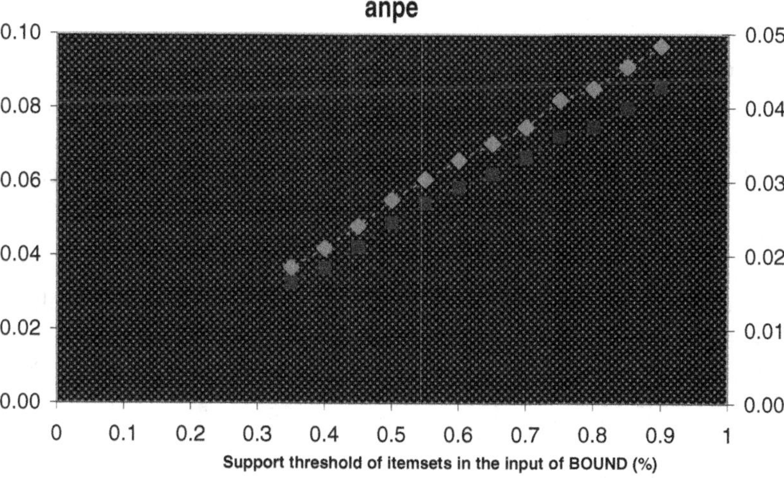

Fig. 1. Average interval size vs. input itemsets' support threshold produced by BOUND on the *connect-4* and *anpe* data sets

In Figure 1 (top) we report the average size of the interval returned by the bounding procedure for different values of min_{supp}. The right-hand scale (diamonds) reports the difference between the ends of the interval, and the left-hand scale (squares) reports the ratio of this difference to the upper limit of the interval. As we can see, a lower min_{supp} results in a better bounding precision. This is due to the increasing number of input itemsets, therefore a richer collection of information about the original data set. However, the cost associated with the computation and the use of these more voluminous collections of summaries also increases.

The support computation of [MT96] potentially involves an exponential number of terms for a single Boolean formula. Typically the computation will involve a significant part of the lattice of itemsets, subsets of the itemset on which we base our random formula. Since our random formulae are based on infrequent itemsets, many terms (often a majority) have unknown supports. However, the interval size we observe is of the same order of magnitude as the support threshold σ, which bounds the error of each unknown support. It seems that the unknown supports tend to cancel out, which appears to be a promising result.

Let us take an example. Consider the support threshold of $\sigma = 15\%$ and that the itemsets on which our random formulae are based have an average support of 2.26%, and never have a support above $\sigma_{max} = 10\%$. Take a single itemset and the corresponding formula; typically, a great number of the itemset's subsets are infrequent, each having support in the $[0, \sigma)$ range. When we compute the support of the formula as in [MT96], naturally most errors will cancel out, but one would not exepct the overall error to be in the $[0, \sigma)$ range; our method yields an average uncertainty of less than 10%.

4.3 Experiments with *anpe*

The *anpe* data set is quite uncorrelated. With its 214 items and over 109 000 rows, it is significantly larger than *connect-4*. Frequent set mining extracts relatively small collections, unless we set a very small min_{supp}. We chose to extract itemsets at these low thresholds. In Figure 1 (bottom) we report the average interval sizes. As previously, we relate the scores to different min_{supp}.

The results look fairly similar to those of the previous experiment. In comparing the graphs it should be noted that the scaling of the axes is different: in this experiment, both the relative and the absolute errors are below 0.1 for all runs. In other words, this less dense data set allowed much greater precision in the resulting intervals.

4.4 Observed Running Times

In our experiments, we first gather summary query answers, to simulate either off-line or on-the-fly collecting of highly processed information. Then, we draw random formulae, as described in Section 4.1. For each random formula, we execute two steps: conversion into an $\mathrm{LP}'(\mathcal{A}, \mathcal{B}, \mathcal{C})$ problem and solving it.

During the experiments, we observed that frequent itemset mining is the most expensive phase, despite the optimization of using an efficient condensed representation of the itemset collection described in [BR01]. For example, for the *connect-4* data set and $min_{supp} = 5\%$ it took more than 3000 seconds. Conversion to $\mathrm{LP}'(\mathcal{A}, \mathcal{B}, \mathcal{C})$ took 4.78 seconds per formula on average, and solving $\mathrm{LP}'(\mathcal{A}, \mathcal{B}, \mathcal{C})$ took only about 3.1 seconds per formula. Thus, the bounding procedure can be quite efficient in practice, after the frequent itemset mining has been performed.

5 Discussion and Future Work

We have considered the problem of bounding the support of Boolean formulae when some aggregate information is available. We showed that the bounding problem can be reduced to a linear programming problem whose size can in the worst case be exponential in the number of attributes. While our result is foremost a theoretical one, we also gave empirical results showing that the bounding method can be effectively used to obtain additional information from frequent itemsets or other summaries.

We emphasize that our aim is to find exact bounds. Another approach would be to approximate the frequency of the query and give some kind of tail bounds for the error of the approximation. The most natural way would be to take a sample from the database and compute all queries on the sample; thus, instead of frequent sets, the sample would serve as the representation of the original data. This kind of a method has been used for computing frequent sets (see [Toi96]). A more sophisticated approximation can be based on frequent sets (or similar summaries) by building a probabilistic model over the variables occurring in the formula. A method using the maximum entropy principle is described in [PMS00]. Like our solution, it suffers from exponential complexity in the number of variables occurring in the query.

Calders and Goethals [Cal02,CG02] have studied a similar problem. They have derived deduction rules for bounding the support of an itemset given the exact supports of all its proper subsets. While the rules are sound and complete for that task, they don't solve our more general problem. In particular, these rules are not applicable when the supports of some subsets are unknown. Thus they cannot derive directly the support of a derivable itemset, but must first bound recursively the supports of all its proper subsets. They deal only with itemsets, i.e., conjunctions of attributes, not arbitrary formulae. The full set of rules is exponentially large, although Calders and Goethals give experimental evidence that a small subset of the rules suffices to give a reasonably good result.

Several open problems remain. One area is obtaining a faster method for the inference problem. With large, redundant summaries such as frequent itemsets, the solution by linear programming is quite slow, and it is in many cases outperformed by the simple "scan the database once and count" method. The method could, however, be useful in cases where the data set is not available or where the set of queries Φ (corresponding to known supports) carries a lot of information condensed in well chosen summaries, orders of magnitude smaller than the data set itself. Thus, the following fundamental issue is interesting.

Problem 1. Given a relation r, an amount Z of storage, and a class of queries Ψ that we wish to perform on r, what should we store in Z (which presumably cannot hold all of r) in order to most effectively answer the queries in Ψ?

Frequent sets are typically redundant collections, and thus are not optimal. In fact, in our experiments we used the smaller collection of disjunction-free sets [BR01], and further gains could be obtained using the Calders–Goethals rules [CG02]. Another interesting representation is the AD-tree [ML98]. In general, if we store in Z the answers to some Boolean queries $\phi_1, \phi_2, \ldots, \phi_N$, the linear programming approach shows the limits of what we can reconstruct. Perhaps a suitable set of formulae would allow an analytical solution, possibly only approximate, of the linear program. The problem of computing frequent sets from data has been extensively studied, and they were used in [MT96], which formed the starting point for our research. But the linear programming framework does not depend on them—it can be used with supports of any formulae.

Another interesting issue is how to relax (if possible) the requirements of Definition 6 if the complete procedure is too slow. We do not want to give unsound answers, but too wide intervals are not necessarily harmful. The simplest incomplete and sound algorithm "return the interval $[0, 1]$" is not useful, but we suspect there might be a reasonably fast compromise between it and the complete linear programming approach.

Problem 2. How close to completeness can a polynomial-time (or linear-time, or randomized polynomial-time) sound solution to BOUND come?

Acknowledgements

Part of the work was done when Artur Bykowski was visiting the Laboratory of Computer and Information Science at Helsinki University of Technology. The *connect-4* data set was provided by researchers at the IBM Almaden research center, and the *anpe* data set was preprocessed and anonymized by Christophe Rigotti at INSA-Lyon. Professor Heikki Mannila presented to us the question about support inference and gave useful comments on an earlier version of this manuscript.

References

[AMS⁺96] Rakesh Agrawal, Heikki Mannila, Ramakrishnan Srikant, Hannu Toivonen, and A. Inkeri Verkamo. Fast discovery of association rules. In Usama M. Fayyad, Gregory Piatetsky-Shapiro, Padhraic Smyth, and Ramasamy Uthurusamy, editors, *Advances in Knowledge Discovery and Data Mining*, chapter 12, pages 307–328. AAAI Press, 1996.

[BBR00] Jean-François Boulicaut, Artur Bykowski, and Christophe Rigotti. Approximation of frequency queries by means of free-sets. In *PKDD'00*, Lecture Notes in Computer Science, Vol. 1910, pages 75–85. Springer, 2000.

[BR01] Artur Bykowski and Christophe Rigotti. A condensed representation to find frequent patterns. In *Proc. of the 20th ACM SIGACT-SIGMOD-SIGART Symposium on Principles of Database Systems (PODS'01)*, Santa Barbara, CA, USA, May 2001. ACM.

[Cal02] Toon Calders. Deducing bounds on the frequency of itemsets. In *EDBT 2002 Workshop on Database Technologies for Data Mining*, 2002.

[CG02] Toon Calders and Bart Goethals. Mining all non-derivable frequent itemsets. In Tapio Elomaa, Heikki Mannila, and Hannu Toivonen, editors, *Proceedings of the 6th European Conference on Principles of Data Mining and Knowledge Discovery*, volume 2431 of *Lecture Notes in Computer Science*, pages 74–85. Springer-Verlag, 2002.

[GZ01] Karam Gouda and Mohammed Javeed Zaki. Efficiently mining maximal frequent itemsets. In *Proc. of the 2001 IEEE International Conference on Data Mining (ICDM'01)*, pages 163–170, San Jose, California, USA, 2001.

[HPY00] Jiawei Han, Jian Pei, and Yiwen Yin. Mining frequent patterns without candidate generation. In Weidong Chen, Jeffrey Naughton, and Philip A. Bernstein, editors, *2000 ACM SIGMOD Intl. Conference on Management of Data*, pages 1–12. ACM Press, May 2000.

[Kar84] N. Karmarkar. A new polynomial-time algorithm for linear programming. In *Proceedings of the sixteenth annual ACM symposium on Theory of computing*, pages 302–311, 1984.

[Kre93] Erwin Kreyszig. *Advanced Engineering Mathematics.* John Wiley Inc., seventh edition, 1993.

[ML98] Andrew Moore and Mary Soon Lee. Cached sufficient statistics for efficient machine learning with large datasets. *Journal of Artificial Intelligence Research*, 8:67–91, 1998.

[MSW96] Jiří Matoušek, Micha Sharir, and Emo Welzl. A subexponential bound for linear programming. *Algorithmica*, 16(4/5):498–516, 1996.

[MT96] Heikki Mannila and Hannu Toivonen. Multiple uses of frequent sets and condensed representations: Extended abstract. In *Proc. of the 2nd International Conference on Knowledge Discovery and Data Mining (KDD'96)*, pages 189–194, Portland, Oregon, USA, August 1996. AAAI Press.

[MTV97] Heikki Mannila, Hannu Toivonen, and A. Inkeri Verkamo. Discovery of frequent episodes in event sequences. *Data Mining and Knowledge Discovery*, 1(3):259–289, 1997.

[PBTL99] Nicolas Pasquier, Yves Bastide, Rafik Taouil, and Lotfi Lakhal. Efficient mining of association rules using closed itemset lattices. *Information Systems*, 24(1):25–46, 1999.

[PMS00] Dmitry Pavlov, Heikki Mannila, and Padhraic Smyth. Probabilistic models for query approximation with large sparse binary datasets. In *Proc. of the 16th Conference in Uncertainty in Artificial Intelligence (UAI'00)*, Stanford, California, USA, 2000.

[Toi96] Hannu Toivonen. Sampling large databases for association rules. In *Proc. of the 22th International Conference on Very Large Data Bases (VLDB'96)*, pages 134–145, Mumbai (Bombay), India, 1996.

[ZRL97] Tian Zhang, Raghu Ramakrishnan, and Miron Livny. BIRCH: A new data clustering algorithm and its applications. *Data Mining and Knowledge Discovery*, 1(2):141–182, 1997.

Condensed Representations for Sets of Mining Queries

Arnaud Giacometti[1],
Dominique Laurent[1], and Cheikh Talibouya Diop[1,2]

[1] LI, Université de Tours, 41000 Blois, FRANCE
{giaco,laurent}@univ-tours.fr
[2] Université Gaston Berger, Saint-Louis, SENEGAL
cdiop@ugb.sn

Abstract. In this paper, we propose a general framework for condensed representations of sets of mining queries. To this end, we adapt the standard notions of maximal, closed and key patterns introduced in previous works, including those dealing with condensed representations. Whereas these previous works concentrate on condensed representations of the answer to a *single* mining query, we consider the more general case of *sets* of mining queries defined by monotonic and anti-monotonic selection predicates.

1 Introduction

In the past decades, the problem of discovery of interesting patterns in large databases has motivated many research efforts. Whereas these works have focussed mainly on the efficiency of the algorithms [1,6,12,16], some other issues have been recently considered, among which the problem of efficient storage of the result of an extraction [4,14,15]. In this paper, we propose a general framework for condensed representations of the answers to a *set* of mining queries. More precisely, we assume that we are given:

1. A set $\mathbf{\Delta}$ of all data sets Δ from which the patterns are to be discovered.
2. A partially ordered set of patterns $\mathbb{L}$, where the partial ordering is denoted by $\preceq$.
3. A set of selection predicates $\mathbb{Q}$, a selection predicate being a boolean function defined over $\mathbb{L} \times \mathbf{\Delta}$.
4. A set of measure functions $\mathbb{F}$, a measure function being a real function defined over $\mathbb{L} \times \mathbf{\Delta}$.

Moreover, given a selection predicate q and a data set Δ in $\mathbf{\Delta}$, we say that a pattern φ in $\mathbb{L}$ is *interesting in Δ with respect to q* if $q(\varphi, \Delta)$ has the value *true*. Any selection predicate is also called a *simple mining query* and the set of interesting patterns in Δ with respect to q, denoted by $sol(q/\Delta)$, is called the *answer of* q in Δ.

R. Meo et al. (Eds.): Database Support for Data Mining Applications, LNAI 2682, pp. 250–269, 2004.

We call *extended mining query* any pair of the form (q, f) where q is in $\mathbb{Q}$ and f is in $\mathbb{F}$. The answer in Δ to an extended mining query (q, f), denoted by $ans(q, f/\Delta)$, is the set of pairs $(\varphi, f(\varphi, \Delta))$ such that φ is in $sol(q/\Delta)$.

In the following example, that will be used as a running example throughout the paper, we illustrate these notions in the classical association rule mining problem of [1].

Running Example 1. *Given a set of items Items, the set of patterns* $\mathbb{L}$ *considered in our approach is the set of all subsets of Items, i.e.,* $\mathbb{L} = 2^{Items}$. *Moreover, the partial ordering over* $\mathbb{L}$ *that we consider is set inclusion: given two patterns* φ *and* φ' *in* $\mathbb{L}$, *we say that* $\varphi \preceq \varphi'$ *if* $\varphi' \subseteq \varphi$.

In this context, a data set Δ *is defined by a set of transactions* Tr *and a function it from* Tr *to* $\mathbb{L}$. *Given a transaction* $x \in Tr$, $it(x)$ *is the set of items in transaction* x. *The support of a pattern is an example of measure function of* $\mathbb{F}$. *More precisely, for every pattern* φ, *the support of* φ *in* Δ, *denoted by* $sup(\varphi, \Delta)$, *is defined by:*

$$sup(\varphi, \Delta) = |\{x \in Tr \mid it(x) \preceq \varphi\}|/|Tr|.$$

Note that, given a minimal support threshold minsup, we can consider the selection predicate q *defined by: for every pattern* $\varphi \in \mathbb{L}$, $q(\varphi, \Delta) = true$ *if* $sup(\varphi, \Delta) \geq minsup$.

In the rest of the paper, we consider the case where the set of items is $Items = \{A, B, C, D, E\}$ *and where the set of transactions is* $Tr = \{1, 2, \ldots, 10\}$. *For the sake of simplicity, sets of items are denoted by the concatenation of their elements, e.g. the set of items* $\{A, B, C\}$ *is denoted by* ABC. *The function it from* Tr *to* $\mathbb{L}$ *that defines the data set* Δ *is represented in the table of Figure 1.*

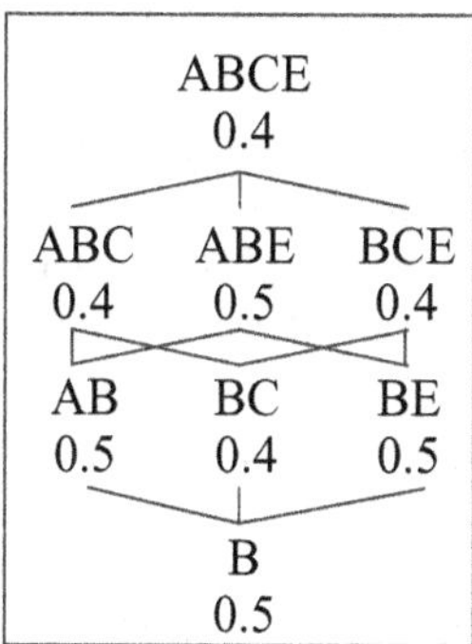

Tr	Set of Items
1	A
2	DE
3	ABCE
4	ABE
5	ABCDE
6	ACD
7	ABCE
8	AE
9	ABCDE
10	CD

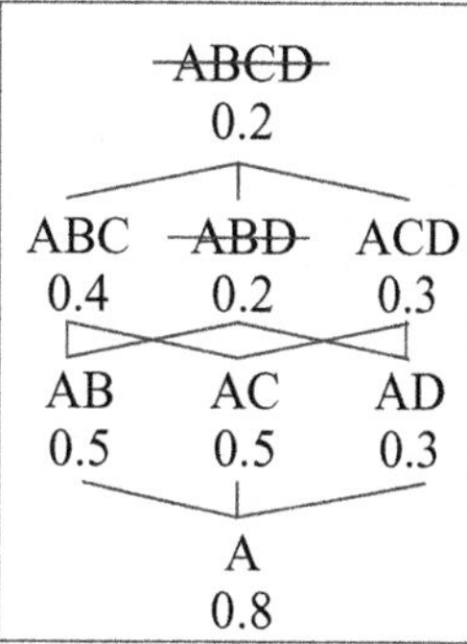

Fig. 1. Example of data set and sub-lattices of interesting patterns

Let m_1, m_2, a_1 *and* a_2 *be selection predicates defined for every pattern* φ *in* $\mathbb{L}$ *by:*

- $m_1(\varphi, \Delta) = true$ *if* $B \subseteq \varphi$, $m_2(\varphi, \Delta) = true$ *if* $A \subseteq \varphi$.
- $a_1(\varphi, \Delta) = \overline{a_1}(\varphi, \Delta) \wedge \tilde{a_1}(\varphi, \Delta)$, *where* $\overline{a_1}(\varphi, \Delta) = true$ *if* $sup(\varphi, \Delta) \geq 0.4$, *and* $\tilde{a_1}(\varphi, \Delta) = true$ *if* $\varphi \subseteq ABCE$.
- $a_2(\varphi, \Delta) = \overline{a_2}(\varphi, \Delta) \wedge \tilde{a_2}(\varphi, \Delta)$ *where* $\overline{a_2}(\varphi, \Delta) = true$ *if* $sup(\varphi, \Delta) \geq 0.3$, *and* $\tilde{a_2}(\varphi, \Delta) = true$ *if* $\varphi \subseteq ABCD$.

$q_1 = m_1 \wedge a_1$ *and* $q_2 = m_2 \wedge a_2$ *are simple mining queries. Moreover, it is easy to see from the table in Figure 1 that:*

$sol(m_1 \wedge a_1/\Delta) = \{B, AB, BC, BE, ABC, ABE, BCE, ABCE\}$
$sol(m_2 \wedge a_2/\Delta) = \{A, AB, AC, AD, ABC, ACD\}$

On the other hand, (q_1, sup) *and* (q_2, sup) *are examples of extended mining queries, and we have:*

$$ans(q_1, sup/\Delta) = \{(B, 0.5), (AB, 0.5), (BC, 0.4), (BE, 0.5), (ABC, 0.4), (ABE, 0.5), (BCE, 0.4), (ABCE, 0.4)\}$$
$$ans(q_2, sup/\Delta) = \{(A, 0.5), (AB, 0.5), (AC, 0.5), (AD, 0.3), (ABC, 0.4), (ACD, 0.3)\}$$

The answers in Δ *of* (q_1, sup) *and* (q_2, sup) *are also represented in Figure 1.* □

In the case of a simple mining query q, we recall that $sol(q/\Delta)$ can be computed *without any access* to Δ if only the maximal and minimal elements of $sol(q/\Delta)$ (with respect to the partial ordering on $\mathbb{L}$) are known [9,12]. Indeed, denoting these sets by $G(q/\Delta)$ and $S(q/\Delta)$, respectively, we know that a pattern φ is in $sol(q/\Delta)$ if and only if there exist $\varphi_g \in G(q/\Delta)$ and $\varphi_s \in S(q/\Delta)$ such that $\varphi_s \preceq \varphi \preceq \varphi_g$. Since $G(q/\Delta) \cup S(q/\Delta) \subseteq sol(q/\Delta)$, we say that $\{G(q/\Delta), S(q/\Delta)\}$ is a *condensed representation* of $sol(q/\Delta)$.

In our Running Example 1, it can be seen that $G(q_1/\Delta) = \{B\}$ and $S(q_1/\Delta) = \{ABCE\}$. Thus, $sol(q_1/\Delta)$ is the set of all itemsets φ such that $B \subseteq \varphi \subseteq ABCE$, and this can be computed independently from Δ.

On the other hand, in the case of extended mining queries, we adapt the notions of closed patterns and of key patterns ([3,4,16]) to our formalism, which allows us to obtain condensed representations of the set $ans(q, f/\Delta)$ (see Section 3.3). For instance, in our Running Example 1, for q_1 and the function sup, it will be seen that the answer $ans(q_1, f/\Delta)$ can be computed *without any access to* Δ, from the three sets $\{B\}$, $\{ABCE\}$, and $\{(ABE, 0.5), (ABCE, 0.4)\}$. In this case, we say that these three sets constitute an *extended condensed representation* of $ans(q_1, sup/\Delta)$.

As the main contribution of this paper, we consider the case of *sets* of mining queries (simple or extended). Noting that the union of condensed representations of different mining queries is *not* a condensed representation of the corresponding set of mining queries ([9]), we extend the notions of maximal, minimal, closed and key patterns to the case of sets of mining queries. Then, we propose condensed representations for such sets, in the sense that, given a set $\mathcal{Q}$ of mining queries, the answers in Δ of the queries in $\mathcal{Q}$ can be computed based *only* on the condensed representation, i.e., without any access to the data set Δ.

In the case of our Running Example 1, consider the set of simple mining queries $\mathcal{Q} = \{q_1, q_2\}$. Then, it will be seen in Section 4 that the sets of pairs

$\{(ABCE, q_1), (ACD, q_2)\}$ and $\{(B, q_1), (A, q_2)\}$ constitute a condensed representation of $sol(q_1/\Delta)$ and $sol(q_2/\Delta)$. We would like to emphasize that in the first set above, the maximal element ABC in $sol(q_2/\Delta)$ does not appear in the given condensed representation. Thus, in condensed representations of sets of mining queries, some maximal or minimal elements with respect to single mining queries can be omitted.

Comparing our approach to that of [6,7], we note that in [6,7] the authors consider conjunctive queries made of monotonic and anti-monotonic primitives, which correspond to what we call simple mining queries. Moreover, it is shown in [6,7] that the answer to one such query can be represented by its minimal and maximal elements only. However, contrary to the present paper, the case of *sets* of queries is not considered.

On the other hand, in [4], the authors also consider conjunctive queries. They use a caching technique to store condensed representations of the answers to these queries together with their supports. In our terminology, this corresponds to extended mining queries. However, in [4], each answer is condensed separately and stored in the cache, whereas our approach allows to benefit from relationships between the queries in order to further condense the answers to the queries.

Thus, our approach can be seen as an extension of [6,7] and [4]. In this paper, however, we do not consider computational aspects, such as the computation and the maintenance of condensed representations.

The paper is organized as follows: In Section 2, we give the formal definitions of the basic concepts of our approach, and in Section 3, mining queries, condensed representations as well as maximal, closed and key patterns are introduced. Section 4 deals with condensed representations of sets of mining queries. In Section 5, we conclude the paper and we propose further research directions based on this work.

2 Basic Definitions

In our formalism, we assume that we are given:

1. A set $\mathbf{\Delta}$ of all data sets from which the patterns are to be discovered. For instance, $\mathbf{\Delta}$ can be thought of as being the set of all instances of a given relation schema.
2. A set of patterns $\mathbb{L}$ and a partial ordering $\preceq$ over $\mathbb{L}$. Given two patterns φ_1, φ_2 in $\mathbb{L}$, we say that φ_1 is more specific than φ_2 (or that φ_2 is more general than φ_1) if we have $\varphi_1 \preceq \varphi_2$.
3. A set of selection predicates $\mathbb{Q}$, a selection predicate $q \in \mathbb{Q}$ being a boolean function defined over $\mathbb{L} \times \mathbf{\Delta}$. Moreover, given a pattern φ in $\mathbb{L}$ and a data set Δ in $\mathbf{\Delta}$, we say that φ is interesting in Δ with respect to q if $q(\varphi, \Delta) = true$.
4. A set of measure functions $\mathbb{F}$, a measure function being a function defined from $\mathbb{L} \times \mathbf{\Delta}$ to $\Re$.

Now, we define when a selection predicate is independent from $\mathbf{\Delta}$.

Definition 1 - Data Independency. *Let q be a selection predicate in $\mathbb{Q}$. q is* data independent *(or* independent *for short) if there exists a function $\widetilde{q}$ from $\mathbb{L}$ to $\{true, false\}$ such that for every data set Δ in $\mathbf{\Delta}$ and every pattern φ in $\mathbb{L}$, $q(\varphi, \Delta) = \widetilde{q}(\varphi)$.*

In our Running Example 1, it is easy to see that the selection predicates m_1, m_2, $\widetilde{a_1}$ and $\widetilde{a_2}$ are independent. In the following, we denote by $\widetilde{\mathbb{Q}}$ the set of all independent selection predicates, and by $\overline{\mathbb{Q}}$ the complement of $\widetilde{\mathbb{Q}}$ in $\mathbb{Q}$, i.e., $\overline{\mathbb{Q}} = \mathbb{Q} \setminus \widetilde{\mathbb{Q}}$.

In this paper, we consider only selection predicates that are monotonic or anti-monotonic, and measure functions that are monotonic increasing.

Definition 2 - Monotonicity. *Let q be a selection predicate.*

- *q is* monotonic *if for every data set Δ in $\mathbf{\Delta}$ and every pair of patterns (φ_1, φ_2) in $\mathbb{L}^2$, we have:*
$$\text{if } \varphi_1 \preceq \varphi_2 \text{ and } q(\varphi_2, \Delta) = true, \text{ then } q(\varphi_1, \Delta) = true.$$
- *q is* anti-monotonic *if for every data set Δ in $\mathbf{\Delta}$ and every pair of patterns (φ_1, φ_2) in $\mathbb{L}^2$, we have:*
$$\text{if } \varphi_1 \preceq \varphi_2 \text{ and } q(\varphi_1, \Delta) = true, \text{ then } q(\varphi_2, \Delta) = true.$$

Let f be a measure function. f is a monotonic increasing *function if for every data set Δ in $\mathbf{\Delta}$ and every pair of patterns (φ_1, φ_2) in $\mathbb{L}^2$, we have:*
$$\text{if } \varphi_1 \preceq \varphi_2, \text{ then } f(\varphi_1, \Delta) \leq f(\varphi_2, \Delta).$$

In our Running Example 1, it is easy to see that the selection predicates m_i $(i = 1, 2)$ are monotonic, whereas the selection predicates $\overline{a_i}$ and $\widetilde{a_i}$ $(i = 1, 2)$ are anti-monotonic. Moreover, the measure function *sup* is an example of monotonic increasing measure function.

In the following, we denote by $\mathbb{A}$ the set of all anti-monotonic selection predicates and by $\mathbb{M}$ the set of all monotonic selection predicates. Moreover, we denote by $\widetilde{\mathbb{A}}$ (respectively $\widetilde{\mathbb{M}}$) the set of all selection predicates in $\mathbb{A}$ (respectively $\mathbb{M}$) that are independent, and by $\overline{\mathbb{A}}$ (respectively $\overline{\mathbb{M}}$) the set of all selection predicates in $\mathbb{A}$ (respectively $\mathbb{M}$) that are not independent. Finally, we denote by $\mathbb{I}$ the set of all monotonic increasing measure functions.

In our approach, selection predicates are compared according to the following definition.

Definition 3 - Selectivity. *Let q_1 and q_2 be two selection predicates. q_1 is* more selective *than q_2, denoted by $q_1 \sqsubseteq q_2$, if for every data set Δ in $\mathbf{\Delta}$ and every pattern φ in $\mathbb{L}$, we have: if $q_1(\varphi, \Delta) = true$, then $q_2(\varphi, \Delta) = true$.*

In the context of our Running Example 1, let α_1 and α_2 be two support thresholds. For $i = 1, 2$, let a_i be the selection predicate defined by: for every pattern φ, $q_i(\varphi, \Delta) = true$ if $sup(\varphi, \Delta) \geq \alpha_i$. It is easy to see that if $\alpha_2 \geq \alpha_1$, then $q_2 \sqsubseteq q_1$.

In the rest of the paper, we consider a *fixed* data set Δ in $\mathbf{\Delta}$. Therefore, for notational convenience, we shall omit Δ in the subsequent definitions and propositions. For instance, referring to the previous two definitions, $q(\varphi, \Delta)$ and $f(\varphi, \Delta)$ will be simply denoted by $q(\varphi)$ and $f(\varphi)$, respectively.

3 Mining Query and Condensed Representations

3.1 Basic Definitions

In our approach, we define two types of mining query.

Definition 4 - Mining Query. *A* simple mining query *is a selection predicate q. Given a data set Δ, the* answer of q in Δ, *denoted by $sol(q)$, is defined by:*

$$sol(q) = \{\varphi \in \mathbb{L} \mid q(\varphi) = true\}.$$

$sol(q)$ denotes the set of all interesting patterns *in $\mathbb{L}$ with respect to q.*

An extended mining query *is a pair (q, f) where q is a selection predicate and f is a measure function. Given a data set Δ, the* answer of (q, f) in Δ, *denoted by $ans(q, f)$, is defined by:*

$$ans(q, f) = \{(\varphi, f(\varphi)) \mid \varphi \in sol(q)\}.$$

Note that an algorithm proposed in [5] can compute directly $sol(q)$ and $ans(q, f)$ if $q = m \wedge a$ with $m \in \mathbb{M}$ and $a \in \mathbb{A}$.

Let $Y = \{(y_1^i, y_2^i, \ldots, y_n^i) \mid i = 1, \ldots, p\}$ be a set of tuples whose first elements are patterns in $\mathbb{L}$. The projection of Y on $\mathbb{L}$, denoted by $\pi_{\mathbb{L}}(Y)$, is defined by: $\pi_{\mathbb{L}}(Y) = \{y_1^1, y_1^2, \ldots, y_1^p\}$. We note that $\pi_{\mathbb{L}}(Y) \subseteq \mathbb{L}$, and that for every $q \in \mathbb{Q}$ and $f \in \mathbb{F}$, $sol(q) = \pi_{\mathbb{L}}(ans(q, f))$.

We now introduce the notion of condensed representation.

Definition 5 - Condensed Representation. *Let $X_1, \ldots, X_K$ be sets of patterns, i.e., $X_k \subseteq \mathbb{L}$ ($k = 1, \ldots, K$). Given a mining query $q \in \mathbb{Q}$ and a data set Δ, $\{X_1, \ldots, X_K\}$ is a* condensed representation *of $sol(q)$, denoted by $X_1, \ldots, X_K \models sol(q)$, if:*

- *$(X_1 \cup \ldots \cup X_K) \subseteq sol(q)$, and*
- *there exists a function F independent from Δ such that: $sol(q) = F(X_1, \ldots, X_K)$.*

Let Y be a set of pairs (φ, α) where φ is a pattern in $\mathbb{L}$ and α is a real. Given an extended mining query $(q, f) \in \mathbb{Q} \times \mathbb{F}$ and a data set Δ, $\{X_1, \ldots, X_K, Y\}$ is an extended condensed representation *of $ans(q, f)$, denoted by $X_1, \ldots, X_K, Y \models_e ans(q, f)$, if:*

- *$(X_1 \cup \ldots \cup X_K \cup \pi_{\mathbb{L}}(Y)) \subseteq \pi_{\mathbb{L}}(ans(q, f))$, and*
- *there exists a function F independent from Δ such that: $ans(q, f) = F(X_1, \ldots, X_K, Y)$.*

Given a simple mining query q and a measure function f, we now consider condensed representations of $sol(q)$ and extended condensed representations of $ans(q, f)$.

3.2 Maximal Patterns

In this paper, we consider only simple mining queries that are defined by conjunction of anti-monotonic and monotonic selection predicates. In this case, the answer of a simple mining query can be represented by its most specific and most general patterns [9,12].

Definition 6. *Let $q = m \wedge a$ be simple mining queries with $m \in \mathbb{M}$ and $a \in \mathbb{A}$.*

- *The set of* most specific patterns in $sol(q)$, *denoted by $S(q)$, is defined by:*
  $S(q) = min_{\preceq}(sol(q)) = \{\varphi \in sol(q) \mid (\nexists \varphi' \in sol(q))(\varphi' \prec \varphi)\}$.
- *The set of* most general patterns in $sol(q)$, *denoted by $G(q)$, is defined by:*
  $G(q) = max_{\preceq}(sol(q)) = \{\varphi \in sol(q) \mid (\nexists \varphi' \in sol(q))(\varphi \prec \varphi')\}$.

The following lemma, whose easy proof is omitted, shows that $sol(q)$ can be computed from $S(q)$ and $G(q)$.

Lemma 1. *Let $q = m \wedge a$ be a simple mining query with $m \in \mathbb{M}$ and $a \in \mathbb{A}$. We have: $sol(q) = \{\varphi \in \mathbb{L} \mid (\exists \varphi_s \in S(q))(\exists \varphi_g \in G(q))(\varphi_s \preceq \varphi \preceq \varphi_g)\}$.*

Therefore, we have the following proposition.

Proposition 1. *Let $q = m \wedge a$ be a simple mining query with $m \in \mathbb{M}$ and $a \in \mathbb{A}$. The set $\{G(q),\ S(q)\}$ is a condensed representation of $sol(q)$, i.e., $G(q),\ S(q) \models sol(q)$.*

Proof: *Let F be the function defined by: $F(X_1, X_2) = \{\varphi \in \mathbb{L} | (\exists \varphi_1 \in X_1)(\exists \varphi_2 \in X_2)(\varphi_1 \preceq \varphi \preceq \varphi_2)\}$. Using Lemma 1, we have $sol(q) = F(S(q), G(q))$. Moreover, F is independent from the data set Δ since $\preceq$ does not depend on Δ. Finally, we have $S(q) \cup G(q) \subseteq sol(q)$, which completes the proof.* □

We point out that algorithms for computing $S(m \wedge a)$ and $G(m \wedge a)$ directly have been proposed recently, e.g. the level-wise version space algorithm in [6].

Example 1. *Let q_1 and q_2 be the simple mining queries as given in our Running Example 1. We recall that: $G(q_1) = \{B\}$, $S(q_1) = \{ABCE\}$, $G(q_2) = \{A\}$, and $S(q_2) = \{ABC, ACD\}$. Applying Proposition 1, we have: $G(q_1),\ S(q_1) \models sol(q_1)$ and $G(q_2),\ S(q_2) \models sol(q_2)$.* □

It is important to note that $G(m) \models sol(m)$, $S(a) \models sol(a)$ and $sol(m \wedge a) = sol(m) \cap sol(a)$. Therefore, $sol(q)$ can be computed from $G(m)$ and $S(a)$. However, the set $\{G(m), S(a)\}$ is *not* always a condensed representation of $sol(q)$, since we can have $sol(q) \subset (G(m) \cup S(a))$. This is in particular the case for a query $q = m \wedge a$ such that $sol(q) = \emptyset$, $sol(m) \neq \emptyset$, and $sol(a) \neq \emptyset$.

On the other hand, in [12], the authors consider what they call the *positive* and the *negative* borders of the answer to a mining query. In our approach, given a simple mining query q, the corresponding positive and negative borders, respectively denoted by $Bd^+(q)$ and $Bd^-(q)$, can be defined as follows:

- $Bd^+(q) = \{S(q), G(q)\}$, where $S(q)$ and $G(q)$ have been defined previously
- $Bd^-(q) = \{S^-(q), G^-(q)\}$, where $S^-(q)$ and $G^-(q)$ are the following sets: $S^-(q) = max_{\preceq}\{\varphi \in sol(m) \mid \varphi \notin sol(a)\}$ and $G^-(q) = min_{\preceq}\{\varphi \in sol(a) \mid \varphi \notin sol(m)\}$.

Therefore, according to Definition 5, the positive border can be seen as a condensed representation of $sol(q)$, whereas the negative border can not. Indeed, although the sets $S^-(q)$ and $G^-(q)$ allow to recompte $sol(q)$ without any access to the data set, the first point of Definition 5 is not satisfied, since neither $S^-(q)$ nor $G^-(q)$ is a subset of $sol(q)$.

We shall not consider the case of negative borders in the rest of the paper, but we note in this respect that (i) storing $Bd^-(q)$ is not optimal in general (since its cardinality can be much greater than that of $sol(q)$), and (ii) $Bd^-(q)$ can be seen in our approach as a condensed representation of the set $sol(q) \cup Bd^-(q)$.

3.3 Closed and Key Patterns

In this section, we give alternative definitions of the notions of closed and key patterns introduced in [3,4,16]. To this end, given a measure function f, we consider the partial ordering $\leq_f$ defined for every pair of patterns (φ, φ') by:

$$\varphi \leq_f \varphi' \text{ if } \varphi \preceq \varphi' \text{ and } f(\varphi) = f(\varphi').$$

Definition 7. *Let q be a mining query in $\mathbb{Q}$ and f be a measure function in $\mathbb{F}$. Let Δ be a data set and φ be a pattern in $\mathbb{L}$.*

- *The set of all* interesting closed patterns *in Δ with respect to q and f, denoted by $SC(q, f)$, is defined by:*
$$SC(q, f) = min_{\leq_f}(sol(q)).$$
- *The set of all* interesting key patterns *in Δ with respect to q and f, denoted by $GK(q, f)$, is defined by:*
$$GK(q, f) = max_{\leq_f}(sol(q)).$$

It can be shown that our notions of interesting closed patterns and interesting key patterns coincide with those of [3,4,16] in the context of classical association rules mining [1].

Moreover, it is easily seen that for every extended mining query (q, f) with $q \in \mathbb{Q}$ and $f \in \mathbb{I}$, we have $S(q) \subseteq SC(q, f)$ and $G(q) \subseteq GK(q, f)$. More precisely, the following lemma holds.

Lemma 2. *Let q be a selection predicate in $\mathbb{Q}$ and f be a monotonic increasing measure function in $\mathbb{I}$. We have:*

$$S(q) = min_{\preceq}(SC(q, f)) \quad and \quad G(q) = max_{\preceq}(GK(q, f)).$$

Proof: *We first show that $S(q) \subseteq min_{\preceq}(SC(q, f))$. Let $\varphi \in S(q)$. There does not exist a pattern $\varphi' \in sol(q)$ such that $\varphi' \prec \varphi$. Therefore, there does not exist a pattern $\varphi' \in sol(q)$ such that $\varphi' \prec \varphi$ and $f(\varphi') = f(\varphi)$, which shows* *that* $\varphi \in SC(q, f)$. *Assume now that* $\varphi \notin min_{\preceq}(SC(q, f))$. *Then, there exists* $\varphi' \in SC(q, f)$ *such that* $\varphi' \prec \varphi$, *which is in contradiction with the hypothesis* $\varphi \in S(q)$. *Hence, we have* $S(q) \subseteq min_{\preceq}(SC(q, f))$.

Now, we show that $min_{\preceq}(SC(q, f)) \subseteq S(q)$. *Let* $\varphi \in min_{\preceq}(SC(q, f))$. *Assume that* $\varphi \notin S(q)$. *Then, there exists* $\varphi' \in S(q)$ *such that* $\varphi' \prec \varphi$. *Since it has been shown above that* $S(q) \subseteq min_{\preceq}(SC(q, f))$, *we have that* $\varphi' \in SC(q, f)$. *This is in contradiction with the hypothesis* $\varphi \in min_{\preceq}(SC(q, f))$. *Hence, we have* $min_{\preceq}(SC(q, f)) \subseteq S(q)$.

Thus the proof that $S(q) = min_{\preceq}(SC(q, f))$ *is complete. In the same way, it can be shown that* $G(q) = max_{\preceq}(GK(q, f))$, *which completes the proof.* □

The following lemma states that given any pattern φ in $sol(q)$, $f(\varphi)$ can be computed based on $SC(q, f)$ or $GK(q, f)$.

Lemma 3. *Let q be a selection predicate in $\mathbb{Q}$ and f be a monotonic increasing measure function in $\mathbb{I}$. For every interesting pattern φ in $sol(q)$, we have:*

- $f(\varphi) = max\{f(\varphi') \mid \varphi' \in SC(q, f) \text{ and } \varphi' \preceq \varphi\}$, *and*
- $f(\varphi) = min\{f(\varphi') \mid \varphi' \in GK(q, f) \text{ and } \varphi \preceq \varphi'\}$

where min and max denote respectively the minimum and maximum functions according to the standard ordering of real numbers.

Proof: *Let* $\varphi \in sol(q)$ *and* $X(\varphi) = \{\varphi' \in sol(q) \mid \varphi' \preceq \varphi \text{ and } f(\varphi') = f(\varphi)\}$. *Since* $\varphi \in X(\varphi)$, *we know that* $Y(\varphi) = min_{\preceq}(X(\varphi))$ *is not empty. Given any* $\varphi'' \in Y(\varphi)$, *assume that* $\varphi'' \notin SC(q, f)$. *Then, there exists* $\varphi' \in sol(q)$ *such that* $\varphi' \prec \varphi''$ *and* $f(\varphi') = f(\varphi'')$, *which shows that* $\varphi' \in X(\varphi)$ *and contradicts the fact that* φ'' *is minimal in* $X(\varphi)$. *Hence, there exists* $\varphi_c \in SC(q, f)$ *such that* $\varphi_c \preceq \varphi$ *and* $f(\varphi_c) = f(\varphi)$.

On the other hand, for every $\varphi' \in SC(q, f)$ *such that* $\varphi' \preceq \varphi$, *we have* $f(\varphi') \leq f(\varphi)$. *Therefore, we have* $f(\varphi) = max\{f(\varphi') \mid \varphi' \in SC(q, f) \text{ and } \varphi' \preceq \varphi\}$. *Since the fact that* $f(\varphi) = min\{f(\varphi') \mid \varphi' \in GK(q, f) \text{ and } \varphi \preceq \varphi'\}$ *can be shown in the same way, the proof is complete.* □

Let (q, f) be an extended mining query. In the following, we denote by $SC^*(q, f)$ and $GK^*(q, f)$ the sets defined by:

- $SC^*(q, f) = \{(\varphi, f(\varphi)) \mid \varphi \in SC(q, f)\}$, and
- $GK^*(q, f) = \{(\varphi, f(\varphi)) \mid \varphi \in GK(q, f)\}$.

The following proposition follows from the previous two lemmas.

Proposition 2. *Let $q = m \wedge a$ be a simple mining query with $m \in \mathbb{M}$, $a \in \mathbb{A}$, and let f be a monotonic increasing measure function in $\mathbb{I}$. The sets $\{S(q), G(q), SC^*(q, f)\}$ and $\{S(q), G(q), GK^*(q, f)\}$ are extended condensed representations of $ans(q, f)$, i.e.,*

$$S(q),\ G(q),\ SC^*(q, f) \models_e ans(q, f) \text{ and } S(q),\ G(q),\ GK^*(q, f) \models_e ans(q, f).$$

PROOF: *Let F be the function defined by:* $F(X_1, X_2, Y) = \{(\varphi, \alpha) \in \mathbb{L}\times\Re \mid (\exists\varphi_1 \in X_1)(\exists\varphi_2 \in X_2)(\varphi_1 \preceq \varphi \preceq \varphi_2)$ *and* $\alpha = max\{\alpha' \mid (\exists\varphi' \in \mathbb{L})((\varphi', \alpha') \in Y \wedge \varphi' \preceq \varphi)\}$. *Using Lemma 1 and Lemma 3, we have* $ans(q, f) = F(S(q), G(q), SC^*(q, f))$. *Moreover, F is independent from the data set Δ since $\preceq$ does not depend on Δ, and* $S(q) \cup G(q) \cup SC(q, f) \subseteq sol(q)$. *Therefore,* $\{S(q),\ G(q),\ SC^*(q, f)\}$ *is an extended condensed representation of* $ans(q, f)$. *Since the fact that* $S(q),\ G(q),\ GK^*(q, f) \models_e ans(q, f)$ *can be shown in the same way, the proof is complete.* □

Example 2. *Let q_1 be the simple mining query as defined in our Running Example 1. We can see that:*

$GK^*(q_1, sup) = \{(B, 0.5),\ (BC, 0.4)\}$ *and*
$SC^*(q_1, sup) = \{(ABE, 0.5), (ABCE, 0.4)\}$.

Recalling that $G(q_1) = \{B\}$ *and* $S(q_1) = \{ABCE\}$, *and using Proposition 2, we obtain that* $S(q_1),\ G(q_1),\ SC^*(q_1, sup) \models_e ans(q_1, sup)$ *and that* $S(q_1),\ G(q_1),\ GK^*(q_1, sup) \models_e ans(q_1, sup)$. □

4 Condensed Representations of Sets of Mining Queries

In this section, we extend the notions of condensed representation and of extended condensed representation to the case of sets of mining queries.

4.1 Definitions

Definition 8 - Set of Mining Queries. *Let* $\mathcal{Q} = \{q_1, \ldots, q_n\}$ *be a set of mining queries. Given a data set Δ, the* answer of $\mathcal{Q}$ in Δ, *denoted by* $sol(\mathcal{Q})$, *is the set defined by:*

$$sol(\mathcal{Q}) = \bigcup_{q \in \mathcal{Q}} \{(\varphi, q) \mid \varphi \in sol(q)\}.$$

Let f be a measure function in $\mathbb{F}$. The answer of $(\mathcal{Q}, f)$ in Δ, *denoted by* $ans(\mathcal{Q}, f)$, *is the set defined by:*

$$ans(\mathcal{Q}, f) = \bigcup_{q \in \mathcal{Q}} \{(\varphi, q, f(\varphi)) \mid \varphi \in sol(q)\}.$$

Definition 9 - Condensed Representation. *Let* $\mathcal{X}_1, \ldots, \mathcal{X}_K$ *be sets of pairs* (φ, q) *where* $\varphi \in \mathbb{L}$ *and* $q \in \mathbb{Q}$. *Given a set of mining queries $\mathcal{Q}$ and a data set Δ,* $\{\mathcal{X}_1, \ldots, \mathcal{X}_K\}$ *is a* condensed representation *of* $sol(\mathcal{Q})$, *denoted by* $\mathcal{X}_1, \ldots, \mathcal{X}_K \models sol(\mathcal{Q})$, *if:*

- $\pi_{\mathbb{L}}(\mathcal{X}_1) \cup \ldots \cup \pi_{\mathbb{L}}(\mathcal{X}_K) \subseteq \pi_{\mathbb{L}}(sol(\mathcal{Q}))$, *and*
- *there exists a function F independent from Δ such that:* $sol(\mathcal{Q}) = F(\mathcal{X}_1, \ldots, \mathcal{X}_K)$.

Let Y be a set of pairs (φ, α) where φ is a pattern in $\mathbb{L}$ and α is a real. Given a set of mining queries $\mathcal{Q}$, a measure function f and a data set Δ, $\{\mathcal{X}_1, \ldots, \mathcal{X}_K, Y\}$ *is an* extended condensed representation *of* $ans(\mathcal{Q}, f)$, *denoted by* $\mathcal{X}_1, \ldots, \mathcal{X}_K, Y \models_e ans(\mathcal{Q}, f)$, *if:*

- $\pi_{\mathbb{L}}(\mathcal{X}_1) \cup \ldots \cup \pi_{\mathbb{L}}(\mathcal{X}_K) \cup \pi_{\mathbb{L}}(Y) \subseteq \pi_{\mathbb{L}}(ans(\mathcal{Q}, f))$, *and*
- *there exists a function* F *independent from* Δ *such that:* $ans(\mathcal{Q}, f) = F(\mathcal{X}_1, \ldots, \mathcal{X}_K, Y)$.

Let $\mathcal{C} = \{\mathcal{Z}_1, \ldots, \mathcal{Z}_K\}$ and $\mathcal{C}' = \{\mathcal{Z}'_1, \ldots, \mathcal{Z}'_K\}$ be two condensed representations (extended or not) having the same cardinality K. We say that $\mathcal{C}$ is *more concise* than $\mathcal{C}'$ if there exists a permutation θ of $\{1, \ldots, K\}$ such that for every $i = 1, \ldots, K$, $\mathcal{Z}_i \subseteq \mathcal{Z}'_{\theta(i)}$.

Given a set of mining queries $\mathcal{Q}$ and a measure function f, we study condensed representations of $sol(\mathcal{Q})$ and extended condensed representations of $ans(\mathcal{Q}, f)$.

4.2 Maximal Patterns

Given a set of mining queries $\mathcal{Q} = \{q_1, \ldots, q_n\}$, it is well known [9] that, although $\{S(q_i),\ G(q_i)\}$ is a condensed representation of $sol(q_i)$, for every $i = 1, \ldots, n$, the set $\{S(q_1) \cup \ldots \cup S(q_n),\ G(q_1) \cup \ldots \cup G(q_n)\}$ is *not* a condensed representation of $sol(q_1) \cup \ldots \cup sol(q_n)$.

However, if for every φ in $S(q_1) \cup \ldots \cup S(q_n)$ or in $G(q_1) \cup \ldots \cup G(q_n)$, we keep track of the query q_i the pattern φ comes from, then $sol(\mathcal{Q})$ and $ans(\mathcal{Q}, f)$ can be condensed. For this reason, we define the sets $\mathcal{S}(\mathcal{Q})$ and $\mathcal{G}(\mathcal{Q})$ as follows:

Definition 10. *Let* $\mathcal{Q} = \{q_1, \ldots, q_n\}$ *be a set of mining queries* $q_i \in \mathbb{Q}$ $(i = 1, \ldots, n)$. *The sets* $\mathcal{S}(\mathcal{Q})$ *and* $\mathcal{G}(\mathcal{Q})$ *are defined by:*

$$\mathcal{S}(\mathcal{Q}) = \bigcup_{q \in \mathcal{Q}} \{(\varphi, q) \mid \varphi \in S(q)\} \quad \textit{and} \quad \mathcal{G}(\mathcal{Q}) = \bigcup_{q \in \mathcal{Q}} \{(\varphi, q) \mid \varphi \in G(q)\}.$$

Given these definitions, we have the following proposition.

Proposition 3. *Let* $\mathcal{Q} = \{q_1, \ldots, q_n\}$ *be a set of mining queries* $q_i = m_i \wedge a_i$ *with* $m_i \in \mathbb{M}$ *and* $a_i \in \mathbb{A}$ $(i = 1, \ldots, n)$. *The set* $\{\mathcal{S}(\mathcal{Q}),\ \mathcal{G}(\mathcal{Q})\}$ *is a condensed representation of* $sol(\mathcal{Q})$, *i.e.,* $\mathcal{S}(\mathcal{Q}),\ \mathcal{G}(\mathcal{Q}) \models sol(\mathcal{Q})$.

Proof: *Let* F *be the function defined by:* $F(\mathcal{X}_1, \mathcal{X}_2) = \{(\varphi, q) \in \mathbb{L} \times \mathbb{Q} \mid (\exists(\varphi_1, q_1) \in \mathcal{X}_1)(\exists(\varphi_2, q_2) \in \mathcal{X}_2)(q_1 = q_2 = q \text{ and } \varphi_1 \preceq \varphi \preceq \varphi_2)\}$. *Based on Lemma 1, we can easily see that* $sol(\mathcal{Q}) = F(\mathcal{S}(\mathcal{Q}), \mathcal{G}(\mathcal{Q}))$. *Moreover,* F *is independent from the data set* Δ *since* $\preceq$ *does not depend on* Δ. *Finally, we have* $\mathcal{S}(\mathcal{Q}) \subseteq sol(\mathcal{Q})$ *and* $\mathcal{G}(\mathcal{Q}) \subseteq sol(\mathcal{Q})$, *which completes the proof.* □

Example 3. *In the context of our Running Example 1, let* $q_3 = m_3 \wedge a_3$ *and* $q_4 = m_4 \wedge a_4$ *where* m_3, m_4, a_3 *and* a_4 *are selection predicates defined for every pattern* $\varphi \in \mathbb{L}$ *by:*

- $m_3(\varphi, \Delta) = true$ *if* $A \subseteq \varphi$, *and* $m_4(\varphi, \Delta) = true$ *if* $AC \subseteq \varphi$,
- $a_3(\varphi, \Delta) = true$ *if* $sup(\varphi, \Delta) \geq 0.4$ *and* $\varphi \subseteq ABC$, *and* $a_4(\varphi, \Delta) = true$ *if* $sup(\varphi, \Delta) \geq 0.3$ *and* $\varphi \subseteq ABCD$.

We note that m_3 and m_4 are monotonic selection predicates such that $m_4 \sqsubseteq m_3$, whereas a_3 and a_4 are anti-monotonic selection predicates such that $a_3 \sqsubseteq a_4$. We can see that $S(q_3) = \{ABC\}$, $S(q_4) = \{ABC, ACD\}$, $G(q_3) = \{A\}$ and $G(q_4) = \{AC\}$. Considering $\mathcal{Q} = \{q_3, q_4\}$, we have:

$$\mathcal{S}(\mathcal{Q}) = \{(ABC, q_3), (ABC, q_4), (ACD, q_4)\} \text{ and } \mathcal{G}(\mathcal{Q}) = \{(A, q_3), (AC, q_4)\}$$

Using Proposition 3, we can see that: $\mathcal{S}(\mathcal{Q})$, $\mathcal{G}(\mathcal{Q}) \models sol(\mathcal{Q})$. □

In what follows, we show how to define condensed representations of $sol(\mathcal{Q})$ that are *more concise* than $\{\mathcal{S}(\mathcal{Q}), \mathcal{G}(\mathcal{Q})\}$.

Let $\mathcal{Q} = \{q_1, \ldots, q_n\}$ be a set of mining queries $q_i = m_i \wedge a_i$ with $m_i \in \mathbb{M}$ and $a_i \in \mathbb{A}$ $(i = 1, \ldots, n)$. We define two partial pre-orderings, denoted by $\leq_{\mathbb{A}}$ and $\leq_{\mathbb{M}}$, as follows: for all (φ_i, q_i) and (φ_j, q_j) in $\mathbb{L} \times \mathcal{Q}$:

$$(\varphi_i, q_i) \leq_{\mathbb{A}} (\varphi_j, q_j) \text{ if } \varphi_i \preceq \varphi_j \text{ and } a_i \sqsubseteq a_j$$
$$(\varphi_i, q_i) \leq_{\mathbb{M}} (\varphi_j, q_j) \text{ if } \varphi_i \preceq \varphi_j \text{ and } m_j \sqsubseteq m_i.$$

Then, we denote by $\Sigma(\mathcal{Q})$ the set of all minimal pairs in $\mathcal{S}(\mathcal{Q})$ with respect to $\leq_{\mathbb{A}}$. Similarly, we denote by $\Gamma(\mathcal{Q})$ the set of all maximal pairs in $\mathcal{G}(\mathcal{Q})$ with respect to $\leq_{\mathbb{M}}$. That is:

$$\Sigma(\mathcal{Q}) = min_{\leq_{\mathbb{A}}}(\mathcal{S}(\mathcal{Q})) \text{ and } \Gamma(\mathcal{Q}) = max_{\leq_{\mathbb{M}}}(\mathcal{G}(\mathcal{Q})).$$

The following lemma states that, for every $q \in \mathcal{Q}$, $sol(q)$ can be computed based on $\Sigma(\mathcal{Q})$ and $\Gamma(\mathcal{Q})$, only.

Lemma 4. *Let $\mathcal{Q} = \{q_1, \ldots, q_n\}$ be a set of mining queries $q_i = m_i \wedge a_i$ with $m_i \in \mathbb{M}$ and $a_i \in \mathbb{A}$ $(i = 1, \ldots, n)$. For every q in $\mathcal{Q}$, we have:*

$$sol(q) = \{\varphi \in \mathbb{L} \mid (\exists(\varphi_i, q_i) \in \Sigma(\mathcal{Q}))((\varphi_i, q_i) \leq_{\mathbb{A}} (\varphi, q)) \text{ and } (\exists(\varphi_j, q_j) \in \Gamma(\mathcal{Q}))((\varphi, q) \leq_{\mathbb{M}} (\varphi_j, q_j))\}.$$

PROOF: *Let $X(q)$ be the set defined by:*

$$X(q) = \{\varphi \in \mathbb{L} \mid (\exists(\varphi_i, q_i) \in \Sigma(\mathcal{Q}))((\varphi_i, q_i) \leq_{\mathbb{A}} (\varphi, q)) \text{ and } (\exists(\varphi_j, q_j) \in \Gamma(\mathcal{Q}))((\varphi, q) \leq_{\mathbb{M}} (\varphi_j, q_j))\}.$$

We first show that $X(q) \subseteq sol(q)$. Let $\varphi \in X(q)$. There exist $(\varphi_i, q_i) \in \Sigma(\mathcal{Q})$ and $(\varphi_j, q_j) \in \Gamma(\mathcal{Q})$ such that $(\varphi_i, q_i) \leq_{\mathbb{A}} (\varphi, q)$ and $(\varphi, q) \leq_{\mathbb{M}} (\varphi_j, q_j)$.

On one hand, we know that $q_i(\varphi_i) = true$. Thus, we have $a_i(\varphi_i) = true$. It follows that $a(\varphi_i) = true$ since $a_i \sqsubseteq a$, and that $a(\varphi) = true$ since $\varphi_i \preceq \varphi$ and a is anti-monotonic.

On the other hand, we know that $q_j(\varphi_j) = true$. Thus, we have $m_j(\varphi_j) = true$. It follows that $m(\varphi_j) = true$ since $m_j \sqsubseteq m$, and that $m(\varphi) = true$ since $\varphi \preceq \varphi_j$ and m is monotonic. Therefore, we have $a(\varphi) = true$ and $m(\varphi) = true$, which shows that $\varphi \in sol(q)$. Hence, we have: $X(q) \subseteq sol(q)$.

Now, we show that $sol(q) \subseteq X(q)$. Let $\varphi \in sol(q)$. There exist $\varphi_s \in S(q)$ and $\varphi_g \in G(q)$ such that $\varphi_s \preceq \varphi \preceq \varphi_g$. Thus, we have $(\varphi_s, q) \in \mathcal{S}(\mathcal{Q})$ and $(\varphi_g, q) \in \mathcal{G}(\mathcal{Q})$.

Given the definitions of $\Sigma(\mathcal{Q})$ and $\Gamma(\mathcal{Q})$, there exist $(\varphi_i, q_i) \in \Sigma(\mathcal{Q})$ and $(\varphi_j, q_j) \in \Gamma(\mathcal{Q})$ such that $(\varphi_i, q_i) \leq_{\mathbb{A}} (\varphi_s, q)$ and $(\varphi_g, q) \leq_{\mathbb{M}} (\varphi_j, q_j)$. Moreover, we have $(\varphi_s, q) \leq_{\mathbb{A}} (\varphi, q)$ since $\varphi_s \preceq \varphi$, and $(\varphi, q) \leq_{\mathbb{M}} (\varphi_g, q)$ since $\varphi \preceq \varphi_g$. Thus, $(\varphi_i, q_i) \leq_{\mathbb{A}} (\varphi, q)$ and $(\varphi, q) \leq_{\mathbb{M}} (\varphi_j, q_j)$, which shows that $\varphi \in X(q)$. Hence, we have $sol(q) \subseteq X(q)$, which completes the proof. □

As a consequence of Lemma 4 above, we have the following theorem:

Theorem 1. *Let $\mathcal{Q} = \{q_1, \ldots, q_n\}$ be a set of mining queries $q_i = m_i \wedge a_i$ with $m_i \in \mathbb{M}$ and $a_i \in \mathbb{A}$ $(i = 1, \ldots, n)$. The set $\{\Sigma(\mathcal{Q}),\ \Gamma(\mathcal{Q})\}$ is a condensed representation of $sol(\mathcal{Q})$, i.e., $\Sigma(\mathcal{Q}),\ \Gamma(\mathcal{Q}) \models sol(\mathcal{Q})$.*

Moreover, $\{\Sigma(\mathcal{Q}),\ \Gamma(\mathcal{Q})\}$ is more concise than $\{\mathcal{S}(\mathcal{Q}),\ \mathcal{G}(\mathcal{Q})\}$.

Proof: *Let F be the function defined by: $F(\mathcal{X}_1, \mathcal{X}_2) = \{(\varphi, q) \in \mathbb{L} \times \mathbb{Q} \mid (\exists(\varphi_1, q_1) \in \mathcal{X}_1)((\varphi_1, q_1) \leq_{\mathbb{A}} (\varphi, q))$ and $(\exists(\varphi_2, q_2) \in \mathcal{X}_2)((\varphi, q) \leq_{\mathbb{M}} (\varphi_2, q_2))\}$. Using Lemma 4, we can easily see that $sol(\mathcal{Q}) = F(\Sigma(\mathcal{Q}), \Gamma(\mathcal{Q}))$. Moreover, F is independent from the data set Δ since $\preceq$ and $\sqsubseteq$ do not depend on Δ.*

It is easily seen that we have $\Sigma(\mathcal{Q}) \subseteq \mathcal{S}(\mathcal{Q}) \subseteq sol(\mathcal{Q})$ and $\Gamma(\mathcal{Q}) \subseteq \mathcal{G}(\mathcal{Q}) \subseteq sol(\mathcal{Q})$. Therefore, $\{\Sigma(\mathcal{Q}),\ \Gamma(\mathcal{Q})\}$ is more concise than $\{\mathcal{S}(\mathcal{Q}),\ \mathcal{G}(\mathcal{Q})\}$ and thus, the proof is complete. □

Example 4. *We recall from Example 3 that we have:*

$\mathcal{S}(\mathcal{Q}) = \{(ABC, q_3), (ABC, q_4), (ACD, q_4)\}$ *and* $\mathcal{G}(\mathcal{Q}) = \{(A, q_3), (AC, q_4)\}$.

Since $a_3 \sqsubseteq a_4$, we have $(ABC, q_3) \leq_{\mathbb{A}} (ABC, q_4)$. On the other hand, (A, q_3) and (AC, q_4) are not comparable with respect to $\leq_{\mathbb{M}}$. It follows that:

$$\Sigma(\mathcal{Q}) = \{(ABC, q_3), (ACD, q_4)\} \quad \text{and} \quad \Gamma(\mathcal{Q}) = \{(A, q_3), (AC, q_4)\}$$

Using Theorem 1, we can see that $\Sigma(\mathcal{Q}),\ \Gamma(\mathcal{Q}) \models \mathcal{S}(\mathcal{Q})$. Moreover, since $\Sigma(\mathcal{Q}) \subset \mathcal{S}(\mathcal{Q})$ and $\Gamma(\mathcal{Q}) \subseteq \mathcal{G}(\mathcal{Q})$, $\{\Sigma(\mathcal{Q}),\ \Gamma(\mathcal{Q})\}$ is more concise than $\{\mathcal{S}(\mathcal{Q}),\ \mathcal{G}(\mathcal{Q})\}$. □

We end this subsection by showing how to optimize the computation of $\Sigma(\mathcal{Q})$ (respectively $\Gamma(\mathcal{Q})$) by stating that two pairs (φ_i, q_i) and (φ_j, q_j) in $\mathcal{S}(\mathcal{Q})$ (respectively $\mathcal{G}(\mathcal{Q})$) cannot be comparable with respect to $\leq_{\mathbb{A}}$ (respectively $\leq_{\mathbb{M}}$) if $\varphi_i \neq \varphi_j$.

Indeed, based on this result, it turns out that the computation of $\Sigma(\mathcal{S}) = min_{\leq_{\mathbb{A}}}(\mathcal{S}(\mathcal{Q}))$ (respectively $\Gamma(\mathcal{S}) = max_{\leq_{\mathbb{M}}}(\mathcal{G}(\mathcal{Q})))$ only requires to compare the pairs of $\mathcal{S}(\mathcal{Q})$ (respectively $\mathcal{G}(\mathcal{Q})$) that contain the same pattern.

Proposition 4. *Let $\mathcal{Q} = \{q_1, \ldots, q_n\}$ be a set of mining queries $q_i = m_i \wedge a_i$ with $m_i \in \mathbb{M}$ and $a_i \in \mathbb{A}$ $(i = 1, \ldots, n)$.*

If (φ_i, q_i) and (φ_j, q_j) are two pairs in $\mathcal{S}(\mathcal{Q})$ (respectively $\mathcal{G}(\mathcal{Q})$) such that $(\varphi_i, q_i) \leq_{\mathbb{A}} (\varphi_j, q_j)$ (respectively such that $(\varphi_i, q_i) \leq_{\mathbb{M}} (\varphi_j, q_j)$), then we have $\varphi_i = \varphi_j$.

Proof: *Let (φ_i, q_i) and (φ_j, q_j) be two pairs in $\mathcal{S}(\mathcal{Q})$ such that $(\varphi_i, q_i) \leq_{\mathbb{A}} (\varphi_j, q_j)$. Since $q_i(\varphi_i) = true$, we have $a_i(\varphi_i) = true$ and $a_j(\varphi_i) = true$ since $a_i \sqsubseteq a_j$. On the other hand, since $q_j(\varphi_j) = true$, $\varphi_i \preceq \varphi_j$ and m_j is monotonic,* *we have $m_j(\varphi_j) = true$ and $m_j(\varphi_i) = true$. Therefore, we have $q_j(\varphi_i) = true$, meaning that $\varphi_i \in sol(q_j)$. Moreover, since φ_j is minimal in $sol(q_j)$ with respect to $\preceq$ and $\varphi_i \preceq \varphi_j$, we necessarily have $\varphi_i = \varphi_j$. It can be shown in the same way that if (φ_i, q_i) and (φ_j, q_j) are two pairs in $\mathcal{G}(\mathcal{Q})$ such that $(\varphi_i, q_i) \leq_{\mathbb{M}} (\varphi_j, q_j)$, then $\varphi_i = \varphi_j$. Thus the proof is complete.* □

4.3 Closed and Key Patterns

In this subsection, we consider the case of extended condensed representations of a set $\mathcal{Q} = \{q_1, \ldots, q_n\}$ of simple mining queries with $q_i \in \mathbb{Q}$ $(i = 1, \ldots, n)$ involving a monotonic increasing measure function f in $\mathbb{I}$. To this end, recalling that $SC(q_i, f)$ is the set of all interesting closed patterns in Δ with respect to q_i and f $(i = 1, \ldots, n)$, we define the sets $\mathcal{SC}(\mathcal{Q}, f)$ and $\mathcal{SC}^*(\mathcal{Q}, f)$ as follows:

$$\mathcal{SC}(\mathcal{Q}, f) = min_{\leq_f}(\bigcup_{q \in \mathcal{Q}} SC(q, f)) \text{ and } \mathcal{SC}^*(\mathcal{Q}, f) = \{(\varphi, f(\varphi)) \mid \varphi \in \mathcal{SC}(\mathcal{Q}, f)\}$$

Example 5. *Let $\mathcal{Q} = \{q_1, q_2\}$ be the set of simple mining queries as defined in our Running Example 1. We have:*

- *$SC(q_1, f) = \{ABCE, ABE\}$ and $SC(q_2, f) = \{ABC, ACD, AB, AC, A\}$,*
- *$\mathcal{SC}(\mathcal{Q}, f) = \{ABCE, ABE, ACD, AC, A\}$ and*
- *$\mathcal{SC}^*(\mathcal{Q}, f) = \{(ABCE, 0.4), (ABE, 0.5), (ACD, 0.3), (AC, 0.5), (A, 0.8)\}$.* □

Based on Lemma 3, we can state the following proposition:

Proposition 5. *Let $\mathcal{Q} = \{q_1, \ldots, q_n\}$ be a set of simple mining queries with $q_i \in \mathbb{Q}$ $(i = 1, \ldots, n)$ and f be a monotonic increasing measure function in $\mathbb{I}$. For every $i = 1, \ldots, n$ and $\varphi \in sol(q_i)$, we have:*

$$f(\varphi) = max\{f(\varphi') \mid \varphi' \in \mathcal{SC}(\mathcal{Q}, f) \text{ and } \varphi' \preceq \varphi\}.$$

PROOF: *Let φ_i in $sol(q_i)$. Using Lemma 3, we know that:*

$$f(\varphi_i) = max\{f(\varphi_i') \mid \varphi_i' \in SC(q_i, f) \text{ and } \varphi_i' \preceq \varphi_i\}$$

Let $\varphi_i' \in SC(q_i, f)$ such that $\varphi_i' \preceq \varphi_i$ and $f(\varphi_i') = f(\varphi_i)$. Given the definition of $\mathcal{SC}(\mathcal{Q}, f)$, there exists $\varphi_j' \in \mathcal{SC}(\mathcal{Q}, f)$ such that $\varphi_j' \leq_f \varphi_i'$, i.e., $\varphi_j' \preceq \varphi_i'$ and $f(\varphi_j') = f(\varphi_i')$. Thus, there exists $\varphi_j' \in \mathcal{SC}(\mathcal{Q}, f)$ such that $\varphi_j' \preceq \varphi_i$ and $f(\varphi_j') = f(\varphi_i)$. Finally, for every $\varphi' \in \mathcal{SC}(\mathcal{Q}, f)$ such that $\varphi' \preceq \varphi_i$, we have $f(\varphi') \leq f(\varphi_i)$ since f is a monotonic increasing function. It follows that: $f(\varphi_i) = f(\varphi_j') = max\{f(\varphi') \mid \varphi_i' \in \mathcal{SC}(\mathcal{Q}, f) \text{ and } \varphi' \preceq \varphi_i\}$ which completes the proof. □

The same idea applies for key patterns. Recalling that $GK(q_i, f)$ is the set of all interesting key patterns in Δ with respect to q_i and f $(i = 1, \ldots, n)$, we define the sets $\mathcal{GK}(\mathcal{Q}, f)$ and $\mathcal{GK}^*(\mathcal{Q}, f)$ by:

$$\mathcal{GK}(\mathcal{Q}, f) = max_{\leq_f}(\bigcup_{q \in \mathcal{Q}} GK(q, f)) \text{ and } \mathcal{GK}^*(\mathcal{Q}, f) = \{(\varphi, f(\varphi)) \mid \varphi \in \mathcal{GK}(\mathcal{Q}, f)\}$$

The following proposition states how to compute $f(\varphi)$ based on the set $\mathcal{GK}(\mathcal{Q}, f)$.

Proposition 6. *Let $\mathcal{Q} = \{q_1, \ldots, q_n\}$ be a set of simple mining queries with $q_i \in \mathbb{Q}$ $(i = 1, \ldots, n)$ and f be a monotonic increasing measure function in $\mathbb{I}$. For every $i = 1, \ldots, n$ and $\varphi \in sol(q_i)$, we have:*

$$f(\varphi) = min\{f(\varphi') \mid \varphi' \in \mathcal{GK}(\mathcal{Q}, f) \text{ and } \varphi \preceq \varphi'\}.$$

PROOF: *The proof uses similar arguments as that of Proposition 5, and thus is omitted.* □

Using propositions 5, 6 and Theorem 1, the following theorem holds.

Theorem 2. *Let $\mathcal{Q} = \{q_1, \ldots, q_n\}$ be a set of mining queries with $q_i = m_i \wedge a_i$ where $m_i \in \mathbb{M}$ and $a_i \in \mathbb{A}$ $(i = 1, \ldots, n)$. Let f be a monotonic increasing measure function in $\mathbb{I}$.*

The sets $\{\Sigma(\mathcal{Q}),\ \Gamma(\mathcal{Q}),\ \mathcal{SC}^*(\mathcal{Q}, f)\}$ and $\{\Sigma(\mathcal{Q}),\ \Gamma(\mathcal{Q}),\ \mathcal{GK}^*(\mathcal{Q}, f)\}$ are extended condensed representations of $ans(\mathcal{Q}, f)$, i.e.,

$\Sigma(\mathcal{Q}),\ \Gamma(\mathcal{Q}),\ \mathcal{SC}^*(\mathcal{Q}, f) \models_e ans(\mathcal{Q}, f)$, *and*
$\Sigma(\mathcal{Q}),\ \Gamma(\mathcal{Q}),\ \mathcal{GK}^*(\mathcal{Q}, f) \models_e ans(\mathcal{Q}, f)$.

PROOF: *Let F be the function defined as follows: for every triple $(\varphi, q, \alpha) \in \mathbb{L} \times \mathbb{Q} \times \Re$, $(\varphi, q, \alpha) \in F(\mathcal{X}_1, \mathcal{X}_2, \mathcal{Y})$ if:*

- *there exists $(\varphi_1, q_1) \in \mathcal{X}_1$ such that $(\varphi_1, q_1) \leq_{\mathbb{A}} (\varphi, q)$, and*
- *there exists $(\varphi_2, q_2) \in \mathcal{X}_2$ such that $(\varphi, q) \leq_{\mathbb{M}} (\varphi_2, q_2)$, and*
- $\alpha = max\{\alpha' \mid (\exists \varphi' \in \mathbb{L})((\varphi', \alpha') \in \mathcal{Y} \text{ and } \varphi' \preceq \varphi)\}$.

Using Theorem 1 and Proposition 5, we can easily see that $ans(\mathcal{Q}, f) = F(\Sigma(\mathcal{Q}),\ \Gamma(\mathcal{Q}),\ \mathcal{SC}^*(\mathcal{Q}, f))$. Moreover, F is independent from the data set Δ since $\preceq$ and $\sqsubseteq$ do not depend on Δ. Finally, we have $\Sigma(\mathcal{Q}) \subseteq \mathcal{S}(\mathcal{Q}) \subseteq sol(\mathcal{Q})$, $\Gamma(\mathcal{Q}) \subseteq \mathcal{G}(\mathcal{Q}) \subseteq sol(\mathcal{Q})$ and $\pi_{\mathbb{L}}(\mathcal{SC}^*(\mathcal{Q}, f)) = \mathcal{SC}(\mathcal{Q}, f) \subseteq \pi_{\mathbb{L}}(sol(\mathcal{Q}))$, which shows that $\Sigma(\mathcal{Q}),\ \Gamma(\mathcal{Q}),\ \mathcal{SC}^*(\mathcal{Q}, f) \models_e ans(\mathcal{Q}, f)$. Using Theorem 1 and Proposition 6, it can be shown in the same way that $\Sigma(\mathcal{Q}),\ \Gamma(\mathcal{Q}),\ \mathcal{GK}^*(\mathcal{Q}, f) \models_e ans(\mathcal{Q}, f)$, thus the proof is complete. □

Example 6. *Let $\mathcal{Q} = \{q_1, q_2\}$ be the set of simple mining queries as defined in our Running Example 1. We recall from examples 1 and 5 that:*

- $S(q_1) = \{ABCE\}$, $S(q_2) = \{ABC, ACD\}$, $G(q_1) = \{B\}$ *and* $G(q_2) = \{A\}$,
- $SC(q_1, f) = \{ABCE, ABE\}$ *and* $SC(q_2) = \{ABC, ACD, AB, AC, A\}$,
- $\mathcal{SC}(\mathcal{Q}, f) = \{ABCE, ABE, ACD, AC, A\}$, *and*
- $\mathcal{SC}^*(\mathcal{Q}, f) = \{(ABCE, 0.4), (ABE, 0.5), (ACD, 0.3), (AC, 0.5), (A, 0.8)\}$.

Therefore, $\mathcal{S}(\mathcal{Q}) = \{(ABCE, q_1), (ABC, q_2), (ACD, q_2)\}$ and $\mathcal{G}(\mathcal{Q}) = \{(B, q_1), (A, q_2)\}$. Since $\mathcal{S}(\mathcal{Q})$ (respectively $\mathcal{G}(\mathcal{Q})$) contains no pairs comparable with respect to $\leq_{\mathbb{A}}$ (respectively $\leq_{\mathbb{M}}$), we have $\Sigma(\mathcal{Q}) = \mathcal{S}(\mathcal{Q})$ (respectively $\Gamma(\mathcal{Q}) = \mathcal{G}(\mathcal{Q})$).

Then using Theorem 2, we know that $\{\Sigma(\mathcal{Q}),\ \Gamma(\mathcal{Q}),\ \mathcal{SC}^*(\mathcal{Q}, f)\}$ is an extended condensed representation of $ans(\mathcal{Q}, f)$, i.e., $\Sigma(\mathcal{Q}),\ \Gamma(\mathcal{Q}),\ \mathcal{SC}^*(\mathcal{Q}, f) \models_e ans(\mathcal{Q}, f)$. Moreover, we note that $\mathcal{SC}^*(\mathcal{Q}, f) \subset SC(q_1, f) \cup SC(q_2, f)$. □

The previous example shows a case where the two condensed representations $\{\mathcal{S}(\mathcal{Q}),\ \mathcal{G}(\mathcal{Q})\}$ and $\{\Sigma(\mathcal{Q}),\ \Gamma(\mathcal{Q})\}$ of $sol(\mathcal{Q})$ are equal. In the next subsection, we show that these condensed representations can be made more concise under additional hypotheses that are satisfied in the traditional case of association rules mining [1].

4.4 Further Improvement

We assume now that every query $q \in \mathcal{Q}$ is of the form $q = \overline{q} \wedge \widetilde{q}$ where $\widetilde{q}$ is an independent selection predicate. Intuively, in order to further condense $\{\Sigma(\mathcal{Q}),\ \Gamma(\mathcal{Q})\}$, we compare queries based on their 'non-independent parts,' since their 'independent parts' can be evaluated without considering the underlying data set.

To this end, given a set of mining queries $\mathcal{Q} = \{q_1, \ldots, q_n\}$ where $q_i = \overline{q_i} \wedge \widetilde{q_i}$ with $\overline{q_i} \in \overline{\mathbb{Q}}$ and $\widetilde{q_i} \in \widetilde{\mathbb{Q}}$, we define two partial pre-orderings, denoted by $\leq_{\overline{\mathbb{A}}}$ and $\leq_{\overline{\mathbb{M}}}$, as follows: for all (φ_i, q_i) and (φ_j, q_j) in $\mathbb{L} \times \mathcal{Q}$:

$$(\varphi_i, q_i) \leq_{\overline{\mathbb{A}}} (\varphi_j, q_j) \text{ if } \varphi_i \preceq \varphi_j \text{ and } \overline{a_i} \sqsubseteq \overline{a_j}$$
$$(\varphi_i, q_i) \leq_{\overline{\mathbb{M}}} (\varphi_j, q_j) \text{ if } \varphi_i \preceq \varphi_j \text{ and } \overline{m_j} \sqsubseteq \overline{m_i}.$$

Then, we introduce the following notations:

$$\overline{\Sigma}(\mathcal{Q}) = min_{\leq_{\overline{\mathbb{A}}}}(\mathcal{S}(\mathcal{Q})) \qquad \text{and} \qquad \overline{\Gamma}(\mathcal{Q}) = max_{\leq_{\overline{\mathbb{M}}}}(\mathcal{G}(\mathcal{Q})).$$

The following lemma states how, for every q in $\mathcal{Q}$, $sol(q)$ can be computed based on $\overline{\Sigma}(\mathcal{Q})$ and $\overline{\Gamma}(\mathcal{Q})$, assuming that the independent part $\widetilde{q}$ of q is known.

Lemma 5. *Let $\mathcal{Q} = \{q_1, \ldots, q_n\}$ be a set of mining queries $q_i = \overline{q_i} \wedge \widetilde{q_i}$ where $\overline{q_i} = \overline{m_i} \wedge \overline{a_i}$ with $\overline{m_i} \in \overline{\mathbb{M}}$, $\overline{a_i} \in \overline{\mathbb{A}}$, and $\widetilde{q_i} = \widetilde{m_i} \wedge \widetilde{a_i}$ with $\widetilde{m_i} \in \widetilde{\mathbb{M}}$, $\widetilde{a_i} \in \widetilde{\mathbb{A}}$ $(i = 1, \ldots, n)$. For every q in $\mathcal{Q}$, we have:*

$$sol(q) = \{\varphi \in sol(\widetilde{q}) \mid (\exists(\varphi_i, q_i) \in \overline{\Sigma}(\mathcal{Q}))((\varphi_i, q_i) \leq_{\overline{\mathbb{A}}} (\varphi, q) \text{ and } (\exists(\varphi_j, q_j) \in \overline{\Gamma}(\mathcal{Q}))((\varphi, q) \leq_{\overline{\mathbb{M}}} (\varphi_j, q_j))\}.$$

PROOF: *Let $\overline{X}(q)$ be the set defined by:*

$$\overline{X}(q) = \{\varphi \in sol(\widetilde{q}) \mid (\exists(\varphi_i, q_i) \in \overline{\Sigma}(\mathcal{Q}))((\varphi_i, q_i) \leq_{\overline{\mathbb{A}}} (\varphi, q) \text{ and } (\exists(\varphi_j, q_j) \in \overline{\Gamma}(\mathcal{Q}))((\varphi, q) \leq_{\overline{\mathbb{M}}} (\varphi_j, q_j))\}.$$

We first show that $\overline{X}(q) \subseteq sol(q)$. Let $\varphi \in \overline{X}(q)$. There exist $(\varphi_i, q_i) \in \overline{\Sigma}(\mathcal{Q})$ and $(\varphi_j, q_j) \in \overline{\Gamma}(\mathcal{Q})$ such that $(\varphi_i, q_i) \leq_{\overline{\mathbb{A}}} (\varphi, q)$ and $(\varphi, q) \leq_{\overline{\mathbb{M}}} (\varphi_j, q_j)$.

On one hand, we know that $q_i(\varphi_i) = true$. Thus, we have $\overline{a_i}(\varphi_i) = true$. It follows that $\overline{a}(\varphi_i) = true$ since $\overline{a_i} \sqsubseteq \overline{a}$, and so, $\overline{a}(\varphi) = true$ since $\varphi_i \preceq \varphi$ and $\overline{a}$ is anti-monotonic.

On the other hand, we know that $q_j(\varphi_j) = true$. Thus, we have $\overline{m_j}(\varphi_j) = true$. It follows that $\overline{m}(\varphi_j) = true$ since $\overline{m_j} \sqsubseteq \overline{m}$, and so, $\overline{m}(\varphi) = true$ since $\varphi \preceq \varphi_j$ and $\overline{m}$ is monotonic. Therefore, we have $\overline{q}(\varphi) = true$. Since $\varphi \in sol(\widetilde{q})$, we have $q(\varphi) = \overline{q}(\varphi) \wedge \widetilde{q}(\varphi) = true$, which shows that $\overline{X}(q) \subseteq sol(q)$.

Now, we show that $sol(q) \subseteq \overline{X}(q)$. Let $\varphi \in sol(q)$. There exist $\varphi_s \in S(q)$ and $\varphi_g \in G(q)$ such that $\varphi_s \preceq \varphi \preceq \varphi_g$. Moreover, we have $(\varphi_s, q) \in \mathcal{S}(\mathcal{Q})$ and $(\varphi_g, q) \in \mathcal{G}(\mathcal{Q})$.

Given the definitions of $\overline{\Sigma}(\mathcal{Q})$ and $\overline{\Gamma}(\mathcal{Q})$, there exist $(\varphi_i, q_i) \in \overline{\Sigma}(\mathcal{Q})$ and $(\varphi_j, q_j) \in \overline{\Gamma}(\mathcal{Q})$ such that $(\varphi_i, q_i) \leq_{\overline{\mathbb{A}}} (\varphi_s, q)$ and $(\varphi_g, q) \leq_{\overline{\mathbb{M}}} (\varphi_j, q_j)$. Moreover, we have $(\varphi_s, q) \leq_{\overline{\mathbb{A}}} (\varphi, q)$ since $\varphi_s \preceq \varphi$, and $(\varphi, q) \leq_{\overline{\mathbb{M}}} (\varphi_g, q)$ since $\varphi \preceq \varphi_g$. Thus, $(\varphi_i, q_i) \leq_{\overline{\mathbb{A}}} (\varphi, q)$ and $(\varphi, q) \leq_{\overline{\mathbb{M}}} (\varphi_j, q_j)$. As $\varphi \in sol(q)$ and as $sol(q) \subseteq sol(\widetilde{q})$, it follows that $\varphi \in \overline{X}(q)$, which entails that $sol(q) \subseteq \overline{X}(q)$. Thus, the proof is complete. □

Based on Lemma 5 above, we can state the following theorem.

Theorem 3. *Let $\mathcal{Q} = \{q_1, \ldots, q_n\}$ be a set of mining queries $q_i = \overline{q_i} \wedge \widetilde{q_i}$ where $\overline{q_i} = \overline{m_i} \wedge \overline{a_i}$ with $\overline{m_i} \in \overline{\mathbb{M}}$, $\overline{a_i} \in \overline{\mathbb{A}}$, and $\widetilde{q_i} = \widetilde{m_i} \wedge \widetilde{a_i}$ with $\widetilde{m_i} \in \widetilde{\mathbb{M}}$, $\widetilde{a_i} \in \widetilde{\mathbb{A}}$ ($i = 1, \ldots, n$). The set $\{\overline{\Sigma}(\mathcal{Q}),\ \overline{\Gamma}(\mathcal{Q})\}$ is a condensed representation of $sol(\mathcal{Q})$, i.e., $\overline{\Sigma}(\mathcal{Q}),\ \overline{\Gamma}(\mathcal{Q}) \models sol(\mathcal{Q})$.*

Proof: *Let us consider the function F defined by:*

$$\begin{aligned} F(\mathcal{X}_1, \mathcal{X}_2) = \{\ & (\varphi, q) \in \mathbb{L} \times \mathbb{Q} \mid \varphi \in sol(\widetilde{q}) \text{ and} \\ & (\exists(\varphi_1, q_1) \in \mathcal{X}_1)((\varphi_1, q_1) \leq_{\overline{\mathbb{A}}} (\varphi, q)) \text{ and} \\ & (\exists(\varphi_2, q_2) \in \mathcal{X}_2)((\varphi, q) \leq_{\overline{\mathbb{M}}} (\varphi_2, q_2))\}) \end{aligned}$$

Using Lemma 5, we can easily see that $sol(\mathcal{Q}) = F(\overline{\Sigma}(\mathcal{Q}), \overline{\Gamma}(\mathcal{Q}))$. Moreover, F is independent from the data set Δ since $\preceq$ and $\sqsubseteq$ do not depend on Δ. Finally, for every pair (φ, q) in $\overline{\Sigma}(\mathcal{Q})$ or $\overline{\Gamma}(\mathcal{Q})$, we know that $(\varphi, q) \in sol(\mathcal{Q})$. Thus, we have $\pi_{\mathbb{L}}(\overline{\Sigma}(\mathcal{Q}) \cup \overline{\Gamma}(\mathcal{Q})) \subseteq \pi_{\mathbb{L}}(sol(\mathcal{Q}))$, which completes the proof. □

Unfortunately, as shown in the following example, the condensed representations $\{\Sigma(\mathcal{Q}),\ \Gamma(\mathcal{Q})\}$ and $\{\overline{\Sigma}(\mathcal{Q}),\ \overline{\Gamma}(\mathcal{Q})\}$ are not comparable in general. Intuitively, this is due to the fact that $(\varphi_1, q_1) \leq_{\mathbb{A}} (\varphi_2, q_2)$ can hold whereas $(\varphi_1, q_1) \leq_{\overline{\mathbb{A}}} (\varphi_2, q_2)$ does not, or conversely.

Example 7. *In the context of our Running Example 1, let $q_5 = m_5 \wedge a_5$ and $q_6 = m_6 \wedge a_6$ where m_5, m_6, a_5 and a_6 are defined for every $\varphi \in \mathbb{L}$ by:*

- $m_5(\varphi, \Delta) = true$ *if* $sup(\varphi, \Delta) \leq 0.8$ *and* $A \subseteq \varphi$,
- $m_6(\varphi, \Delta) = true$ *if* $sup(\varphi, \Delta) \leq 0.9$ *and* $AC \subseteq \varphi$,
- $a_5(\varphi, \Delta) = true$ *if* $sup(\varphi, \Delta) \geq sup(AB, \Delta)$ *and* $\varphi \subseteq AC$,
- $a_6(\varphi, \Delta) = true$ *if* $sup(\varphi, \Delta) \geq sup(AC, \Delta)$ *and* $\varphi \subseteq ABC$.

We note that m_5 and m_6 are monotonic, whereas a_5 and a_6 are anti-monotonic. Moreover, we can see that $S(q_5) = \{AC\}$, $S(q_6) = \{AC\}$, $G(q_5) = \{A\}$ and $G(q_6) = \{AC\}$.

Now, considering $\mathcal{Q} = \{q_5, q_6\}$, we have: $\mathcal{S}(\mathcal{Q}) = \{(AC, q_5), (AC, q_6)\}$ and $\mathcal{G}(\mathcal{Q}) = \{(A, q_5),\ (AC, q_6)\}$. Moreover, we have $(AC, q_5) <_{\mathbb{A}} (AC, q_6)$, whereas (A, q_5) and (AC, q_6) are not comparable with respect to $\leq_{\mathbb{M}}$. Therefore, we have:

$$\Sigma(\mathcal{Q}) = \{(AC, q_5)\} \text{ and } \Gamma(\mathcal{Q}) = \{(A, q_5), (AC, q_6)\}.$$

On the other hand, $(AC, q_6) <_{\overline{\mathbb{M}}} (A, q_5)$, whereas (AC, q_5) and (AC, q_6) are not comparable with respect to $\leq_{\overline{\mathbb{A}}}$. Therefore, we have:

$$\overline{\Sigma}(\mathcal{Q}) = \{(AC, q_5), (AC, q_6)\} \text{ and } \overline{\Gamma}(\mathcal{Q}) = \{(A, q_5)\}.$$

Hence, we have $\Sigma(\mathcal{Q}) \subset \overline{\Sigma}(\mathcal{Q})$ and $\overline{\Gamma}(\mathcal{Q}) \subset \Gamma(\mathcal{Q})$, which shows that $\{\Sigma(\mathcal{Q}),\ \Gamma(\mathcal{Q})\}$ and $\{\overline{\Sigma}(\mathcal{Q}),\ \overline{\Gamma}(\mathcal{Q})\}$ are not comparable. □

The following lemma states a sufficient condition when $\{\overline{\Sigma}(\mathcal{Q}),\ \overline{\Gamma}(\mathcal{Q})\}$ is more concise than $\{\Sigma(\mathcal{Q}),\ \Gamma(\mathcal{Q})\}$. Intuitively, according to this condition, the anti-monotonic (respectively monotonic) queries to be considered must satisfy the fact that if two queries are comparable, then their dependent part are comparable as well.

Lemma 6. *Let $\mathcal{Q} = \{q_1, \ldots, q_n\}$ be a set of mining queries $q_i = \overline{q_i} \wedge \widetilde{q_i}$ where $\overline{q_i} = \overline{m_i} \wedge \overline{a_i}$ with $\overline{m_i} \in \overline{\mathbb{M}}$, $\overline{a_i} \in \overline{\mathbb{A}}$, and $\widetilde{q_i} = \widetilde{m_i} \wedge \widetilde{a_i}$ with $\widetilde{m_i} \in \widetilde{\mathbb{M}}$, $\widetilde{a_i} \in \widetilde{\mathbb{A}}$ $(i = 1, \ldots, n)$.*

If for every $(a_i, a_j) \in \mathbb{A}^2$ such that $a_i \sqsubseteq a_j$, we have $\overline{a_i} \sqsubseteq \overline{a_j}$, and for every $(m_i, m_j) \in \mathbb{M}^2$ such that $m_i \sqsubseteq m_j$, we have $\overline{m_i} \sqsubseteq \overline{m_j}$, then $\{\overline{\Sigma}(\mathcal{Q}),\ \overline{\Gamma}(\mathcal{Q})\}$ is more concise than $\{\Sigma(\mathcal{Q}),\ \Gamma(\mathcal{Q})\}$.

PROOF: *Assume that for every $(a_i, a_j) \in \mathbb{A}^2$ such that $a_i \sqsubseteq a_j$, we have $\overline{a_i} \sqsubseteq \overline{a_j}$. Then, for all pairs (φ_1, q_1) and (φ_2, q_2) in $\mathcal{S}(\mathcal{Q})$ such that $(\varphi_1, q_1) \leq_{\mathbb{A}} (\varphi_2, q_2)$, we also have $(\varphi_1, q_1) \leq_{\overline{\mathbb{A}}} (\varphi_2, q_2)$. Hence, we have $\overline{\Sigma}(\mathcal{Q}) \subseteq \Sigma(\mathcal{Q})$. In the same way, we can see that if for every $(m_i, m_j) \in \mathbb{M}^2$ such that $m_i \sqsubseteq m_j$, we have $\overline{m_i} \sqsubseteq \overline{m_j}$, then $\overline{\Gamma}(\mathcal{Q}) \subseteq \Gamma(\mathcal{Q})$. Thus, the proof is complete.* □

In what follows, we identify a case where the previous lemma applies. This case makes use of the notion of *dense* measure function, defined by:

Definition 11. *Let f be a measure function defined over $\Lambda \subseteq \Re$. We say that f is* dense *in Λ with respect to $\mathbb{L}$, if for every pair $(\lambda_1, \lambda_2) \in \Lambda^2$ such that $\lambda_1 < \lambda_2$ and every pattern $\varphi \in \mathbb{L}$, there exists a data set Δ such that $\lambda_1 < f(\varphi, \Delta) < \lambda_2$.*

Then, we have the following.

Proposition 7. *Let f be a increasing measure function defined from $\mathbb{L} \times \mathbf{\Delta}$ over $\Lambda \subseteq \Re$ such that f is dense in Λ with respect to $\mathbb{L}$.*

Let $\overline{\mathbb{Q}}_f = \overline{\mathbb{A}}_f \cup \overline{\mathbb{M}}_f$ where $\overline{\mathbb{A}}_f = \{\overline{a}_\lambda \mid \lambda \in \Lambda\}$ and $\overline{\mathbb{M}}_f = \{\overline{m}_\lambda \mid \lambda \in \Lambda\}$ are two sets of selection predicates defined by: for every data set Δ and every pattern $\varphi \in \mathbb{L}$, $\overline{a}_\lambda(\varphi, \Delta) = true$ if $f(\varphi, \Delta) \geq \lambda$, and $\overline{m}_\lambda(\varphi, \Delta) = true$ if $f(\varphi, \Delta) \leq \lambda$.

Let $\widetilde{\mathbb{A}}$ and $\widetilde{\mathbb{M}}$ be two sets of independent selection predicates such that for every $\widetilde{a}$ in $\widetilde{\mathbb{A}}$ (respectively $\widetilde{m} \in \widetilde{\mathbb{M}}$), $\widetilde{a}$ is anti-monotonic (respectively $\widetilde{m}$ is monotonic) and $sol(\widetilde{a}) \neq \emptyset$ (respectively $sol(\widetilde{m}) \neq \emptyset$).

Let $\mathcal{Q} = \{q_1, \ldots, q_n\}$ be a set of mining queries $q_i = \overline{q_i} \wedge \widetilde{q_i}$ where $\overline{q_i} = \overline{m_i} \wedge \overline{a_i}$ with $\overline{m_i} \in \overline{\mathbb{M}}_f$, $\overline{a_i} \in \overline{\mathbb{A}}_f$, and $\widetilde{q_i} = \widetilde{m_i} \wedge \widetilde{a_i}$ with $\widetilde{m_i} \in \widetilde{\mathbb{M}}$, $\widetilde{a_i} \in \widetilde{\mathbb{A}}$ $(i = 1, \ldots, n)$. Then, $\{\overline{\Sigma}(\mathcal{Q}),\ \overline{\Gamma}(\mathcal{Q})\}$ is more concise than $\{\Sigma(\mathcal{Q}),\ \Gamma(\mathcal{Q})\}$.

PROOF: *Using the notation of the proposition, based on Lemma 6, we have to show that for every $i, j = \{1, \ldots, n\}$, if $a_i \sqsubseteq a_j$, then $\overline{a_i} \sqsubseteq \overline{a_j}$ and that if $m_i \sqsubseteq m_j$, then $\overline{m_i} \sqsubseteq \overline{m_j}$.*

Assuming that $a_i \sqsubseteq a_j$ and $\overline{a_i} \not\sqsubseteq \overline{a_j}$ implies that there exist two reals λ_i and λ_j such that for every pattern $\varphi \in \mathbb{L}$ and every data set Δ, $\overline{a}_i(\varphi, \Delta) = true$ if $f(\varphi, \Delta) \geq \lambda_i$ and $\overline{a}_j(\varphi, \Delta) = true$ if $f(\varphi, \Delta) \geq \lambda_j$. If $\overline{a_i} \not\sqsubseteq \overline{a_j}$, we necessarily have $\lambda_i < \lambda_j$.

Moreover, given a pattern $\varphi \in sol(\widetilde{a_i})$, there exists a data set Δ such that $\lambda_i < f(\varphi, \Delta) < \lambda_j$. Then, we have $\varphi \in sol(a_i/\Delta)$ and $\varphi \notin sol(a_j/\Delta)$, which contradicts the hypothesis $a_i \sqsubseteq a_j$.

Using similar arguments as above, it can shown that if $m_i \sqsubseteq m_j$, then $\overline{m_i} \sqsubseteq \overline{m_j}$, which completes the proof. □

Now, we note that the previous proposition applies in the traditional case of association rules where $\mathbb{L} = 2^{Items} \setminus \{\emptyset, Items\}$ and the measure function is the function *sup*. Indeed, it is easy to see that the function *sup* is dense in $[0,1]$ with respect to $\mathbb{L} = 2^{Items} \setminus \{\emptyset, Items\}$.

The following example shows how Proposition 7 applies in the context of our Running Example 1.

Example 8. *Let $\mathcal{Q} = \{q_1, q_2\}$ be the set of simple mining queries $q_i = m_i \wedge a_i$ where m_i and a_i $(i = 1, 2)$ are defined in our Running Example 1.*

We recall from Example 6 that $\mathcal{S}(\mathcal{Q}) = \{(ABCE, q_1), (ABC, q_2), (ACD, q_2)\}$ and $\mathcal{G}(\mathcal{Q}) = \{(B, q_1), (A, q_2)\}$. Moreover, we also recall that $\Sigma(\mathcal{Q}) = \mathcal{S}(\mathcal{Q})$ and $\Gamma(\mathcal{Q}) = \mathcal{G}(\mathcal{Q})$.

Since $ABC \subseteq ABCE$ $(ABCE \preceq ABC)$ and $\overline{a_1} \sqsubseteq \overline{a_2}$, we have $(ABCE, q_1) \leq_{\overline{\mathbb{A}}} (ABC, q_2)$. Thus, the pair (ABC, q_2) does not belong to $\overline{\Sigma}(\mathcal{Q})$. Hence, we have

$$\overline{\Sigma}(\mathcal{Q}) = \{(ABCE, q_1), (ACD, q_2)\}.$$

Then, since the pairs (B, q_1) and (A, q_2) are not comparable with respect to $\leq_{\overline{\mathbb{M}}}$, we have $\overline{\Gamma}(\mathcal{Q}) = \overline{\mathcal{G}}(\mathcal{Q})$. In conclusion, using Theorem 3, we can see that $\overline{\Sigma}(\mathcal{Q})$, $\overline{\Gamma}(\mathcal{Q}) \models sol(\mathcal{Q})$. Moreover, since $\overline{\Sigma}(\mathcal{Q}) \subset \Sigma(\mathcal{Q})$ and $\overline{\Gamma}(\mathcal{Q}) \subseteq \Gamma(\mathcal{Q})$, it is easy to see that $\{\overline{\Sigma}(\mathcal{Q}),\ \overline{\Gamma}(\mathcal{Q})\}$ is more concise than $\{\Sigma(\mathcal{Q}),\ \Gamma(\mathcal{Q})\}$. □

Finally, regarding extended condensed representations, we can easily prove the following theorem, based on propositions 5 and 6 and on Theorem 3.

Theorem 4. *Let f be a monotonic increasing measure function in $\mathbb{I}$ and $\mathcal{Q} = \{q_1, \ldots, q_n\}$ be a set of mining queries $q_i = \overline{q_i} \wedge \widetilde{q_i}$ where $\overline{q_i} = \overline{m_i} \wedge \overline{a_i}$ with $\overline{m_i} \in \overline{\mathbb{M}}$, $\overline{a_i} \in \overline{\mathbb{A}}$, and $\widetilde{q_i} = \widetilde{m_i} \wedge \widetilde{a_i}$ with $\widetilde{m_i} \in \widetilde{\mathbb{M}}$, $\widetilde{a_i} \in \widetilde{\mathbb{A}}$ $(i = 1, \ldots, n)$. The sets $\{\overline{\Sigma}(\mathcal{Q}),\ \overline{\Gamma}(\mathcal{Q}),\ \mathcal{SC}^*(\mathcal{Q}, f)\}$ and $\{\overline{\Sigma}(\mathcal{Q}),\ \overline{\Gamma}(\mathcal{Q}),\ \mathcal{GK}^*(\mathcal{Q}, f)\}$ are extended condensed representations of $ans(\mathcal{Q}, f)$, i.e., $\overline{\Sigma}(\mathcal{Q}),\ \overline{\Gamma}(\mathcal{Q}),\ \mathcal{SC}^*(\mathcal{Q}, f) \models_e ans(\mathcal{Q}, f)$ and $\overline{\Sigma}(\mathcal{Q}),\ \overline{\Gamma}(\mathcal{Q}),\ \mathcal{GK}^*(\mathcal{Q}, f) \models_e ans(\mathcal{Q}, f)$.*

5 Conclusion

In this paper, we have considered the problem of defining condensed representations of sets of mining queries. To this end, we have first studied the case of a single mining query and we have extended previous works on version spaces by [9] so as to take into account the presence of measure functions in the query. This has been done based on the well known notions of closed and key patterns ([3,16]). Then, we have seen how to extend this approach to *sets* of mining queries. The main idea in this extension is that, in order to obtain condensed representations in this case, when storing a pattern, one must keep track of the query the pattern comes from.

Based on this work, we are currently investigating how condensed representations can be used to optimize the iterative computation of the answer of mining queries. This problem has been studied for standard association rules [2,10,13,14] and multi-dimensional association rules [8,15]. In our framework, this problem can be stated as follows: given a data set Δ, a set $\mathcal{Q} = \{q_1, \ldots, q_n\}$ of mining queries and a new extended mining query (q, f):

1. How to optimize the computation of $ans(q, f)$ using the extended condensed representations of $ans(\mathcal{Q}, f)$?

2. How to efficiently modify the extended condensed representation of $ans(\mathcal{Q}, f)$ so as to obtain an extended condensed representation of $ans(\mathcal{Q} \cup \{q\}, f)$?

Moreover, it is clear that some tests are necessary to compare the various condensed representations proposed in this paper. To this end, we are implementing our approach in the context of our previous work [8], where mining queries are composed through relational operators. We also investigate how our approach can be used to optimize the iterative computation of iceberg cubes [11].

References

1. R. Agrawal, H. Mannila, R. Srikant, H. Toivonen, A.I. Verkamo (1996). *Fast Discovery of Association Rules.* In Advances in Knowledge Discovery and Data Mining, pp 309–328, AAAI-MIT Press.
2. E. Baralis and G. Psaila (1999). *Incremental Refinement of Mining Queries.* In Proc. of DAWAK'99, pp. 173–182, Florence.
3. Y. Bastide, R. Taouil, N. Pasquier, G. Stumme and L. Lakhal (2000). *Mining Frequent Patterns with Counting Inference.* SIGKDD Explorations, 2(2), p. 66–75.
4. J.-F. Boulicaut, A. Bykowski and C. Rigotti (2000). *Approximation of Frequency Queries by Means of Free-Sets.* In Proc. of PKDD'00, LNCS vol. 1910, pp. 75–85, Springer-Verlag.
5. J.-F. Boulicaut (2001). *Habilitation thesis* (French). INSA-Lyon, France.
6. L. De Raedt and S. Kramer (2001). *The Levelwise Version Space Algorithm and its Application to Molecular Fragment Finding.* In Proc. of IJCAI'01, pp. 853–862.
7. L. De Raedt (2002). *Query execution and optimization for inductive databases.* In Proc. of International Workshop DTDM'02, In conjunction with EDBT 2002, pp. 19–28 (Extended Abstract), Praha, CZ.
8. C.T. Diop, A. Giacometti, D. Laurent and N. Spyratos (2002). *Composition of Mining Contexts for Efficient Extraction of Association Rules.* In Proc. of the EDBT'02, LNCS vol. 2287, pp. 106–123, Springer-Verlag.
9. H. Hirsh (1994). *Generalizing Version Spaces.* Machine Learning, Vol. 17(1), pp. 5–46, Kluwer Academic Publishers.
10. B. Jeudy, J-F. Boulicaut (2002). *Using condensed representations for interactive association rule mining.* In Proc. of ECML/PKDD 2002, Helsinki, LNAI vol. 2431, pp. 225–236, Springer-Verlag.
11. M. Laporte, N. Novelli, R. Cicchetti, L. Lakhal (2002). *Computing Full and Iceberg Datacubes Using Partitions.* In Proc. of ISMIS'2002, LNAI vol. 2366, pp. 244–254, Springer-Verlag.
12. H. Mannila, H. Toivonen (1997). *Levelwise Search and Borders of Theories in Knowledge Discovery.* Techn. Rep. C-1997-8, University of Helsinki.
13. T. Morzy, M. Wojciechowski and M. Zakrzewicz (2000). *Materialized Data Mining Views.* In Proc. of PKDD'2000, LNCS vol. 1910, pp. 65–74, Springer-Verlag.
14. B. Nag, P. Deshpande and D.J. DeWitt (1999). *Using a Knowledge Cache for Interactive Discovery of Association Rules.* In Proc. of KDD'99, pp. 244–253, San Diego, USA.
15. B. Nag, P. Deshpande and D.J. DeWitt (2001). *Caching for Multi-dimensional Data Mining Queries.* In Proc. of SCI'2001, Orlando, Florida.
16. N. Pasquier, Y. Bastide, R. Taouil and L. Lakhal (1999). *Efficient Mining of Association Rules using Closed Itemsets Lattices.* Information Systems, Vol. 24(1), pp. 25–46, Elsevier Publishers.

One-Sided Instance-Based Boundary Sets

Evgueni N. Smirnov, Ida G. Sprinkhuizen-Kuyper, and H. Jaap van den Herik

IKAT, Department of Computer Science, Universiteit Maastricht,
P.O.BOX 616, 6200 MD Maastricht, The Netherlands
{smirnov, kuyper, herik}@cs.unimaas.nl

Abstract. Instance retraction is a difficult problem for concept learning by version spaces. This chapter introduces a family of version-space representations called one-sided instance-based boundary sets. They are correct and efficiently computable representations for admissible concept languages. Compared to other representations, they are the most efficient useful[1] version-space representations for instance retraction.

1 Introduction

Currently, there is a renewed interest in version spaces caused by their applicability in inductive databases [5,6]. This chapter considers version spaces when the inductive query constraints are instances of a concept to be learned; i.e., the task is essentially a concept-learning task. In this context we study two important problems of inductive databases: the problem of efficiently representing version spaces and the problem of efficiency of version spaces for instance retraction.

Mitchell defined version spaces as sets of concept descriptions that are consistent with training data [7,8]. Version-space learning is an incremental process:

- *If an instance i is added*, the version space is revised so that it consists of all the concept descriptions consistent with the processed training data *plus* i.
- *If an instance i is retracted*, the version space is revised so that it consists of all the concept descriptions consistent with the processed training data *minus* i.

For the learning processes version spaces are represented. The standard representation is the boundary-set representation [7,8]. It is correct for the class of admissible concept languages [7], but its size can grow exponentially in the size of training data [1]. To overcome this problem alternative version-space representations were introduced [2,3,4,9,10,11,12,13]. They extended the classes of concept languages for which version spaces are efficiently computable.

However a remaining problem for most version-space representations is that they are inefficient for instance retraction, since they lack a structure that determines the influence of an individual training instance. Hence, if a training instance is retracted, the representations are recomputed [11]. To avoid this problem two version-space representations were proposed. The first one is the

[1] In this chapter, the notion useful has a technical meaning defined in subsection 2.2.

R. Meo et al. (Eds.): Database Support for Data Mining Applications, LNAI 2682, pp. 270–288, 2004.

training-instance representation [3]. By its definition it is efficient for instance retraction. However, the representation has only a theoretical value, since the classification of each instance requires search in the concept language using all the training data. The second representation is instance-based boundary sets (IBBS) [11,12]. It is correct and efficiently computable for the class of admissible concept languages. The instance-retraction algorithm of the IBBS is efficient and it does not recompute the representation. At the moment the IBBS is the most efficient useful version-space representation for instance retraction.

In this chapter we address the question whether it is possible to design new version-space representations that are even more efficient than the IBBS in terms of computability and instance retraction. To answer the question we introduce a family of version-space representations called one-sided instance-based boundary sets. The family consists of two dual representations: instance-based maximal boundary sets (IBMBS) and instance-based minimal boundary sets (IBmBS). Without loss of generality we consider in detail only the IBMBS representation.

The course of the chapter is as follows. In section 2 we formalise the necessary basic notions. They are used in section 3 to define the IBMBS representation. There, we prove that the representation is correct for the class of admissible concept languages and derive the conditions for finiteness. Section 4 presents four IBMBS algorithms for instance addition, instance retraction, version-space collapse and instance classification. It is shown that the IBMBS can be used for instance classification in the presence of noisy training data. In sections 5 and 6 we provide an analysis and an evaluation of usefulness of the IBMBS. The dual representation of the IBMBS, instance-based minimal boundary sets (IBmBS), is touched upon in section 7. We compare the new representations with relevant work in section 8. Finally, in section 9 conclusions are given.

2 Formalisation

This section formalises the necessary basic notions. In subsection 2.1 we formulate the concept-learning task. Then, in subsection 2.2, we introduce version spaces as a solution of the task and we consider the notion of version-space representations together with their characteristics of usefulness. In this context we present the class of admissible concept languages in subsection 2.3.

2.1 The Concept-Learning Task

Concept learning assumes the presence of a universe of all the instances [11].

Notation 1. The universe of all the instances is denoted by I.

Definition 2 (Concept). *A concept $\mathfrak{c}$ is subset of I: $\mathfrak{c} \subseteq I$.*

Given a concept, there exist two types of instance sets.

Definition 3 (Set of Positive/Negative Instances). *A set $I^+ \subseteq I$ is a set of positive instances of a concept $\mathfrak{c} \subseteq I$ if and only if $I^+ \subseteq \mathfrak{c}$. A set $I^- \subseteq I$ is a set of negative instances of a concept $\mathfrak{c} \subseteq I$ if and only if $I^- \cap \mathfrak{c} = \emptyset$.*

The set of all the concepts defined on the universe I is the power set $P(I)$. To represent concepts from $P(I)$ we introduce a language.

Definition 4 (Concept Language). *The concept language Lc is a set of descriptions c.*

To associate a description $c \in Lc$ with a concept $\mathfrak{c} \in P(I)$ that c represents, we define a function $\mathcal{R}_{\mathfrak{c}}$.

Definition 5. *The function $\mathcal{R}_{\mathfrak{c}} : Lc \rightarrow P(I)$ is an injective function that maps a concept description $c \in Lc$ to a concept $\mathfrak{c} \in P(I)$.*

Since $\mathcal{R}_{\mathfrak{c}}$ is a function, no two distinct concepts in $P(I)$ can be represented by the same description in Lc. Since $\mathcal{R}_{\mathfrak{c}}$ is injective, no two distinct descriptions in Lc can represent the same concept in $P(I)$.

Instances are related to concepts by the membership relation. The relation is projected into a cover relation between instances and concept descriptions [7].

Definition 6 (Cover Relation M).
$M : Lc \times I \rightarrow Boolean$
defined by: $M(c, i) \leftrightarrow i \in \mathcal{R}_{\mathfrak{c}}(c)$.

The cover relation M holds for a description $c \in Lc$ and an instance $i \in I$ if and only if i is a member of the concept $\mathcal{R}_{\mathfrak{c}}(c)$. If the relation M holds for $c \in Lc$ and $i \in I$, we say that c covers i; otherwise, we say that c does not cover i.

After the introduction of the elements of the concept-learning task we formulate the task itself according to [7,8,11]. *Given a universe I of all the instances, a concept language Lc, a cover relation M, and the training sets I^+ and I^- of a target concept, the task is to find descriptions of the target concept in Lc.*

2.2 Version Spaces

A version space is a solution of the concept-learning task. It is a set of all the concept descriptions that are consistent with the training sets I^+ and I^- [7,8]. A description $c \in Lc$ is consistent with the sets I^+ and I^- if and only if c covers each instance $p \in I^+$ and does not cover any instance $n \in I^-$. Below we give a formal definition of version spaces.

Definition 7 (Version Space). *Given the training sets I^+ and I^- of a target concept, the version space $VS(I^+, I^-)$ is defined as follows:*

$$VS(I^+, I^-) = \{c \in Lc \mid (\forall p \in I^+)M(c, p) \wedge (\forall n \in I^-)\neg M(c, n)\}.$$

To learn version spaces we need a representation. A version-space representation is a structure that contains "information needed to reconstruct every description in the version space" [7]. It has four possible characteristics: (1) compactness; (2) finiteness; (3) efficient computability [2,11]; and (4) efficiency of the algorithms for instance addition, instance retraction, version-space collapse and instance classification. To encapsulate these characteristics we introduce the notion of a useful version-space representation.

Definition 8 (Useful Version-Space Representation). *A version-space representation is useful if and only if it is compact, finite, efficiently computable and has efficient algorithms for instance addition, instance retraction, version-space collapse, and instance classification.*

2.3 Admissible Concept Languages

The key to find a compact version-space representation is to observe that concept languages can be ordered. This can be done by a partially-ordering relation "more general". The relation is taken from [7,8] and is defined below.

Definition 9 (Relation "More General" ($\geq$)).

$$(\forall c_1, c_2 \in Lc)((c_1 \geq c_2) \leftrightarrow (\forall i \in I)(M(c_1, i) \leftarrow M(c_2, i))).$$

A description $c_1 \in Lc$ is more general than a description $c_2 \in Lc$ ($c_1 \geq c_2$) if and only if for each instance $i \in I$ if c_2 covers i, then c_1 covers i as well. If a description $c_1 \in Lc$ is more general than a description $c_2 \in Lc$ we say that c_1 is a generalisation of c_2 and c_2 is a specialisation of c_1.

If the relation "$\geq$" is defined on a concept language Lc, then Lc is partially-ordered. For defining version-space representations one class of partially-ordered languages was extensively used, viz. the class of admissible concept languages [7,11]. It is introduced after the definition of minimal and maximal sets of a partially-ordered set (cf. [7]).

Definition 10 (Minimal/Maximal Set). *If C is a partially-ordered set, then:*

$$MIN(C) = \{c \in C | (\forall c' \in C)((c \geq c') \rightarrow (c' = c))\}$$
$$MAX(C) = \{c \in C | (\forall c' \in C)((c' \geq c) \rightarrow (c' = c))\}.$$

A partially-ordered concept language Lc is admissible if each subset of Lc is bounded. A partially-ordered set is bounded if all the elements of the set are between its minimal and maximal elements. Below we give a formal definition.

Definition 11 (Admissible Concept Language). *A partially-ordered concept language Lc is admissible if and only if for every nonempty subset $C \subseteq Lc$:*

$$C \subseteq \{c \in Lc | (\exists s \in MIN(C))(c \geq s) \wedge (\exists g \in MAX(C))(g \geq c)\}.$$

Given a version space $VS(I^+, I^-)$ in an admissible concept language, the maximal set of $VS(I^+, I^-)$ is known as the maximal boundary set of $VS(I^+, I^-)$ and the minimal set of $VS(I^+, I^-)$ as the minimal boundary set of $VS(I^+, I^-)$.

Notation 12. The maximal boundary set and the minimal boundary set of version space $VS(I^+, I^-)$ are denoted by $G(I^+, I^-)$ and $S(I^+, I^-)$, respectively.

3 Instance-Based Maximal Boundary Sets

Below we introduce instance-based maximal boundary sets (IBMBS) as a new version-space representation. The correctness of the representation is proven for the class of admissible concept languages. The IBMBS are shown to be compact and their conditions for finiteness are derived.

3.1 Definition and Correctness

The IBMBS representation consists of the set of positive training instances and a family of maximal boundary sets indexed by negative training instances. Below the representation is formally defined.

Definition 13 (Instance-Based Maximal Boundary Sets). *Consider an admissible concept language Lc and training sets $I^+ \subseteq I$ and $I^- \subseteq I$ so that $I^- \neq \emptyset$. Then the instance-based-maximal-boundary-set representation of a version space $VS(I^+, I^-)$ is the ordered pair $\langle I^+, \{G(I^+, \{n\})\}_{n \in I^-} \rangle$.*

The IBMBS are "instance-based" since each of their elements corresponds to particular training instances. The IBMBS are "maximal boundary sets" since each of their elements in $\{G(I^+, \{n\})\}_{n \in I^-}$ is a maximal boundary set. The IBMBS are a one-sided version space representation since its first part is the set I^+; i.e., this part does not contain any boundary-set element.

To prove that the IBMBS are a correct version-space representation we give theorems 14 and 15 from [11]. Theorem 14 states that if a description $c \in Lc$ is more specific than at least one element of each maximal boundary set $G(I^+, \{n\})$ for all $n \in I^-$, then c is consistent with the set I^- of negative instances.

Theorem 14. *If the concept language Lc is admissible, then:*

$$(\forall c \in Lc)((\forall n \in I^-)(\exists g \in G(I^+, \{n\}))(g \geq c) \rightarrow (\forall n \in I^-)\neg M(c, n)).$$

Theorem 15 states that a version space $VS(I_1^+, I_1^-)$ is a subset of a version space $VS(I_2^+, I_2^-)$ if and only if every description in $VS(I_1^+, I_1^-)$ is consistent with the sets I_2^+ and I_2^-.

Theorem 15. $VS(I_1^+, I_1^-) \subseteq VS(I_2^+, I_2^-) \leftrightarrow$
$$(\forall c \in VS(I_1^+, I_1^-))((\forall p \in I_2^+)M(c, p) \wedge (\forall n \in I_2^-)\neg M(c, n)).$$

Theorem 16 (Correctness of IBMBS). *Consider a version space $VS(I^+, I^-)$ represented by IBMBS: $\langle I^+, \{G(I^+, \{n\})\}_{n \in I^-} \rangle$. If the concept language Lc is admissible, then:*

$$(\forall c \in Lc)(c \in VS(I^+, I^-) \leftrightarrow ((\forall p \in I^+)M(c, p) \wedge (\forall n \in I^-)(\exists g \in G(I^+, \{n\}))(g \geq c))).$$

Proof. ($\rightarrow$) Consider an arbitrarily chosen description $c \in VS(I^+, I^-)$. By theorem 15 $(\forall n \in I^-)(VS(I^+, I^-) \subseteq VS(I^+, \{n\}))$. Thus, $(\forall n \in I^-)(c \in VS(I^+, \{n\}))$. Since Lc is admissible, according to definition 11 for each $VS(I^+, \{n\})$ we have:

$$(\forall n \in I^-)(\exists g \in G(I^+, \{n\}))(g \geq c). \quad (1)$$

Since $c \in VS(I^+, I^-)$, according to definition 7:

$$(\forall p \in I^+)M(c, p). \quad (2)$$

From (1) and (2) the first part of theorem is proven.
($\leftarrow$) Let $c \in Lc$ be arbitrarily chosen so that:

$$(\forall p \in I^+)M(c,p). \tag{3}$$
$$(\forall n \in I^-)(\exists g \in G(I^-, \{n\}))(g \geq c). \tag{4}$$

By theorem 14 formula (4) implies:

$$(\forall n \in I^-)\neg M(c,n). \tag{5}$$

Thus, $c \in Lc$, (3), and (5) imply $c \in VS(I^+, I^-)$ according to definition 7. □

Given the IBMBS of a version space $VS(I^+, I^-)$ and an admissible concept language, theorem 16 states that the concept descriptions in $VS(I^+, I^-)$ are exactly those that (1) cover all the positive instances in I^+, and (2) are more specific than at least one element of each maximal boundary set $G(I^+, \{n\})$. This means that the size of IBMBS is not tied to the number of descriptions in the version space. Thus, the IBMBS are a compact version-space representation.

Example 1. Let the instance universe I and the concept language Lc be 1-CNF languages with 8 attributes. The domain of the k-th attribute in I is $\{0,1\}$ and in Lc is $\{0,1,?\}$, where the symbol "?" indicates that any value is acceptable. The procedure of the cover relation M returns true for a concept description $c \in Lc$ and an instance $i \in I$ if and only if for each attribute the values of c and i are equal or the value of c equals "?". In this context we consider a concept-learning task with the set I^+ consisting of one positive instance: $i_1^+ = \langle 1,1,1,1,1,1,1,1\rangle$ and the set I^- consisting of three negative instances: $i_2^- = \langle 0,0,1,1,1,1,1,1\rangle$, $i_3^- = \langle 1,1,0,0,1,1,1,1\rangle$ and $i_4^- = \langle 1,1,1,1,0,0,1,1\rangle$. For this task, the IBMBS: $\langle I^+, \{G(I^+,\{n\})\}_{n\in I^-}\rangle$ of the version space $VS(I^+, I^-)$ consist of four sets:

$$\begin{aligned}
I^+ &= \{\langle 1,1,1,1,1,1,1,1\rangle\},\\
G(I^+,\{i_2^-\}) &= \{\langle 1,?,?,?,?,?,?,?\rangle, \langle ?,1,?,?,?,?,?,?\rangle\},\\
G(I^+,\{i_3^-\}) &= \{\langle ?,?,1,?,?,?,?,?\rangle, \langle ?,?,?,1,?,?,?,?\rangle\},\\
G(I^+,\{i_4^-\}) &= \{\langle ?,?,?,?,1,?,?,?\rangle, \langle ?,?,?,?,?,1,?,?\rangle\}.
\end{aligned}$$
□

3.2 Finiteness

Since the IBMBS are compact, it is important to determine when they are finite. We introduce constraints on the training sets and the concept language. We show that they are sufficient and necessary conditions for the finiteness of the IBMBS.

We start with the constraints on the training sets.

Constraint 17. The training sets I^+ and I^- are finite.

Constraint 17 implies that the number of the maximal boundary sets $G(I^+, \{n\})$ is finite. Hence, IBMBS are finite in this case if each set $G(I^+, \{n\})$ is finite. To guarantee this property we introduce a constraint on the concept language.

Constraint 18. The maximal boundary set $G(\emptyset, \{n\})$ is finite for all $n \in I$.

To explain how constraint 18 affects each maximal boundary set $G(I^+, \{n\})$ theorem 19 is taken from [7]. The theorem states that the set $G(I^+ \cup \{i\}, I^-)$ is equal to the set of those elements of the set $G(I^+, I^-)$ that cover the instance i.

Theorem 19. *Consider training sets $I^+, I^- \subseteq I$. If the concept language Lc is admissible, then for all $p \in I$:*

$$G(I^+ \cup \{p\}, I^-) = \{g \in G(I^+, I^-) | M(g, p)\}.$$

An important consequence of theorem 19 is given in corollary 20 below.

Corollary 20. *Consider sets $I_1^+, I_2^+ \subseteq I$ so that $I_2^+ \subseteq I_1^+$. Then for all $n \in I$:*

$$G(I_1^+, \{n\}) \subseteq G(I_2^+, \{n\}).$$

Using corollary 20 we formulate the following theorem.

Theorem 21. *The maximal boundary set $G(I^+, \{n\})$ is finite for all $n \in I$ and $I^+ \subseteq I$ if and only if constraint 18 holds.*

Combining constraints 17 and 18, and using theorem 21 we finish the section by formulating the theorem of the IBMBS being finite.

Theorem 22. *The IBMBS are finite if and only if constraints 17 and 18 hold.*

4 Algorithms of the IBMBS

This section introduces four algorithms of the IBMBS. The instance-addition algorithm is given in subsection 4.1; the instance-retraction algorithm is given in subsection 4.2; the algorithm for version-space collapse is given in subsection 4.3; and the instance-classification algorithm and its extension for noisy training data are given in subsection 4.4.

4.1 Instance-Addition Algorithm

The instance-addition algorithm of the IBMBS revises the representation given a new training instance. It is correct for the class of admissible concept languages. The algorithm consists of two parts for handling positive and negative training instances. They are based on theorem 23 and theorem 16, respectively.

Theorem 23. *Consider a version space $VS(I^+, I^-)$ represented by IBMBS: $\langle I^+, \{G(I^+, \{n\})\}_{n \in I^-} \rangle$, and a version space $VS(I^+ \cup \{i\}, I^-)$ represented by IBMBS: $\langle I^+ \cup \{i\}, \{G(I^+ \cup \{i\}, \{n\})\}_{n \in I^-} \rangle$. If the concept language Lc is admissible, then:*

$$G(I^+ \cup \{i\}, \{n\}) = \{g \in G(I^+, \{n\}) \,|\, M(g, i)\} \text{ for all } n \in I^-.$$

Proof. The theorem follows from theorem 19. □

Instance-Addition **Algorithm**
Input: i: a new training instance.
$\langle\{I^+, \{G(I^+, \{n\})\}_{n\in I^-}\rangle$: IBMBS of $VS(I^+, I^-)$.
Output:
$\langle I^+ \cup \{i\}, \{G(I^+ \cup \{i\}, \{n\})\}_{n\in I^-}\rangle$: IBMBS of $VS(I^+ \cup \{i\}, I^-)$ if i is positive.
$\langle I^+, \{G(I^+, \{n\})\}_{n\in I^- \cup \{i\}}\rangle$: IBMBS of $VS(I^+, I^- \cup \{i\})$ if i is negative.
Precondition: Lc is admissible.

if instance i is positive **then**
  for $n \in I^-$ **do**
    $G(I^+ \cup \{i\}, \{n\}) = \{g \in G(I^+, \{n\}) \mid M(g, i)\}$
  return $\langle I^+ \cup \{i\}, \{G(I^+ \cup \{i\}, \{n\})\}_{n\in I^-}\rangle$
if instance i is negative **then**
  Generate the set $G(\emptyset, \{i\})$
  $G(I^+, \{i\}) = \{g \in G(\emptyset, \{i\}) | (\forall p \in I^+) M(g, p)\}$
  return $\langle\{I^+, \{G(I^+, \{n\})\}_{n\in I^- \cup \{i\}}\rangle$.

Fig. 1. The Instance-Addition Algorithm

The instance-addition algorithm can be described as follows (see in figure 1). If a new positive training instance i is given, the algorithm forms the maximal boundary sets $G(I^+ \cup \{i\}, \{n\})$ for all $n \in I^-$. Each set $G(I^+ \cup \{i\}, \{n\})$ is formed from those elements of the corresponding set $G(I^+, \{n\})$ that cover the instance i. The resulting IBMBS of the version space $VS(I^+ \cup \{i\}, I^-)$ are formed from the set $I^+ \cup \{i\}$ and the maximal boundary sets $G(I^+ \cup \{i\}, \{n\})$ for all $n \in I^-$.

If the instance i is negative, the algorithm first forms the maximal boundary set $G(I^+, \{i\})$ in two steps. In the first step it generates the maximal boundary set $G(\emptyset, \{i\})$. In the second step the algorithm forms $G(I^+, \{i\})$ from the elements of $G(\emptyset, \{i\})$ that cover all the instances in I^+ (see theorem 19). Then, the resulting IBMBS of the version space $VS(I^+, I^- \cup \{i\})$ are formed from the set I^+ and the maximal boundary sets $G(I^+, \{n\})$ for all $n \in I^- \cup \{i\}$.

Example 2. Let us illustrate the instance-addition algorithm given the IBMBS from example 1. Assume that we have a new negative training instance $i_5^- = \langle 1, 1, 1, 1, 1, 1, 0, 0\rangle$. The algorithm first generates the maximal boundary set $G(I^+, \{i_5^-\}) = \{\langle ?, ?, ?, ?, ?, ?, 1, ?\rangle, \langle ?, ?, ?, ?, ?, ?, ?, 1\rangle\}$. Then, it adds $G(I^+, \{i_5^-\})$ to the IBMBS. The resulting IBMBS consist of five sets:

$$\begin{aligned} I^+ &= \{\langle 1, 1, 1, 1, 1, 1, 1, 1\rangle\}, \\ G(I^+, \{i_2^-\}) &= \{\langle 1, ?, ?, ?, ?, ?, ?, ?\rangle, \langle ?, 1, ?, ?, ?, ?, ?, ?\rangle\}, \\ G(I^+, \{i_3^-\}) &= \{\langle ?, ?, 1, ?, ?, ?, ?, ?\rangle, \langle ?, ?, ?, 1, ?, ?, ?, ?\rangle\}, \\ G(I^+, \{i_4^-\}) &= \{\langle ?, ?, ?, ?, 1, ?, ?, ?\rangle, \langle ?, ?, ?, ?, ?, 1, ?, ?\rangle\}, \\ G(I^+, \{i_5^-\}) &= \{\langle ?, ?, ?, ?, ?, ?, 1, ?\rangle, \langle ?, ?, ?, ?, ?, ?, ?, 1\rangle\}. \end{aligned}$$

Assume now that we have a new positive instance $i_6^+ = \langle 1, 0, 1, 0, 1, 0, 1, 0\rangle$. The algorithm forms for each $n \in I^-$ the maximal boundary set $G(I^+ \cup \{i_6^+\}, \{n\})$

from the elements of the set $G(I^+, \{n\})$ that cover the instance i_6^+. It adds the instance i_6^+ to the training set I^+. The resulting IBMBS consist of five sets:

$$I^+ = \{\langle 1,1,1,1,1,1,1,1\rangle, \langle 1,0,1,0,1,0,1,0\rangle\},$$
$$G(I^+, \{i_2^-\}) = \{\langle 1,?,?,?,?,?,?,?\rangle\}, G(I^+, \{i_3^-\}) = \{\langle ?,?,1,?,?,?,?,?\rangle\},$$
$$G(I^+, \{i_4^-\}) = \{\langle ?,?,?,?,1,?,?,?\rangle\}, G(I^+, \{i_5^-\}) = \{\langle ?,?,?,?,?,?,1,?\rangle\}. \quad \square$$

4.2 Instance-Retraction Algorithm

The instance-retraction algorithm of the IBMBS revises the representation when an instance is removed from one of the training sets. It is correct for the class of admissible concept languages when the property G holds [11].

Definition 24 (Property G). *An admissible concept language is said to have property G if for all $n_1, n_2 \in I$:*

$$\{g \in G(\emptyset, \{n_1\}) | \neg M(g, n_2)\} = \{g \in G(\emptyset, \{n_2\}) | \neg M(g, n_1)\}.$$

An admissible concept language has the property G if for all $n_1, n_2 \in I$ the subset of the elements of the set $G(\emptyset, \{n_1\})$, that do not cover the instance n_2, equals the subset of the elements of the set $G(\emptyset, \{n_2\})$, that do not cover the instance n_1. A simple consequence of the property G is given in a corollary below.

Corollary 25. *If the property G holds, then for all $n_1, n_2 \in I$, and all $I^+ \subseteq I$:*

$$\{g \in G(I^+, \{n_1\}) | \neg M(g, n_2)\} = \{g \in G(I^+, \{n_2\}) | \neg M(g, n_1)\}.$$

The instance-retraction algorithm consists of two parts for handling positive and negative instances. They are based on theorems 26 and 16, respectively.

Theorem 26. *Consider a version space $VS(I^+, I^-)$ represented by IBMBS: $\langle I^+, \{G(I^+, \{n\})\}_{n \in I^-}\rangle$, and a second version space $VS(I^+ \setminus \{i\}, I^-)$ represented by IBMBS: $\langle I^+ \setminus \{i\}, \{G(I^+ \setminus \{i\}, \{n\})\}_{n \in I^-}\rangle$, where $i \in I^+$. If the concept language Lc is admissible and the property G holds, then:*

$$G(I^+ \setminus \{i\}, \{n\}) = G(I^+, \{n\}) \cup \{g \in G(I^+ \setminus \{i\}, \{i\}) | \neg M(g, n)\} \text{ for all } n \in I^-.$$

Proof. For each $n \in I^-$:

$$G(I^+ \setminus \{i\}, \{n\}) = \{g \in G(I^+ \setminus \{i\}, \{n\}) | M(g, i)\} \cup \{g \in G(I^+ \setminus \{i\}, \{n\}) | \neg M(g, i)\}.$$

According to theorem 19:

$$\{g \in G(I^+ \setminus \{i\}, \{n\}) | M(g, i)\} = G(I^+, \{n\})$$

and according to corollary 25:

$$\{g \in G(I^+ \setminus \{i\}, \{n\}) | \neg M(g, i)\} = \{g \in G(I^+ \setminus \{i\}, \{i\}) | \neg M(g, n)\}.$$

Thus,

$$G(I^+ \setminus \{i\}, \{n\}) = G(I^+, \{n\}) \cup \{g \in G(I^+ \setminus \{i\}, \{i\}) | \neg M(g, n)\}. \quad \square$$

Instance-Retraction **Algorithm**
Input: i: a training instance in $I^+ \cup I^-$.
$\langle\{I^+, \{G(I^+, \{n\})\}_{n\in I^-}\rangle$: IBMBS of $VS(I^+, I^-)$.
Output: $\langle\{I^+ \setminus \{i\}, \{G(I^+ \setminus \{i\}, \{n\})\}_{n\in I^-}\rangle$: IBMBS of $VS(I^+ \setminus \{i\}, I^-)$ if $i \in I^+$.
$\langle\{I^+, \{G(I^+, \{n\})\}_{n\in I^- \setminus\{i\}}\rangle$: IBMBS of $VS(I^+, I^- \setminus \{i\})$ if $i \in I^-$.
Precondition: Lc is admissible, property G holds, and $|I^-| > 1$ if $i \in I^-$.

if $i \in I^+$ **then**
  Generate the set $G(\emptyset, \{i\})$
  $G(I^+ \setminus \{i\}, \{i\}) = \{g \in G(\emptyset, \{i\}) | (\forall p \in I^+ \setminus \{i\}) M(g, p)\}$
  for $n \in I^-$ **do**
    $G(I^+ \setminus \{i\}, \{n\}) = G(I^+, \{n\}) \cup \{g \in G(I^+ \setminus \{i\}, \{i\}) \,|\, \neg M(g, n)\}$
  return $\langle I^+ \setminus \{i\}, \{G(I^+ \setminus \{i\}, \{n\})\}_{n\in I^-}\rangle$
if $i \in I^-$ **then**
  return $\langle I^+, \{G(I^+, \{n\})\}_{n\in I^- \setminus\{i\}}\rangle$.

Fig. 2. The Instance-Retraction Algorithm

The instance-retraction algorithm can be described as follows (see figure 2). If an instance i is removed from the set I^+, the algorithm executes two steps following theorem 26. In the first step it forms the maximal boundary set $G(I^+ \setminus \{i\}, \{i\})$. This is done by first generating the maximal boundary set $G(\emptyset, \{i\})$ and then by removing those elements of $G(\emptyset, \{i\})$ that do not cover at least one instance in $I^+ \setminus \{i\}$ (see theorem 19). In the second step the algorithm forms the maximal boundary set $G(I^+ \setminus \{i\}, \{n\})$ for each $n \in I^-$. The set $G(I^+ \setminus \{i\}, \{n\})$ is formed as a union of the corresponding sets $G(I^+, \{n\})$ and $\{g \in G(I^+ \setminus \{i\}, \{i\}) \,|\, \neg M(g, n)\}$. The resulting IBMBS of the version space $VS(I^+ \setminus \{i\}, I^-)$ are formed from the set $I^+ \setminus \{i\}$ and the maximal boundary sets $G(I^+ \setminus \{i\}, \{n\})$ for all $n \in I^-$.

If the instance i is removed from the set I^-, the algorithm forms the resulting IBMBS of the version space $VS(I^+, I^- \setminus \{i\})$ from the set I^+ and the maximal boundary sets $G(I^+, \{n\})$ for all $n \in I^- \setminus \{i\}$.

Example 3. Let us illustrate the instance-retraction algorithm given the last IBMBS from example 2. Note that the property G holds for the concept language used. Assume that we have to retract the positive training instance $i_6^+ = \langle 1, 0, 1, 0, 1, 0, 1, 0\rangle$. The algorithm forms the boundary set $G(I^+ \setminus \{i_6^+\}, \{i_6^+\}) = \{\langle ?, 1, ?, ?, ?, ?, ?, ?\rangle, \langle ?, ?, ?, 1, ?, ?, ?, ?\rangle, \langle ?, ?, ?, ?, ?, 1, ?, ?\rangle, \langle ?, ?, ?, ?, ?, ?, ?, 1\rangle\}$. Then, it forms for each $n \in I^-$ the maximal boundary set $G(I^+ \setminus \{i_6^+\}, \{n\})$ as a union of the sets $G(I^+, \{n\})$ and $\{g \in G(I^+ \setminus \{i_6^+\}, \{i_6^+\}) \,|\, \neg M(g, n)\}$. The instance i_6^+ is excluded from the training set I^+ and the resulting IBMBS coincide with the first IBMBS from example 2.

Assume now that we have to retract the negative training instance $i_5^- = \langle 1, 1, 1, 1, 1, 1, 0, 0\rangle$. The algorithm excludes: (1) the instance from the training set I^-, and (2) the maximal boundary set $G(I^+, \{i_5\})$ from the IBMBS. The resulting IBMBS coincide with those from example 1. □

4.3 Algorithm for Version-Space Collapse

The algorithm for version-space collapse checks whether a version space represented by IBMBS is empty. It is proposed for the class of admissible concept languages when the intersection-preserving property (IP) holds [11].

Definition 27 (Intersection-Preserving Property (IP)). *An admissible concept language is said to have the intersection-preserving property if for each nonempty set $C \subseteq Lc$ there exists a description $c \in Lc$ so that:*

$$(\forall i \in I)((\forall c' \in C)M(c', i) \leftrightarrow M(c, i)).$$

An admissible concept language Lc exhibits the property IP when for each nonempty subset $C \subseteq Lc$ there exists a description $c \in Lc$ so that an instance $i \in I$ is covered by all the descriptions $c' \in C$ if and only if i is covered by c. The property is introduced because it guarantees that if the training set I^- is not empty, the version space $VS(I^+, I^-)$ is not empty if and only if for each $n \in I^-$ the version space $VS(I^+, \{n\})$ is not empty (see theorem 28 taken from [11]).

Theorem 28. *Consider an admissible concept language Lc such that the property IP holds. If the set I^- is nonempty, then:*

$$(VS(I^+, I^-) \neq \emptyset) \leftrightarrow (\forall n \in I^-)(VS(I^+, \{n\}) \neq \emptyset).$$

To check a version space $VS(I^+, I^-)$ for collapse, by theorem 28 we can check for collapse of the version spaces $VS(I^+, \{n\})$ for $n \in I^-$. Since $VS(I^+, \{n\})$ are given by maximal boundary sets $G(I^+, \{n\})$ in the IBMBS of $VS(I^+, I^-)$, we give a relation between the sets $G(I^+, \{n\})$ and version spaces $VS(I^+, \{n\})$ [11].

Theorem 29. $(VS(I^+, I^-) \neq \emptyset) \leftrightarrow (G(I^+, I^-) \neq \emptyset)$.

Theorems 28 and 29 imply corollary 30 below. It states that if the property IP holds and the training set I^- is nonempty, the version space $VS(I^+, I^-)$ is nonempty if and only if for each $n \in I^-$ the set $G(I^+, \{n\})$ is nonempty.

Corollary 30. *Consider an admissible concept language Lc such that the property IP holds. If the set I^- is nonempty, then:*

$$(VS(I^+, I^-) \neq \emptyset) \leftrightarrow (\forall n \in I^-)(G(I^+, \{n\}) \neq \emptyset).$$

The version-space collapse algorithm is given in figure 3. If a version space $VS(I^+, I^-)$, given by IBMBS, is checked for collapse, the algorithm visits the maximal boundary sets $G(I^+, \{n\})$ for $n \in I^-$. If none of the sets $G(I^+, \{n\})$ is empty, by corollary 30 $VS(I^+, I^-)$ is not empty and the algorithm returns false. Otherwise, by corollary 30 $VS(I^+, I^-)$ is empty and the algorithm returns true.

Example 4. Let us illustrate the algorithm for version-space collapse given the IBMBS from example 1. Note that property IP holds for the concept language used. The algorithm checks the maximal boundary sets $G(I^+, \{i_2^-\})$, $G(I^+, \{i_3^-\})$, and $G(I^+, \{i_4^-\})$. Since none of them is empty, the algorithm returns false; i.e., the version space is nonempty. □

VS-Collapse **Algorithm**
Input: $\langle\{I^+, \{G(I^+, \{n\})\}_{n\in I^-}\rangle$: IBMBS of $VS(I^+, I^-)$.
Output: true if $VS(I^+, I^-) = \emptyset$.
false if $VS(I^+, I^-) \neq \emptyset$.
Precondition: Lc is admissible and the property IP holds.

```
for n ∈ I⁻ do
    if G(I⁺, {n}) = ∅ then
        return true
return false.
```

Fig. 3. The Algorithm for Version-Space Collapse

4.4 Instance-Classification Algorithm

Instance classification with version spaces is realised by the unanimous-voting rule: an instance is classified if and only if all the descriptions in a version space agree on a classification of the instance [7,8]. The rule can be implemented using theorems 31 and 32 taken from [11]. Theorem 31 states that all the descriptions of a version space $VS(I^+, I^-)$ do cover an instance $i \in I$ if and only if the version space $VS(I^+, I^- \cup \{i\})$ is empty. Theorem 32 states that all the descriptions of a version space $VS(I^+, I^-)$ do not cover an instance $i \in I$ if and only if the version space $VS(I^+ \cup \{i\}, I^-)$ is empty.

Theorem 31. $(\forall i \in I)((\forall c \in VS(I^+, I^-))M(c, i) \leftrightarrow (VS(I^+, I^- \cup \{i\}) = \emptyset))$.

Theorem 32. $(\forall i \in I)((\forall c \in VS(I^+, I^-))\neg M(c, i) \leftrightarrow (VS(I^+ \cup \{i\}, I^-) = \emptyset))$.

The instance-classification algorithm of the IBMBS realises the unanimous-voting rule for the class of admissible concept languages if the property IP holds. The positive instance classification is based on theorem 31, and the negative instance classification is based on theorem 33. Theorem 33 is used instead of theorem 32 for efficiency reasons. It states that if the concept language is admissible and the property IP holds, then all the descriptions of a version space $VS(I^+, I^-)$ do not cover an instance $i \in I$ if and only if there exists a version space $VS(I^+, \{n\})$ of which all the descriptions do not cover the instance i.

Theorem 33. *Consider an admissible concept language Lc such that the property IP holds. If the set I^- is nonempty, then:*

$$(\forall i \in I)((\forall c \in VS(I^+, I^-))\neg M(c, i) \leftrightarrow (\exists n \in I^-)(\forall c \in VS(I^+, \{n\}))\neg M(c, i)).$$

Proof. Consider an arbitrary $i \in I$. Then:

$(\forall c \in VS(I^+, I^-))\neg M(c, i)$ iff (theorem 32)
$VS(I^+ \cup \{i\}, I^-) = \emptyset$ iff (theorem 28)
$(\exists n \in I^-) VS(I^+ \cup \{i\}, \{n\}) = \emptyset$ iff (theorem 32)
$(\exists n \in I^-)(\forall c \in VS(I^+, \{n\}))\neg M(c, i)$ □

By theorem 33 a negative instance classification can be obtained by the version spaces $VS(I^+, \{n\})$. Since $VS(I^+, \{n\})$ are given with the maximal boundary sets $G(I^+, \{n\})$ in the IBMBS of $VS(I^+, I^-)$, we show how to use these sets for the classification using theorem 34 from [11].

Theorem 34. $(\forall i \in I)((\forall c \in VS(I^+, I^-))\neg M(c,i) \leftrightarrow (\forall g \in G(I^+, I^-))\neg M(g,i))$.

Theorems 33 and 34 imply corollary 35 below. Corollary 35 states that if an admissible concept language has the property IP, then none of the descriptions of a version space $VS(I^+, I^-)$ covers an instance $i \in I$ if and only if there exists a set $G(I^+, \{n\})$ of which all the descriptions do not cover the instance i.

Corollary 35. *Consider an admissible concept language Lc such that the property IP holds. If the set I^- is nonempty, then:*

$$(\forall i \in I)((\forall c \in VS(I^+, I^-))\neg M(c,i) \leftrightarrow (\exists n \in I^-)(\forall g \in G(I^+, \{n\}))\neg M(g,i)).$$

The instance-classification algorithm of the IBMBS is shown in figure 4. Given a nonempty version space $VS(I^+, I^-)$, it classifies an instance $i \in I$ in two steps. In the first step the algorithm forms the IBMBS of the version space $VS(I^+, I^- \cup \{i\})$ using the instance-addition algorithm applied on the IBMBS of $VS(I^+, I^-)$ with the instance i labeled as negative. If $VS(I^+, I^- \cup \{i\})$ is empty, by theorem 31 all the descriptions in $VS(I^+, I^-)$ cover the instance. Hence, the instance i is positive and the algorithm returns "+". If $VS(I^+, I^- \cup \{i\})$ is not empty, during the second step the algorithm visits the sets $G(I^+, \{n\})$ for $n \in I^-$. If none of the elements of one of these sets covers the instance i, by corollary 35 all the descriptions in $VS(I^+, I^-)$ do not cover the instance. Thus, the instance i is negative and the algorithm returns "–". Otherwise, the algorithm returns "?".

Instance-Classification **Algorithm**

Input: i: an instance to be classified.
$\langle I^+, \{G(I^+, \{n\})\}_{n \in I^-}\rangle$: IBMBS of $VS(I^+, I^-)$.

Output: "+" if $(\forall c \in VS(I^+, I^-))M(c,i)$.
"–" if $(\forall c \in VS(I^+, I^-))\neg M(c,i)$.
"?" otherwise.

Precondition: Lc is admissible, the property IP holds, and $VS(I^+, I^-) \neq \emptyset$.

label i as negative
$\langle\{I^+, \{G(I^+, \{n\})\}_{n \in I^- \cup \{i\}}\rangle$ = *Instance-Addition*$(i, \langle\{I^+, \{G(I^+, \{n\})\}_{n \in I^-}\rangle)$
if *VS-Collapse*$(\langle I^+, \{G(I^+, \{n\})\}_{n \in I^- \cup \{i\}}\rangle)$ **then**
  return "+"
for $n \in I^-$ **do**
  if $(\forall g \in G(I^+, \{n\}))\neg M(g,i)$ **then**
    return "–"
return "?".

Fig. 4. The Instance-Classification Algorithm

Example 5. Let us illustrate the classification algorithm given the IBMBS from example 1. Assume that we have to classify instance $i = \langle 1, 1, 1, 1, 1, 1, 0, 0 \rangle$. In the first step the algorithm updates the IBMBS with the instance i considered as negative. The resulting IBMBS coincide with the first IBMBS from example 2 and represent a nonempty version space. In the second step the algorithm determines that all the elements of the maximal boundary sets $G(I^+, \{i_2^-\})$, $G(I^+, \{i_3^-\})$, and $G(I^+, \{i_4^-\})$ do cover the instance. Thus, the algorithm returns "?"; i.e., the instance classification cannot be determined. □

The instance classification with the IBMBS can be extended to situations when the training instances are noisy. The key idea is to use flexible matching between instances and concept descriptions. Below we describe two procedures, based on flexible matching, for positive and negative classification, respectively.

The positive classification procedure, given an instance i to be classified, first forms the maximal boundary set $G(\emptyset, \{i\})$. Then for each positive training instance $p \in I^+$ it determines the number of descriptions $g \in G(\emptyset, \{i\})$ that do not cover the instance. If at least P_p positive training instances are not covered by at least P_g descriptions $g \in G(\emptyset, \{i\})$, the instance i is classified as positive, where P_p and P_g are parameters of flexible matching.

The negative classification procedure is similar to that given in [10]. Given an instance i to be classified, it determines for each negative training instance $n \in I^-$ the number of descriptions $g \in G(I^+, \{n\})$ that do not cover the instance i. If there exist at least N_n maximal boundary sets $G(I^+, \{n\})$ of which at least N_g descriptions do not cover the instance i, then the instance is classified as negative, where N_n and N_g are parameters of flexible matching.

5 Analysis

This section analyses the IBMBS. Subsection 5.1 gives a worst-case complexity analysis of the IBMBS and the algorithms presented. Subsection 5.2 uses the results of the analysis to determine (1) whether the IBMBS algorithms are efficient, and (2) whether the IBMBS are efficiently computable.

5.1 The Worst-Case Complexity Analysis

The worst-case complexity analysis is made in terms of the computational characteristics of admissible concept languages. The characteristics are chosen so that they do not depend on the size of the training data. They are given below:

Γ_n: the maximal size of the maximal boundary set $G(\emptyset, \{n\})$ for all $n \in I$;
$t_n^{\uparrow}$: the maximal time for generating the set $G(\emptyset, \{n\})$ for all $n \in I$;
Σ_p: the maximal size of the minimal boundary set $S(\{p\}, \emptyset)$ for all $p \in I$;
$t_p^{\downarrow}$: the maximal time for generating the set $S(\{p\}, \emptyset)$ for all $p \in I$ [2];
t_m: the maximal time of the operator of the relation $M(c, i)$ for all $c \in Lc, i \in I$.

[2] The computational characteristics Σ_p and $t_p^{\downarrow}$ are given for the complexity analysis of the instance-based minimal boundary sets presented in section 7.

The condition for the worst-case complexity analysis is that the size of the maximal boundary sets $G(I^+, \{n\})$ is equal to the size Γ_n for all $n \in I, I^+ \subseteq I$.

Space Complexity

The worst-case space complexity of the IBMBS is $|I^+|$ plus the worst-case space complexity $O(|I^-|\Gamma_n)$ of the G-part. Thus, it is $O(|I^+| + |I^-|\Gamma_n)$.

Time Complexities

The Instance-Addition Algorithm. The worst-case time complexity of the algorithm part for processing one positive instance is $O(|I^-|\Gamma_n t_m)$. The factor $|I^-|$ arises because we have $|I^-|$ maximal boundary sets $G(I^+ \cup \{i\}, \{n\})$. The factor $\Gamma_n t_m$ arises because in order to form each maximal boundary set $G(I^+ \cup \{i\}, \{n\})$ we test Γ_n elements of the set $G(I^+, \{n\})$ whether they cover the instance.

The worst-case time complexity of the algorithm part for processing one negative instance is $O(t_n^{\uparrow} + |I^+|\Gamma_n t_m)$. The term $O(t_n^{\uparrow})$ arises because the maximal boundary set $G(\emptyset, \{i\})$ is generated. The term $O(|I^+|\Gamma_n t_m)$ arises because the maximal boundary set $G(I^+, \{i\})$ is generated from Γ_n elements of the set $G(\emptyset, \{i\})$ that are tested to cover all the positive instances in the set I^+.

The Instance-Retraction Algorithm. The worst-case time complexity of the algorithm part for processing one positive instance is the sum $O(t_n^{\uparrow} + |I^+|\Gamma_n t_m) + O(|I^-|\Gamma_n t_m)$. The term $O(t_n^{\uparrow} + |I^+|\Gamma_n t_m)$ is the worst-case time complexity for generating the maximal boundary set $G(I^+ \setminus \{i\}, \{i\})$. (The sub-term $t_n^{\uparrow}$ arises because the set $G(\emptyset, \{i\})$ is generated. The sub-term $|I^+|\Gamma_n t_m$ arises because the size of the set $G(\emptyset, \{i\})$ is Γ_n and each element of $G(\emptyset, \{i\})$ is checked whether it covers all the positive instances in the set $I^+ \setminus \{i\}$.) The second term $O(|I^-|\Gamma_n t_m)$ is the time complexity for constructing the maximal boundary sets $G(I^+ \setminus \{i\}, \{n\})$ for all $n \in I^-$. (The factor $|I^-|$ arises because we have $|I^-|$ sets $G(I^+ \setminus \{i\}, \{n\})$. The factor $\Gamma_n t_m$ arises because formation of each set $G(I^+ \cup \{i\}, \{n\})$ requires Γ_n elements of the set $G(I^+ \setminus \{i\}, \{i\})$ to be tested not to cover the corresponding instance n.) Thus, the worst-case time complexity of this part of the algorithm is $O(t_n^{\uparrow} + |I^+|\Gamma_n t_m) + O(|I^-|\Gamma_n t_m) = O(t_n^{\uparrow} + (|I^+| + |I^-|)\Gamma_n t_m)$.

The worst-case time complexity of the algorithm part for processing one negative instance is $O(1)$ because its maximal boundary set is removed only.

The Algorithm for Version-Space Collapse. The worst-case time complexity of the algorithm is $O(|I^-|)$. The term $|I^-|$ arises because in the worst case $|I^-|$ maximal boundary sets $G(I^+, \{n\})$ are checked whether they are empty.

The Instance-Classification Algorithm. The instance-classification algorithm consists of two parts. The first part is the positive instance-classification part. Its worst-case time complexity is the sum $O(t_n^{\uparrow} + |I^+|\Gamma_n t_m) + O(|I^-|)$. (The first term is the worst-case time complexity of the instance-addition algorithm given the instance i to be classified labeled as negative. The second term is the worst-case time complexity of the algorithm for version-space collapse.) The second part is the negative instance-classification part. Its worst-case time complexity is $O(|I^-|\Gamma_n t_m)$. The factor $|I^-|$ arises because we have $|I^-|$ maximal boundary sets $G(I^+, \{n\})$. The factor $\Gamma_n t_m$ arises because the elements of each maximal boundary set $G(I^+, \{n\})$ are tested not to cover the instance i. Thus, the worst-case time complexity of the instance-classification algorithm is: $O(|I^-|)+O(t_n^{\uparrow}+|I^+|\Gamma_n t_m)+O(|I^-|)+O(|I^-|\Gamma_n t_m) = O(t_n^{\uparrow}+(|I^+|+|I^-|)\Gamma_n t_m)$.

The IBMBS complexities are summarised in table 1.

Table 1. Worst-Case Complexities of the IBMBS and their Algorithms

Space:	$O(\|I^+\| + \|I^-\|\Gamma_n)$
Time	
Instance-Addition Algorithm ($\oplus$ instance):	$O(\|I^-\|\Gamma_n t_m)$
Instance-Addition Algorithm ($\ominus$ instance):	$O(t_n^{\uparrow} + \|I^+\|\Gamma_n t_m)$
Instance-Retraction Algorithm ($\oplus$ instance):	$O(t_n^{\uparrow} + (\|I^+\| + \|I^-\|)\Gamma_n t_m)$
Instance-Retraction Algorithm ($\ominus$ instance):	$O(1)$
Version-Space Collapse Algorithm:	$O(\|I^-\|)$
Instance-Classification Algorithm:	$O(t_n^{\uparrow} + (\|I^+\| + \|I^-\|)\Gamma_n t_m)$

5.2 IBMBS and Efficiency

To determine whether the algorithms of the IBMBS are efficient we employ a rule proposed in [2]: an algorithm of a version-space representation is efficient for a concept language if the worst-case time complexity of the algorithm is polynomial in the computational features of the language and the size of the input. From the previous subsection we know that the worst-case time complexities of the IBMBS algorithms are polynomial in the computational characteristics $t_n^{\uparrow}$, Γ_n, t_m, and the sizes $|I^+|$ and $|I^-|$. In this context we note that the upper bound of the size of the input of the algorithms is the size of the IBMBS; i.e., $|I^+|$ plus $|I^-|\Gamma_n$. Thus, we conclude that the IBMBS algorithms for instance addition, instance retraction, version-space collapse and instance classification are efficient for admissible concept languages. In addition, we emphasise that the instance-retraction algorithm does not recompute the IBMBS.

To determine whether the IBMBS are efficiently computable we employ a second rule proposed in [2]: a version-space representation is efficiently computable for a concept language if in the worst case its size is polynomial in the computational features of the language and the sizes of the training sets, and the representation has an efficient instance-addition algorithm. ¿From the previous subsection we know that the worst-case space complexity of the IBMBS is polynomial in the computational characteristic Γ_n, and the sizes $|I^+|$ and $|I^-|$. Since the IBMBS instance-addition algorithm is efficient we conclude that the IBMBS are efficiently computable for admissible concept languages.

6 Usefulness of the IBMBS

This section evaluates the usefulness of the IBMBS. For this purpose we summarise the IBMBS employing the characteristics of a useful version-space representation (definition 8). The characteristics are compactness, finiteness, efficient computability, and efficiency of the IBMBS algorithms.

We showed that the IBMBS are a correct and compact version-space representation for admissible concept languages (section 3). They are finite if the training sets are finite and the maximal boundary set $G(\emptyset, \{n\})$ is finite for all $n \in I$. The IBMBS are efficiently computable and have efficient algorithms for instance addition, instance retraction, version-space collapse and instance classification for admissible concept languages (sections 4 and 5). The only restrictions are that the instance retraction algorithm requires the property G while the algorithms for version-space collapse and instance classification require the property IP on the concept language used.

From this summary we conclude according to definition 8 that the IBMBS are a useful version-space representation for the class of admissible concept languages if the training sets are finite, the maximal boundary set $G(\emptyset, \{n\})$ is finite for all $n \in I$, and the property G as well as the property IP hold.

7 Instance-Based Minimal Boundary Sets

Instance-based minimal boundary sets (IBmBS) and their algorithms can be derived by duality from the previous sections[3]. Therefore, we refrain from providing details. The IBmBS complexities are given in table 2.

Table 2. Worst-Case Complexities of the IBmBS and their Algorithms

Space:	$O(\|I^+\|\Sigma_p + \|I^-\|)$
Time	
Instance-Addition Algorithm ($\oplus$ instance):	$O(t_p^\downarrow + \|I^-\|\Sigma_p t_m)$
Instance-Addition Algorithm ($\ominus$ instance):	$O(\|I^+\|\Sigma_p t_m)$
Instance-Retraction Algorithm ($\oplus$ instance):	$O(1)$
Instance-Retraction Algorithm ($\ominus$ instance):	$O(t_p^\downarrow + (\|I^+\| + \|I^-\|)\Sigma_p t_m)$
Version-Space Collapse Algorithm:	$O(\|I^+\|)$
Instance-Classification Algorithm:	$O(t_p^\downarrow + (\|I^+\| + \|I^-\|)\Sigma_p t_m)$

8 Comparison with Relevant Work

Below we compare the IBMBS and the IBmBS with the training-instance representation [3] and the instance-based boundary-set representation [11,12], i.e., with version-space representations that are efficient for instance retraction. The comparison is made using the characteristics of useful version-space representations (definition 8).

The training-instance representation is a version-space representation that consists of the sets of positive and negative training instances. By definition the

[3] We note that the dual of the property G is the property S, and the dual of the intersection-preserving property (IP) is the union-preserving property (UP) [11].

representation is compact. Obviously, the conditions for finiteness of the training-instance representation are a subset of the conditions for finiteness of the IBMBS (IBmBS). An analogous conclusion can be derived when the representations are compared with respect to efficient computability. The training-instance representation allows much more efficient algorithms for instance addition and instance retraction. This advantage comes with a price: the instance-classification algorithm determines the classification of each instance by a search in the concept language using all the training data and the instance [2]. Thus, the training-instance representation has only a theoretical value. This contrasts with the IBMBS and the IBmBS: their instance-classification algorithms are not based on search and this is one of the factors of their usefulness.

The instance-based boundary-set representation (IBBS) is a useful version-space representation that consists of a family of minimal boundary sets indexed by positive training instances and a family of maximal boundary sets indexed by negative training instances [11,12]. The representation is correct and compact for admissible concept languages. It is possible to prove that the conditions for finiteness of the IBBS are a superset of the conditions for finiteness of the IBMBS (IBmBS). In order to compare the representations in terms of efficiency we examine the IBBS worst-case complexities given in table 3. An analysis of these complexities shows that each of them is equal to the sum of the corresponding complexities of the IBMBS and IBmBS. Thus, the IBMBS and the IBmBS have two advantages: (1) they are more efficiently computable than the IBBS, and (2) the IBMBS and the IBmBS algorithms for instance addition, instance retraction, version-space collapse, and instance classification are more efficient. Moreover, the applicability of the IBBS instance-retraction algorithm is more restricted: the algorithm can be applied only if both properties S and G hold.

Table 3. Worst-Case Complexities of the IBBS and their Algorithms

Space:	$O(\|I^+\|\Sigma_p + \|I^-\|\Gamma_n)$
Time	
Instance-Addition Algorithm ($\oplus$ instance):	$O(t_p^{\downarrow} + \|I^-\|(\Sigma_p + \Gamma_n)t_m)$
Instance-Addition Algorithm ($\ominus$ instance):	$O(t_n^{\uparrow} + \|I^+\|(\Sigma_p + \Gamma_n)t_m)$
Instance-Retraction Algorithm ($\oplus$ instance):	$O(t_n^{\uparrow} + (\|I^+\| + \|I^-\|)\Gamma_n t_m)$
Instance-Retraction Algorithm ($\ominus$ instance):	$O(t_p^{\downarrow} + (\|I^+\| + \|I^-\|)\Sigma_p t_m)$
Version-Space Collapse Algorithm:	$O(\|I^+\| + \|I^-\|)$
Instance-Classification Algorithm:	$O(t_p^{\downarrow} + t_n^{\uparrow} + (\|I^+\| + \|I^-\|)(\Sigma_p + \Gamma_n)t_m)$

9 Conclusion

This chapter introduced a family of useful version-space representations called one-sided instance-based boundary sets (IBMBS and IBmBS). We showed that these representations are correct and compact for the class of admissible concept languages. This allowed us to derive the conditions for finiteness. In addition, we demonstrated that the one-sided instance-based boundary sets are efficiently computable and have efficient algorithms for instance addition, instance retraction, version-space collapse and instance classification for the class of admissible concept languages. We compared the one-sided instance-based boundary sets with other existing version-space representations that are efficient for instance retraction. From the comparison we conclude that the one-sided instance-based boundary sets are at the moment the most efficient useful version-space representations for instance retraction. So, our research question from section 1 has been answered positively.

References

1. Haussler, D.: Quantifying Inductive Bias: AI Learning Algorithms and Valiants Learning Framework. Artificial Intelligence **36** (1988) 177–221
2. Hirsh, H.: Polynomial-Time Learning with Version Spaces. In: Proceedings of the Tenth National Conference on Artificial Intelligence, AAAI Press, Menlo Park, CA (1992) 117–122
3. Hirsh, H., Mishra, N., Pitt, L.: Version Spaces without Boundary Sets. In: Proceedings of the Fourteenth National Conference on Artificial Intelligence, AAAI Press, Menlo Park, CA (1997) 491–496
4. Idemstam-Almquist, P.: Demand Networks: An Alternative Representation of Version Spaces. Master's Thesis, Department of Computer Science and Systems Sciences, Stockholm University, Stockholm, Sweden (1990)
5. De Raedt, L.: Query Evaluation and Optimisation for Inductive Databases using Version Spaces, In: Online Proceedings of the International Workshop on Database Technologies for Data Mining, Prague, Czech Republic (2002)
6. De Raedt, L., Jaeger, M., Lee, S., Mannila, H.: A Theory of Inductive Query Answering, In: Proceedings of the 2002 IEEE International Conference on Data Mining, IEEE Publishing, (2002) 123–128
7. Mitchell, T.: Version Spaces: An Approach to Concept Learning. Ph.D. Thesis, Electrical Engineering Department, Stanford Univeristy, Stanford, CA (1978)
8. Mitchell, T.: Machine Learning. McGraw-Hill, New York, NY (1997)
9. Sablon, G., DeRaedt, L., Bruynooghe, L.: Iterative Versionspaces. Artificial Intelligence **69** (1994) 393–410
10. Sebag, M., Rouveirol. C.: Resource-bounded Relational Reasoning: Induction and Deduction through Stochastic Matching. Machine Learning **38** (2000) 41–62
11. Smirnov, E.N.: Conjunctive and Disjunctive Version Spaces with Instance-Based Boundary Sets. Ph.D. Thesis, Department of Computer Science, Universiteit Maastricht, Maastricht, The Netherlands (2001)
12. Smirnov, E.N., Braspenning, P.J.: Version Space Learning with Instance-Based Boundary Sets. In: Proceedings of The Thirteenth European Conference on Artificial Intelligence. Jonh Willey and Sons, Chichester, UK (1998) 460–464
13. Smith, B.D., Rosenbloom, P.S.: Incremental Non-Backtracking Focusing: A Polynomially Bounded Generalization Algorithm for Version Spaces. In: Proceedings of the Eight National Conference on Artificial Intelligence, MIT Press, MA (1990) 848–853

Domain Structures in Filtering Irrelevant Frequent Patterns

Kimmo Hätönen and Mika Klemettinen

Nokia Research Center, P.O.Box 407, FIN-00045 Nokia Group, Finland
{kimmo.hatonen, mika.klemettinen} @nokia.com

Abstract. Events are used to monitor many types of processes in several technical domains. Computers and efficient electronic communication networks make it very easy to increase the accuracy and amount of logged details. While the size of logs is growing, the collection and analysis of them are becoming harder all the time. Frequent episodes offer one possible method to structure and find information hidden in logs. Unfortunately, as events reflecting simultaneous independent processes are stored to central monitoring points, signs of several unrelated phenomena get mixed with each other. This makes the algorithm searching for frequent episodes to produce accidental and irrelevant results. As a solution to this problem, we introduce here a notion of domain constraints that are based on distance measures, which can be defined in terms of domain structure and used taxonomies. We also show how these constraints can be used to prune irrelevant event combinations.

1 Introduction

Episode rules are a powerful way to search and describe patterns in event sequences. Even though the algorithms are fast, they easily find too much new information. The resulting problem, sometimes referred to as second order data mining, has been recognized quite early (e.g. [8]). In the case of telecommuncation network alarm and system log analysis, for example, two different kinds of problems should be emphasized. In the pool of discovered rules — among the interesting ones — there are (1) proper rules that are of no interest and (2) rules that have been generated between items or sets of items that can not have any interdependence with each other.

In this paper we concentrate on the second problem: how to minimize the effect of accidental event combinations in analysis, especially in telecommunication network log analysis. The telecommunication networks produce large amounts of different types of events that are logged in central monitoring points. These events include different log entries reflecting normal operation of the network as well as alarm information about faults and problems that occur.

In the network there are all the time plenty of independent processes going on. These processes emit alarms, when they get disturbed by faults. It often happens that many independent processes get simultaneously affected by a fault and they all start to alarm, not necessarily about the fault itself, but about its

R. Meo et al. (Eds.): Database Support for Data Mining Applications, LNAI 2682, pp. 289–305, 2004.

secondary reflections. Thus, alarms and log entries, which are sent, actually carry second-hand information about the incident. They do not necessarily identify the primary fault at all.

Alarms that network processes emit are collected to the centralized monitoring points. This makes the analysis even more difficult, because at each monitoring point, the symptoms and reflection of separated problems are merged into one information flow. The combined flow also contains noisy information caused by natural phenomena like thunderstorms or by normal maintenance operations.

A starting point for a network analyst in a fault condition is always localization and isolation of the fault, i.e., finding the area where the problem is located and identification of all network elements that are affected by the fault. Localization and isolation is based on the assumption that it is probable that the fault itself is local although its reflections are widespread. In this situation alarms coming from the same network element or its direct neighbors are related to one reflection of the fault. After the localization has been done it is easier to do the actual identification of the fault.

Episode and association rule based techniques have been used in semi-automatic knowledge acquisition from alarm data in order to collect the required knowledge for knowledge based systems like alarm correlators [5,7,6]. Episode rules [9,10] describe temporal proximity and temporal ordering of recurrent combinations of alarms in a given alarm database. Association rules [1,2], in turn, describe the properties of individual alarms without taking the temporal relationships of the alarms into account. Given such rules holding in an alarm database, a fault management expert is able to verify whether the rules are useful or not. Some of the rules may reflect known causal connections, and some may be irrelevant, while some rules give new insight to the behavior of the network elements. Selected rules can be used as a basis of correlation patterns for alarm correlation systems.

Based on our experience, simple-minded use of discovery algorithms poses problems with the amount of generated rules and their relevance. In the KDD process [3], it is often reasonable or even necessary to constrain the discovery using background knowledge. If no constraints are applied, the discovered result set of, say, episode rules might become huge and contain mostly trivial, uninteresting or even impossible rules.

The basic methods for finding episodes use event distances in time but ignore all the other knowledge about the domain. There is, however, plenty of other useful domain knowledge, e.g., topological information, available from a technical domain like telecommunications networks. Probably the most common information is some kind of part-of hierarchy between domain objects. Also different taxonomies and definitions of control hierarchy and material flows between objects are usually available. We suggest that this background knowledge can be used as a basis for domain specific *distance measures*. These distance measures can then be used to prune out accidental event occurences that are caused by simultaneous but independent phenomena in the network.

This kind of approach to incorporate domain knowledge into the mining process by using distance constraints is general. It is not limited only to a task of finding correlating event sequences but it can be applied to mining of different types of patterns. It can also be used in mining market basket type of data, especially if there are taxonomies available.

There are some earlier studies that discuss the problem of incorporating constraints to the data mining algorithms. Most of the approaches (e.g. [11,4]) concentrate on selecting interesting items or item combinations outside the analysis process. The analysis is then focused so that only those patterns describing the selected items are searched for. Zaki [12] studies the relationships of single events in multiple event sequences. He introduces templates that are used to restrict supporting occurences only to those that fulfill given time interval restriction.

In this article, we introduce methods to apply domain knowledge in searching for episodes. First, in Section 2 we present different approaches that take the structure of the domain into consideration. In Section 3 we give empirical evidence from real-life telecommunication alarm datasets supporting the use of the presented methods. In Section 4 we generalize the notion of distances and show how several distance constraints can be combined. Finally, Section 5 contains general discussion, conclusions, and issues for future work.

2 Application of Domain Knowledge

Episode rules and *episodes* are a modification of the concept of *association rules* and *frequent sets*, applied to sequential data. In this paper we adopt the definition given by Mannila and Toivonen [9] and concentrate on reducing the number of invalid episodes, which has a straightforward impact on the number and quality of resulting episode rules. We have also adopted the basic algorithm they present.

2.1 Data

The data set S used in the episode calculation is usually a set of tuples of the form (t, E), where t is an event occurrence time and E is an event type. An event type can be either a type number or some other field of the event, e.g., a message text attached to it. Episodes are calculated using these event types.

In reality, this data provides only a small fraction of the available information about the events and the underlying network. For instance, by looking at the network configuration — both the topological structure and the taxonomies of event types — it is possible to prune impossible relationships between events, i.e., to reduce the number of episodes found.

To make it possible to use this type of knowledge about the structure of the domain, the data format presented above has to be expanded by adding the infomation about the source of the event. Thus, the data set will be a collection of triples of the form (t, s, E), where s is a sender or source of the event. For example, in the alarm database tests presented below, events are of the form $(time, NEId, alarm\ type)$, where $time$ gives the beginning time of an alarm,

NEId the unique id of the network element that has sent the alarm and *alarm type* the type of the alarm.

2.2 Domain Model

In addition to the data, it is often possible to have a model $\mathcal{M} = (N, F, R, D)$ about the domain and especially about its structure. Here $N = \{ne_0, \ldots, ne_n\}$ is a set of objects of the domain, F is a set of facts defining the structure of the domain, i.e., the relationships that connect the objects together, R is a set of relations that define how it is possible to deduce new relationships, and D is a set of distance functions $d_i(source(e_j), source(e_k)) = d_i(ne_n, ne_m)$, where $e_j, e_k \in S$ and $ne_n, ne_m \in N$. These distance functions return the distances of sources of two events in the defined structure.

In the telecommunication network data sets used in experiments of this article, for example, the model $\mathcal{M}$ consists of N that is a set of network elements, a configuration table F that is a binary relation and defines the control relations between network elements, a set of relations R that show how different ancestor or sibling relations can be deduced for a node and a distance function $d_0(source(e_i), source(e_j))$ that returns an integer which corresponds to a number of steps that must be taken in the control hierarchy in order to travel from the network element that sent the event e_i to another one that sent the event e_j.

The model $\mathcal{M}$ can be used to restrict the generation of P, a set of frequent episodes. For this, it is possible to use a *domain constraint* $C = c_0, \ldots, c_i$ to give maximum thresholds for corresponding distance functions $d_i()$. During the computation of episode candidate supports a support will be increased only if it is true for all required distances d_i that $d_i(source(e_j), source(e_k)) \leq c_i$ for all pairs of events e_j, e_k in a possible candidate occurrence.

On the left in Figure 1 there are two nodes that send alarms, which are collected to the same database. However, the nodes are situated far from each other in the network topology; i.e., the node N_1 is not directly connected to the node N_2 but there are several steps between them. If the distance function $d_0()$ has been defined as a number of node-to-node steps between network nodes, it returns a value that is larger than one. In the real world situation, it is, for example, most likely that if the $d_i(N_1, N_2)$ grows much larger than 1 then the probability that events, which either of the nodes sends, could correlate with events from the other one, decreases[1].

A list of event types and their occurrence times are given on the right in Figure 1. If the frequency threshold is 2 and the window in which the events can correlate is of length 3, then all event types, i.e., A, B, C, D, are frequent. Thus, there are six candidates (AB, AC, AD, BC, BD, and CD) that can be frequent.

If the distance constraint c_0 for distance $d_0()$ is not required for occurrence validation, then four candidates appear to be frequent, namely AB, AC, BC,

[1] Of course, exceptions exist, but they are usually quite rare. And if this phenomenon is very rare, it would be pruned anyhow because of the frequency threshold.

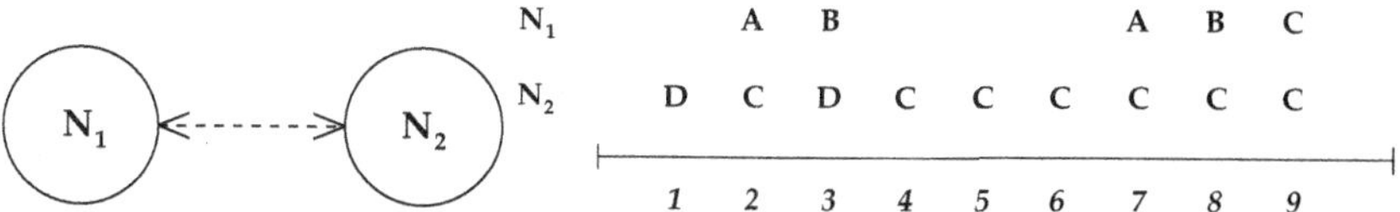

Fig. 1. A simple network with two nodes not directly connected to each other (on the left) and a set of sent alarms (on the right)

and CD. The other two occur only once in the data set. The distance constraint d_0 can be enforced by setting $c_0 = 0$, i.e., by demanding that to be valid, all events of an occurrence have to come from the same source. Thus, the number of frequent sets decreases to two. Then only candidates AB and CD occur twice inside a given time window and events in all occurrences of them are sent by the same node.

2.3 Implementation Alternatives

A straightforward way to apply distance constraints would be to prevent occasional occurrences of a candidate from increasing the candidate's support. This can be done during the frequent episode computation in the algorithm, which we have adopted [9]. When the supports for candidates are computed, we check every possible occurrence of a candidate against a constraint, to see whether an occurrence is appropriate. If the occurence is valid, then we increase the corresponding support counter.

There are, however, some drawbacks in this approach. Probably the most serious one is, that in the worst case it might take exponential time to check all possible instances of all possible episode candidates as has been discussed in [9].

Let us consider, for example, the alarm flow given in Figure 1 and there the candidate AC.[2] There are eight possible occurrences of the candidate: $\{(A_{1,2}, C_{2,2}), (A_{1,2}, C_{2,4}), (A_{1,7}, C_{2,7}), (A_{1,7}, C_{2,6}), (A_{1,7}, C_{2,8}), (A_{1,7}, C_{2,5}), (A_{1,7}, C_{1,9}), (A_{1,7}, C_{2,9})\}$. Seven out of these are improper with respect to the defined distance constraint $d_0(e_i, e_j) \leq c_0$, where $c_0 = 1$. The only proper occurrence is (A_{17}, C_{19}), where $d_0(N_1, N_1) = 0$. Altogether, there are 21 possible occurrences of candidates to be checked against the domain constraint and out of these only 6 are proper ones.

In the straightforward approach explained above, there are still plenty of unnecessarily composed impossible candidates. The events in these faulty candidates are introduced, for example, by sources that emit different types of events at the same time but are located so that used distance constraints would not allow events emitted by them to support any candidate episodes. All these candidates have to be checked against the data. A lot of work can be avoided if it is possible to find *nonoverlapping subsets* from the data.

---

[2] In the following we denote an alarm A sent by node N_1 at time moment 2 by using the notation $A_{1,2}$. All other alarms are denoted respectively.

Definition 1. *Two subsets S_k and S_l of the dataset S are nonoverlapping iff*

1. *there is no such event e that $e \in S_k$ and $e \in S_l$ and*
2. *there are no such events $e_i \in S_k$ and $e_j \in S_l$ for which $d_m(source(e_i), source(e_j)) < c_m$ for any distance function $d_m \in D$ and distance constraint $c_m \in C$.*

In other words, nonoverlapping subsets of the data do not contain the same events and it is impossible for any reason that the events in separate nonoverlapping subsets could interfere with each other, i.e., be included in proper occurrences of episode candidates. There can, however, be pairs of events in a subset, which together can not support a candidate episode.

Nonoverlapping subsets of events can be found, if the distance constraints, which check the candidate occurrences, can be used to introduce nonoverlapping sets of sources. Thus, these event subsets can be separated from each other and frequent episode sets can be computed from each of them and summed up to a set of frequent episodes of the combined data set. The separation can be done with a straightforward algorithm (see Figure 2) in linear time with respect to the data size but in doubled space. The worst case is that there are separate network elements sending each of the alarms. Normally, however, this is not the case.

1. $SRC = \bigcup source(e), e \in S$;
2. Divide SRC to source subsets SRC_k, such that $\forall ne_i, ne_j \in SRC_k : d_m(ne_i, ne_j) < c_m$, for all $d_m \in D$ and $c_m \in C$, and there is no such network element $ne_i \in SRC_k$ that $ne_i \in SRC_l$, if $l \neq k$.
3. **forall** source subsets SRC_k **do**
4.   Initialize empty partition S_{SRC_k}
5. **forall** events $e \in S$ **do**
6.   Include e to $S_{source\ subset(source(e))}$
7. **return** multiset of SRC_k;

Fig. 2. Algorithm for finding nonoverlapping subsets

The partitioning approach explained above can not be applied if there are no ways to separate nonoverlapping subsets from the data.

A nonoverlapping subset S_k is said to be dense iff for every pair of events $e_i \in S_k$ and $e_j \in S_k$, $d_m(source(e_i), source(e_j)) < c_m$ for all distance functions $d_m \in D$ and distance constraint $c_m \in C$. In such a case, all event combinations may support a candidate episode. This is an optimal case of partitioning, in which there is no need to check candidate occurrences againts distance constraints.

Let us add two new nodes to our network. The nodes N_3 and N_4 are added as shown in Figure 3. The distance function is the same as above: $d_0(ne_i, ne_j)$ returns the number of steps from the node ne_i to the node ne_j. The domain constraint sets the allowed number of steps to one, i.e, $c_0 = 1$. Thus, the constraint requires that in order to be proper, an occurrence of a candidate has to contain only such events, which are sent by neighboring nodes of the network. For example, events sent by N_2 and N_4 can not be in the same proper candidate occurrence.

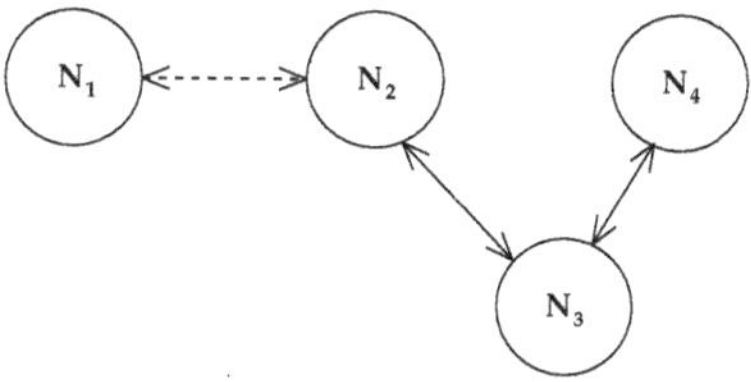

Fig. 3. A network with four nodes. Solid lines represent direct connections and dashed line connection that is not direct

It is possible to separate two nonoverlapping subsets of data from the alarm set. This is possible since events sent by N_1 can not interfere with events sent by any other node in the network. The other part of the network contains a nonoverlapping subset of sources that can not be further divided to nonoverlapping dense subsets. The events sent by node N_3 can interfere with events sent by either of the nodes N_2 and N_4, although events sent by N_2 and N_4 can not interfere with each other since they are not direct neighbors. The subsets are shown in Figure 4.

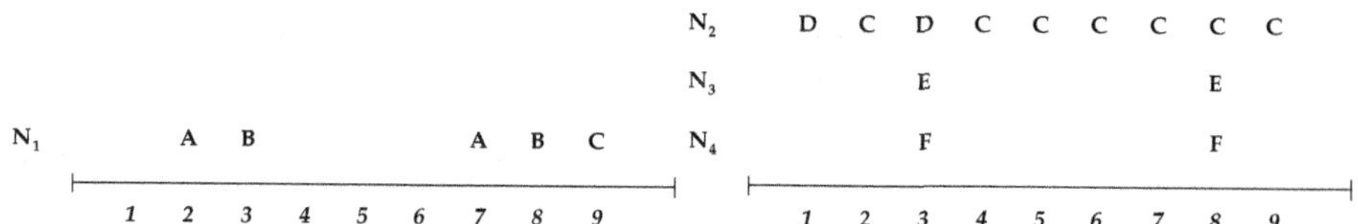

Fig. 4. Alarms sent by nodes in two nonoverlapping subsets

Separation to nonoverlapping subsets of data can be used to improve the effectiveness of the straightforward approach. When the separation has been done, each of the partitions is smaller than the original data set. Thus, there are fewer occurrences of possible candidates and the evaluation times of the restrictive constraints are reduced.

While computing the frequencies of the episodes from the data partition that include only events sent by node N_1, there is no need to make any kind of domain constraint evaluation. This is because all three requirements set for a nonoverlapping dense subset hold in this partition. In the other partition the domain constraint has to be evaluated. If we evaluate it during the frequent set computation, the constraint has to be evaluated with 17 occurrences out of which 7 occurences are proper. If it, on the other hand, is evaluated afterwards with occurences of only those episodes, which are frequent if the domain constraint is omitted, then there will be 14 occurrences to be evaluated. This is because out of six candidates (CD, CE, CF, DE, DF, EF) only four (CD, CE, CF, EF) are frequent without the domain constraint.

3 Experiments with Domain Knowledge

3.1 A Structure of the Network

A GSM network is organised as functional groups of network elements. In each group, there is one base station controller (BSC) that controls the internal behavior of the group and interface to other groups. The group has an internal tree form structure as is shown in Figure 5. These groups are called BSC groups below.

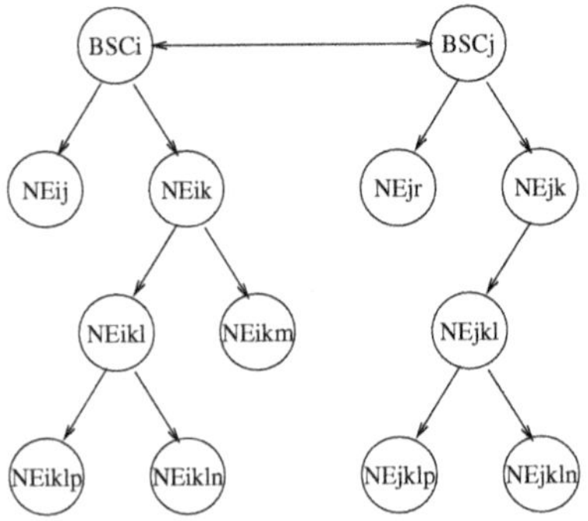

Fig. 5. A network with two BSC groups

It is more probable that the alarms emitted by the network elements controlled by the same BSC are related than those alarms coming from separated groups. Therefore the distance measurement d_0 that is used with the following alarm data sets, is defined as a number of archs in control hierarchy that are between the network elements sending the alarms. For example, $d_0(NE_{jkl}, BSC_j) = d_0(NE_{jklp}, NE_{jkln}) = 2$ and $d_0(NE_{jkl}, BSC_i) = 3$.

3.2 Telecom Event and Alarm Data Sets

The first data set, A, contains 55690 events with 295 different event types. They were emitted by 19 BSC groups located around a large geographical area. The period that is covered by the data set is 15 days. Each of groups has an internal structure like the one shown in Figure 5. Events are sent by network elements within the groups. The distance measure is defined as explained above.

In the search phase, we used time and distance measure $d_0()$ to constrain episode occurrences. Only those episode occurrencies were searched for, which were coming during one hour from network elements within distance c_0, where $c_0 = 0, 1, 2, 3, 4$ and 5.

The second data set, B, contains 25000 alarms consisting of 115 types and emitted by 28 BSC groups during 38 days. The distance measure was the same as with the first data set.

3.3 Restrictive Power

The restrictive power of the control hierarchy distance was tested by setting the window size of the episode algorithm to a large value and by changing the limiting distance that was used as domain constraint c_0. The test was run separately with the first two data sets, A and B. With the first data set, the support was set to 1 and with the second to 2. For the first set, only the episodes of size two were composed.

Table 1. Summary of the restriction test results

	Maximum Distance						
Episode size	$c_0 = 0$	$c_0 = 1$	$c_0 = 2$	$c_0 = 3$	$c_0 = 4$	$c_0 = 5$	$c_0 = \infty$
Test set A							
1	295	295	295	295	295	295	295
2	2447	3411	6858	10345	11432	13428	14468
Test set B							
1	115	115	115	115	115	115	115
2	383	893	1730	2706	3860	4464	5069
3	933	3338	12979	33442	77196	104543	140007
4	1618	8265	63827	–	–	–	–
5	2007	15052	–	–	–	–	–
6	1813	21337	–	–	–	–	–
7	...	...	–	–	–	–	–

From Table 1 we can see that while the distance restriction is loosened, the size of the resulting set of frequent episodes is growing quite rapidly. However, it keeps well under the corresponding counter of the situation where the restriction is not applied at all.

The results shown here actually support the common rule of thumb that only those alarms or events that were emitted by the direct neighbors in the control hierarchy are meaningful. When the distance constraint is changed from 1 to 2, the amount of the frequent candidate sets increases rapidly. Candidates that were generated with larger distances than 1 are more vulnerable for accidental occurrences. The bigger amounts of frequent episodes made also the computation times to grow rapidly and therefore the larger frequent episodes were omitted.

We used also a third data set, C, that is an application log of a large software system. It gave us similar results as sets A and B.

3.4 Quality Improvements

The quality of frequent sets was evaluated by studying the largest frequent episodes found. From each set, it was checked whether it contained only related event types. Evaluation was based on an assumption that a frequent episode is of good quality if it contains only such event types, which can be caused by the related functions under some circumstances.

The results with all of the test sets were encouraging. There were much less clearly unrelated combinations of events in all the frequent episode collections. Especially with the data set C, when the distance restriction was not applied, all the larger covering sets were combinations of unrelated events. On the other hand, when the restriction was used there were fewer frequent sets and only meaningful combinations.

With the test sets A and B the situation was somewhat similar. However, there were also frequent episodes that consisted of unrelated event types, but the amount of them was quite small.

4 Generalization of the Approach

In a complex environment like telecommunication network the role of the domain knowledge becomes important. Especially, information about different structures in the network is needed in order to understand the data that the network provides. For example, Figure 6 shows a small imaginary network that is constructed in a small town. The town is built around two large water areas.

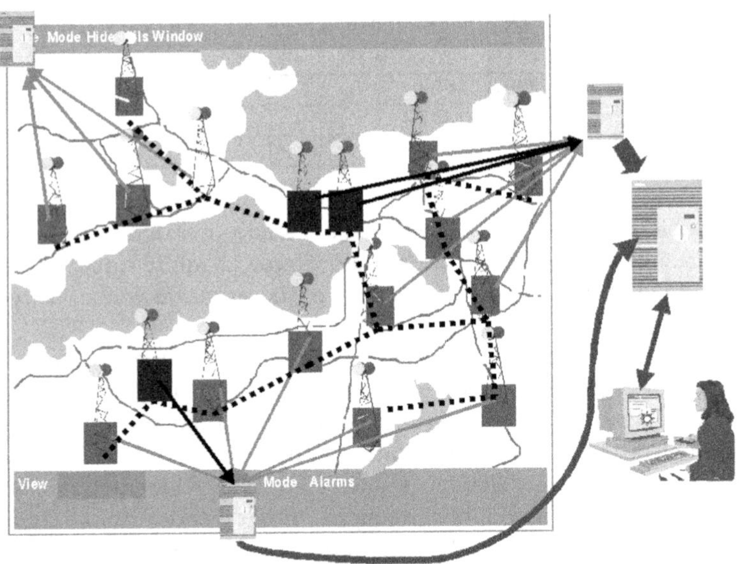

Fig. 6. A network in an urban area

Cells in the network are grouped under three controllers so that each group would be as homogenous as possible and that there would be only a few handovers from one group to another. In the network there are three alarming cells, marked with dark color. All alarms that they emit are collected to a centralized monitoring unit, where they are analysed by a human expert.

When an expert looks at a map with the alarming units, there are several possible relationships that he must consider before he can make his decision about what is the reason for the alarms. The first thing to consider is the geographical distance of the alarming cells: are they close to each other? If they are, as is the case with the alarming cell couple in between the two gulfs, would it be possible that they actually share the reason for their behavior? On the other hand, as is the case between the couple and the third alarming cell below the gulf, would it be possible that the cells are interfering with each other? This might be possible since open water area might carry signals quite far away.

In addition to the geographical relationships, there might be other functional connections between cells. In Figure 6 there is, for example, a transmission network depicted with black dotted lines. A structure of a transmission network is not necessarily reflecting at all in the organization of the radio network. They might even be operated by different operators. However, if a transmission line is broken, it immediately affects not only to its both end points but also to all the traffic that has been routed through it. Therefore, a single wire cut might make half of the network to alarm vigorously.

4.1 Distance Types

In a general case a data set S can be seen as a collection of events e_i so that each event has attached properties, which include event type E and its occurrence time t. Traditionally event time and type have been seen more important as the other properties such as cancellation time and severity of the event. However, this is not necessarily always the case. Any one of the properties can be used as a target for pattern mining. Depending on the application and the data set, some other properties might be more informative.

In earlier analysis [5,9,10] the time has been used as the only property that separates possible event patterns from each others. In this article also domain structures have been introduced for the same purpose. In general, there might be several different types of constraining distance measures that can be used. The most important ones are property distances like distance in time, domain distances like the distances introduced in this article, and characteristics distances like frequency of the type in the data set. The fourth type of constraints are the ones that are deduced from the previous mining results or that are set by the user, in order to reduce the amount of results and to focus on the most interesting phenomena.

Property distances are distance measures that are defined using event properties. Such a property can be, for example, beginning or cancellation time of an event. These can be computed without seeing anything else than a data set or part of it. The distance between beginning or cancellation times of two events might be computed without knowing anything else about the domain.

Domain distances are defined by using additional domain knowledge, for example, a control structure of a telecommunication network. In order to be able to compute distances over these structures, additional information of the domain in a form of a model $\mathcal{M}$ is needed as was discussed in Chapter 2.2. There a distance d_0 of two events was computed by counting control hierarchy steps between them: $d_0(source(e_i), source(e_j))$.

Characteristics distance is defined by the local statistical characteristicts that can be computed from the data set S. For example, frequencies of event types or average activity times of its instances can differ from each others so much that it is obvious that they can not relate to each other. Another such a characteristic that can separate event types, is the distribution over time. An event type can occur all over the data set from its beginning to its end. On the other hand, another event type might be as numerous in the data but it might be concentrated in a short peak.

4.2 Generalized Support

Values of an event property $prop_i$ form a *value set* V_{prop_i}. Each value v_j in a set can be understood as a node. In many cases, it is possible to define a graph between these nodes. For example, values of event property *beginning time* form a graph that is directed and where a total order between the nodes can be defined. A control hierarchy of a network, on the other hand, forms a partially ordered directed graph, while geographical locations of network elements can be understood as an undirected graph.

The notion of support can be generalized so that different types of value graphs can be used. The original support was defined as a number of time windows, where an instance of a pattern occurs. Thus – as is shown in Figure 7 –

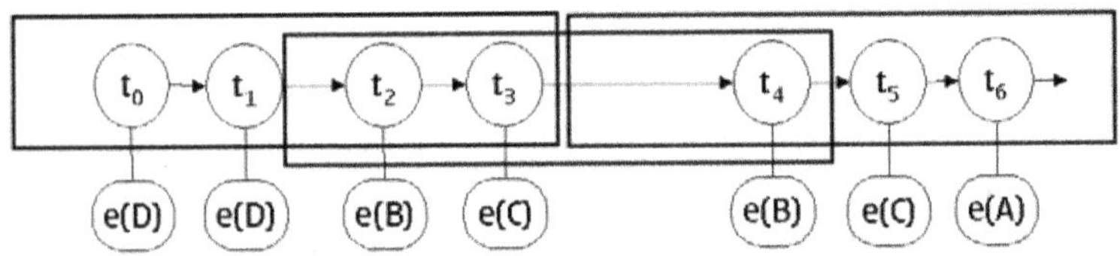

Fig. 7. A time window that has traversed over a graph of time points. At each time point there has been one event

a time window was moved over the time value graph. In each position of the window all the events, whose beginning times were inside the window, were interpreted to occur in the window. Correspondingly, it is possible to use other value graphs for this purpose. For example, in a control topology a window can be defined to contain all network elements that are not more than maximum number of steps away from the starting node. The window can then be moved stepwise from the root node towards the leaves so that each branch is visited once. In Figure 8, windows of size one step in a control topology, are shown. When defined this way, a control topology support of a pattern tells in how many different network element branches a certain event type combination occurs. This information is very useful, for example, in order to evaluate the reason for a malfunction.

A special case for the generalized support are such property value sets, in which it is impossible to define any kind of graph between the values. Such a value set actually gives a natural partitioning of the data set. In such a value set only those events whose values of a given property are equal, can support a pattern.

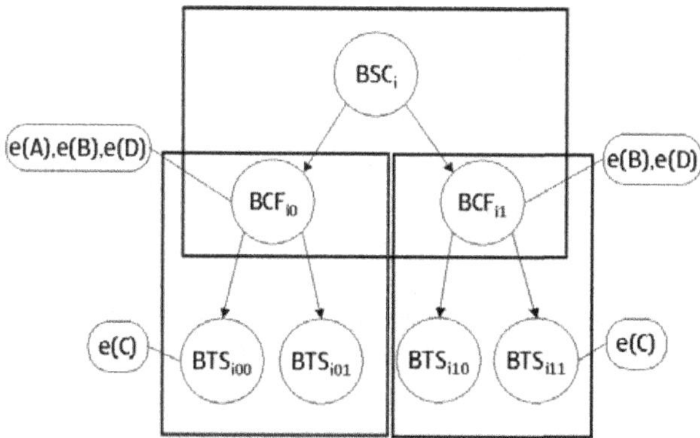

Fig. 8. A topology window that has traversed over a topology graph. Sets of events have been attached to the elements that have sent them

It is possible to use value graphs of different properties to prune out impossible or uninteresting candidates. For example, an alarm combination that occurs only in one network element and its direct children, migh not be as interesting as a combination that occurs in several network element branches. So there might be a threshold s_{prop_i} for each property $prop_i$. This leads to a very interesting result: to be covering a pattern p must have $support_{prop_i}(p) > s_{prop_i}$ for all $i = 0..n$, where n is the number of properties used for pruning. Because of this, in order to get an exact answer to a search for valid patterns, we must first check each possible occurrence of a pattern and if it is valid in respect to all the constraints, only after that one should update all the support counters of the pattern.

Fortunately it is also possible to compute an approximation of the set of valid patterns. This can be done by using each support threshold to compute a corresponding set of covering patterns, where only that support has been checked. The resulting sets of covering patterns are called as *margin sets*. In order to be covering in all respects, it must apply for a pattern p_i that $support_{prop_j}(p_i) > s_{prop_j}$ for each $j = 0, ..., n$. In other words, the pattern has to have all of its property supports greater than the corresponding thresholds, i.e., it has to belong to all of these margin sets. It is in many cases much faster to compute all these margin sets separately and take an intersection of the results – a so called *approximation set* – than to check each occurrence of all the candidate patterns in all respects.

This makes it possible to optimize the pattern computation by selecting the order in which the supports are computed so that those thresholds that are most powerful in pruning candidates or are easiest to compute are computed first.

The problem here is, of course, that there might be such patterns in the approximation set of covering patterns, whose property supports are greater than corresponding thresholds, but whose occurrences actually are not valid. Therefore, if an exact answer is needed, the occurrences of the resulting patterns must be validated against the data set. Fortunately, the amount of occurrences to be validated is much smaller than if the validation would have been done in all respects at once.

4.3 Tests with Generalized Support

With our test set B (see Chapter 3.2) we compared the pruning power of time window support and topological support. The topological support was based on a window that was sliding from the root node to element branches. The results of the experiment are given in Figure 9. It shows how the number of covering pairs changes while the time support threshold is increased. The constant value in the figure is the number of covering pairs that were computed while the topological support was set to 1. The curve Intersection gives the amount of covering pairs that were left to the approximation set. The approximation set was formed by taking intersection of pairs found with pure time support and pairs found with topology support. The curve Integrated gives the amount of covering sets, which were covering with respect to both support thresholds and whose all occurrences were checked against distance constraints. Already the approximation set improves accuracy of the answer by cutting the size of the resulting set to half compared to the time support margin set. Also the topology support margin set gives better results if compared to the low time support threshold.

When the topology support threshold is further increased, the size of the approximation set decreases. This can be seen, for example, in Figure 10. In this domain it makes quite a lot of sense to do this – especially when the information need is to find such alarm combinations that occur often in the network and which contain the same set of alarms every time. If we are using only the

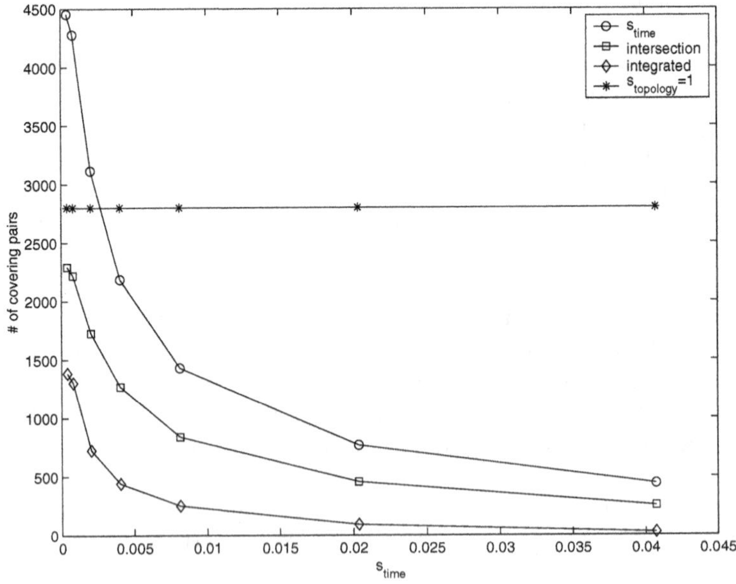

Fig. 9. Comparison of sizes of marginal sets, approximation set and exact results

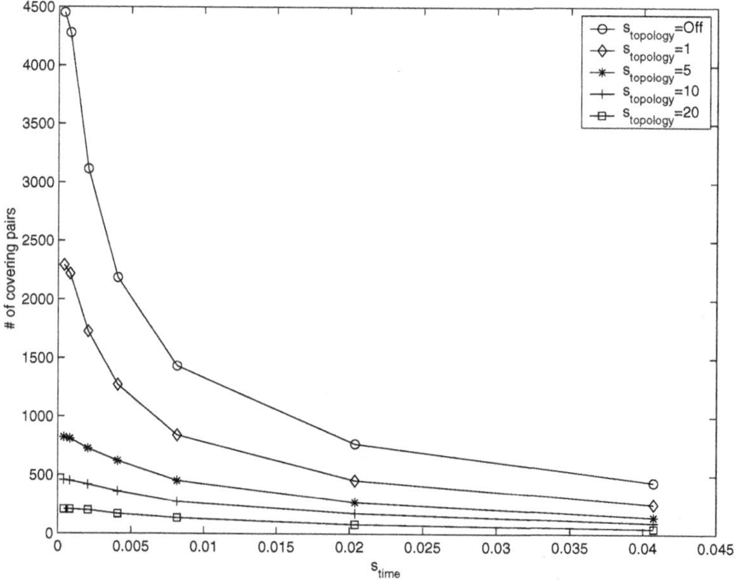

Fig. 10. Improved pruning results with a topology support

time support to prune candidates, it integrates away all such occurrences of the patterns that are coming from different sources but at the same time point. It gives us much more focused results when we can use relatively low time support together with a relatively low topological support and still keep the result set focused.

In windowing approach it varies how different graph types emphasize events attached to different parts of the graph. For example, in a tree structured graph like the one used in our experiments, where the depth of graph is relatively low, the nodes in the middle get more weight than leaves or roots. On the other hand, if the leaves contain a lot of the same types of events, the situation is balanced. In any case, the root and its direct descendants migh be discriminated. This is because the root might be covered only by a few windows while there are lots of windows covering middle parts of the branches or the similar leaves. However, this is not a problem in our domain, where the events sent by low level network elements usually form the majority of events in the data set. They are the first ones to be correlated or combined and removed from the data.

5 Conclusions

Domain knowledge and especially different distance measures between domain objects can effectively be used in pruning impossible frequent episodes and other types of patterns. They can be used in two ways: they might provide efficient natural partitionings to the data. These partitions can then be used to minimize the algorithm execution times. On the other hand, and what can be considered even more important, with them one can prune unnatural and impossible combinations of events away from the result set and thus help to ease the burden of a human expert, who is analysing the telecommunications network event logs.

In the near future, we will formalize the notion of general support. We will also continue experiments on different datasets and domains. We will study more closely the time complexity as well as the usability of the approximation set.

References

1. Rakesh Agrawal, Tomasz Imielinski, and Arun Swami. Mining association rules between sets of items in large databases. In P. Buneman and S. Jajodia, editors, *Proceedings of ACM SIGMOD Conference on Management of Data (SIGMOD'93)*, pages 207–216, Washington, D.C., USA, May 1993. ACM.
2. Rakesh Agrawal, Heikki Mannila, Ramakrishnan Srikant, Hannu Toivonen, and A. Inkeri Verkamo. Fast discovery of association rules. In Usama M. Fayyad, Gregory Piatetsky-Shapiro, Padhraic Smyth, and Ramasamy Uthurusamy, editors, *Advances in Knowledge Discovery and Data Mining*, pages 307–328. AAAI Press, Menlo Park, CA, 1996.
3. Usama M. Fayyad, Gregory Piatetsky-Shapiro, and Padhraic Smyth. From data mining to knowledge discovery: An overview. In Usama M. Fayyad, Gregory Piatetsky-Shapiro, Padhraic Smyth, and Ramasamy Uthurusamy, editors, *Advances in Knowledge Discovery and Data Mining*, pages 1–34. AAAI Press, Menlo Park, CA, 1996.
4. Minos N. Garofalakis, Rajeev Rastogi, and Kyuseok Shim. SPIRIT: Sequential pattern mining with regular expression constraints. In *Proc. of the 25th International Conference on Very Large Data Bases*, pages 223–234, Edinburgh, Scotland, 1999.
5. Kimmo Hätönen, Mika Klemettinen, Heikki Mannila, Pirjo Ronkainen, and Hannu Toivonen. TASA: Telecommunication alarm sequence analyzer, or "How to enjoy faults in your network". In *Proceedings of the 1996 IEEE Network Operations and Management Symposium (NOMS'96)*, pages 520–529, Kyoto, Japan, April 1996. IEEE.
6. Gabriel Jakobson and Mark Weissman. Real-time telecommunication network management: Extending event correlation with temporal constraints. In *Integrated Network Management IV*, pages 290–301. Chapman & Hall, London, 1995.
7. Gabriel Jakobson and Mark D. Weissman. Alarm correlation. *IEEE Network*, 7(6):52–59, November 1993.
8. Mika Klemettinen, Heikki Mannila, Pirjo Ronkainen, Hannu Toivonen, and A. Inkeri Verkamo. Finding interesting rules from large sets of discovered association rules. In *Proceedings of the Third International Conference on Information and Knowledge Management (CIKM'94)*, pages 401–407, Gaithersburg, MD, November 1994. ACM.
9. Heikki Mannila and Hannu Toivonen. Discovering generalized episodes using minimal occurrences. In *Proceedings of the Second International Conference on Knowledge Discovery and Data Mining (KDD'96)*, pages 146–151, Portland, Oregon, August 1996. AAAI Press.

10. Heikki Mannila, Hannu Toivonen, and A. Inkeri Verkamo. Discovery of frequent episodes in event sequences. *Data Mining and Knowledge Discovery*, 1(3):259–289, 1997.
11. Raymond T. Ng, Laks V. S. Lakshmanan, Jiawei Han, and Alex Pang. Exploratory mining and pruning optimizations of constrained associations rules. In *Proc. of the 1998 ACM SIGMOD International Conference on Management of Data*, pages 13–24, Seattle, Washington, USA, 1998.
12. Mohammed J. Zaki. Sequence mining in categorical domains: Incorporating constraints. In *Proc. of the 2000 ACM CIKM International Conference on Information and Knowledge Management*, pages 422–429, McLean, VA, USA, 2000.

Integrity Constraints over Association Rules*

Artur Bykowski[1], Thomas Daurel[1,2], Nicolas Méger[1], and Christophe Rigotti[1]

[1] Laboratoire d'Informatique de Recherche
en Image et Systèmes d'information (LIRIS)
Bâtiment Blaise Pascal, INSA Lyon, 69621 Villeurbanne Cedex, France
[2] Etudes et Productions Schlumberger 1, rue Henri Becquerel,
92142 Clamart Cedex, France

Abstract. In this paper, we propose to investigate the notion of integrity constraints in inductive databases. We advocate that integrity constraints can be used in this context as an abstract concept to encompass common data mining tasks such as the detection of corrupted data or of patterns that contradict the expert beliefs. To illustrate this possibility we propose a form of constraints called *association map constraints* to specify authorized confidence variations among the association rules. These constraints are easy to read and thus can be used to write clear specifications. We also present experiments showing that their satisfaction can be tested in practice.

1 Introduction

Integrity constraints are a central notion in databases used primarily to ensure data consistency. It has shown to be a fruitful and useful concept, with important additional benefits to guide very different aspects such as design, implementation and also query optimization (see [21,1] for an overview).

Basically, integrity constraints are an abstract specification of the possible contents of the database with respect to our current knowledge of the data domain. They have been deeply investigated in the context of relational databases as well as object-oriented databases, according to various objectives (e.g., specifications, efficient checking).

Recently, the concept of inductive database (IDB) has emerged [13,15,8], promoting the vision that a database dedicated to data mining contains not only data (e.g., customer transactions) but also all patterns that hold in the data (e.g., association rules [2]). Ideally, the user of an IDB can query data and patterns within a single language and can also express operations involving both data and patterns. The collection of patterns may be several orders of magnitude larger than the set of data itself, and thus cannot be materialized in general in the IDB. However from the user point of view, each pattern that holds should be considered as available for querying.

* This research is partially funded by the European Commission IST Programme - Accompanying Measures, AEGIS project (IST-2000-26450).

R. Meo et al. (Eds.): Database Support for Data Mining Applications, LNAI 2682, pp. 306–323, 2004.

We advocate in this paper that integrity constraints are also a very promising concept in IDB. Like in the classical database frameworks, they can be used for specifying data consistency and for rejecting inconsistent updates. However, in the context of IDB, integrity constraints raise new interesting and challenging issues, if we consider that they can also be applied to the patterns that hold in the data. Regarding this view, they can be used to specify what knowledge the designer (or an expert) considers to be reasonable to find in the data. Then, the violation of a *pattern integrity constraint* may be seen as an evidence of various phenomena. For example, if we have an IDB containing alarm logs with a daily insertion of a batch of new logs, then if after such an update one of the pattern integrity constraints is no longer satisfied this may highlight that this set of log records has not been properly cleaned. In this case the IDB engine can abort and undo the insertion, and let the IDB administrator (or a user) check these new logs. Another useful possibility is to consider that the designer/expert specifies intensionally with pattern integrity constraints the set of patterns that in her/his opinion could be found in the data. This provides a way to delimit the acceptable laws that could hold with respect to the knowledge that the designer/expert has about the domain. In this case an integrity violation can be assimilated to the occurrence of an unexpected phenomenon and the patterns violating that constraint can be considered as subjectively interesting piece of information for the designer/expert. Obviously, in the context of a multi-user IDB, such integrity constraints on patterns can be customized by each user, so that she/he can add more specific constraints than the ones set by the designer, to reflect her/his own belief and background knowledge.

The detection of corrupted data and the identification of new interesting knowledge among the extracted patterns are common tasks in data mining (see for example the classification of actions proposed by [19] when beliefs are contradicted). The idea, that we want to point out in this paper, is that large parts of these processes can be incorporated nicely in the IDB framework by means of integrity constraint specification and checking.

Of course, most forms of integrity constraints proposed previously in the database domain can be reused to specify the contents of an IDB in terms of tuples or objects (e.g., functional dependency, class inclusion hierarchy). And these constraints can be used directly to specify the data that are admissible in an IDB but also to specify the admissible patterns themselves, when these patterns are encoded as tuples or objects. So, at first sight, we can imagine to choose one of the very expressive languages already proposed in the literature (e.g., using a data manipulation language itself [21]) and use it to specify a large class of constraints over the patterns. The drawback of this approach is that it does not take into account the tradeoff between expressivity and computational complexity of constraint checking in the context of IDB.

For example using a Datalog like language with a polynomial evaluation complexity (w.r.t. the number of tuples in the database) to express constraints may be reasonable in a relational database, but will be in general not applicable in practice for IDB. The reason is that the number of patterns *stored* in an IDB (in a materialized way or not) is for common families of patterns inherently exponential with respect to the pattern domain parameters. For instance, if we consider patterns called *frequent itemsets*, they are defined w.r.t. a set of binary attributes $\mathcal{A}$, and a frequent itemset may be any subset of $\mathcal{A}$, leading in the worst case to a collection of $2^{|\mathcal{A}|}$ patterns. Even if this number can remain reasonable in some practical cases (e.g., using a high frequency threshold on a sparse data set), when we set more difficult conditions (e.g., a lower frequency), all practitioners have had to deal with the problem of the exponential growth of the number of frequent itemsets extracted. The same problem can be illustrated on other commonly used patterns (e.g., association rules [2], frequent Datalog patterns [11]). So, we cannot expect to be able to apply a general integrity checking process (even one ensuring a polynomial evaluation complexity) on this set of patterns of exponential size.

The situation can be even worse since in most cases, in an IDB these patterns are not fully materialized, and thus some extra (in general non-polynomial) computation is needed to enumerate them.

In the context of IDB we propose to investigate the notion of integrity constraint for patterns, by taking advantage of the following observation. In IDB each pattern is an expression of a specific pattern domain with its own semantics and thus could come with its specific family of integrity constraints, offering an acceptable tradeoff between expressivity and evaluation cost.

In the rest of the paper we focus on a very common pattern called *association rule* [2] and we propose a dedicated form of integrity constraints called *association map constraints.*

Association rules were proposed to represent dependencies between the occurrences of items in customer transactions. Originally, the form of these rules was $A_1, A_2, A_3, \ldots \Rightarrow B$ where $A_1, A_2, A_3, \ldots$ and B denote items. The left hand side is called the *antecedent*, and the right hand side the *consequent*[1]. A confidence measure is defined for these rules. The value of the confidence could be considered as the conditional probability of having the consequent in a transaction when we have all items of the antecedent. Another quality measure, called relative support, is generally associated to the rules. A 10% relative support for a rule means that 10% of the observed transactions support the rule, i.e., the items (antecedent and consequent) could be observed together in 10% of the transactions. It should be noticed that mining association rules is not restricted to basket data analysis, and has been applied on many kinds of data sets after an appropriated encoding with Boolean variables (e.g., [20]). Association rules have received a lot of attention and several algorithms (e.g.,[18,3,12]) have been designed to extract them for given confidence and support thresholds.

An association map is an abstract specification of the set of association rules that could hold in the data according to our current knowledge of the data do-

[1] Consequents made of several items are also considered in the literature. The notion of association map constraint proposed in this paper can be adapted easily to this other form.

main. Association maps are a good candidate of dedicated forms of integrity constraints since they are very concise and readable in the following sense. Firstly, a small association map is sufficient to constrain a huge collection of association rules. Secondly, an association map has a strong hierarchical structure enabling quick intuitive browsing while its semantics remain very simple. And finally, another interesting property of association maps for their use as integrity constraints is that their satisfaction can be checked in a reasonably efficient way in practice.

The rest of this paper is organized as follows. In Section 2 we informally present the notion of association map constraint. More formal definitions and an algorithm to compute association maps are given in Section 3. In Section 4 we describe experiments showing that these constraints can be checked efficiently in practice even in difficult cases. We review related work and conclude with a summary in Section 5.

2 Informal Presentation

In this section we introduce in an informal way the notion of integrity constraint based on association map for IDB.

The key idea behind association map is to represent what should be the confidence variation if a particular item is added to or removed from the antecedent of a rule. Let us take a toy example where each transaction in the data set of the IDB describes one person involved in a car crash (her/his characteristic, the context, the damages).

We suppose that the designer has some knowledge in the car crash domain and wants to use it as integrity constraints over the association rules she/he thinks that could reasonably hold in the IDB. We make the hypothesis that there is a wide variety of such knowledge that can be expressed as the variation of rule confidence w.r.t. the presence/absence of a particular attribute in the left-hand side of the rule. For example, consider that for car crashes the expert thinks that the use of an airbag reduces the probability of severe injury, except for persons that wear glasses. This opinion can be seen as a constraint (denoted IC_1 below) on the variation of the confidence of rules concluding on *severe injury*, w.r.t. a *variation criterion* which is the presence/absence of an airbag.

Consider the following association rules that hold (among others) in the current instance of our IDB. For each rule we indicate the corresponding confidence, and one can easily see that this set of rules does not contradict the integrity constraint IC_1 set by the designer.

$\emptyset \Rightarrow$	$severe\, injury$ 20%
$airbag \Rightarrow$	$severe\, injury$ 10%
$driver \Rightarrow$	$severe\, injury$ 18%
$driver, airbag \Rightarrow$	$severe\, injury$ 10%
$wear\, glasses \Rightarrow$	$severe\, injury$ 15%
$wear\, glasses, airbag \Rightarrow$	$severe\, injury$ 20%
$wear\, glasses, driver \Rightarrow$	$severe\, injury$ 20%
$wear\, glasses, driver, airbag \Rightarrow$	$severe\, injury$ 25%

An association map is simply an explicit synthetic representation of these confidence variations in terms of effects of the presence/absence of the variation criterion in rule antecedents. A map is defined for a given consequent (e.g., *severe injury*), for a particular item used as variation criterion (e.g., *airbag*) and for a given support threshold, but without any confidence threshold. It contains a set of *regions*, where each region is characterized by an homogeneous effect on rule confidence when we add the variation criterion to the rule antecedent. The effect is called *positive* (resp. *negative*) if the addition of the variation criterion results in an increase (resp. a decrease) of the rule confidence. A region is delimited by a lower bound (w.r.t. set inclusion) which is a rule antecedent (a set of items) called *base*. Upward, a region is delimited by a *border* composed of the rule antecedents that are the minimal supersets of the base where the effect changes, from positive to negative or from negative to positive (neutral effects are not considered as real changes). And finally, all elements in a border of a region can be themselves the bases of new regions. Additionally, it should be noticed that rules having a support lower than the given threshold are not represented by the map.

The constraint IC_1 can be expressed as the association map depicted on Figure 1. If we consider all possible rule antecedents (excluding items *severe injury* and *airbag* that represent the rule consequent and the variation criterion), these antecedents can be organized in a lattice (w.r.t. set inclusion). This lattice is depicted using dashed lines on Figure 1. The bases and the border elements are simply particular elements of this lattice that delimitate the regions of homogeneous effects. For constraint IC_1 we have two such regions. The first having for base the empty set and as border $\{wear\_glasses\}$, and in which the effect is negative. And the second, with base $\{wear\_glasses\}$ and border $\{wear\_glasses, driver\}$ where the effect is positive. On the graphical representation, the space of all supersets of a base is sketched by a conic shape. The map presented on Figure 1 can be read as follows: If we add *airbag* to the antecedent of $\emptyset \Rightarrow severe\ injury$ then the confidence decreases, and this holds for all rules excepted if the antecedent contains $wear\_glasses$ in which case the confidence increases.

Suppose that a new set of transactions representing data related to pregnant women is inserted in the IDB, and that now we have the additional rules[2]:

$$
\begin{array}{lr}
pregnant \Rightarrow & severe\,injury\ 30\% \\
pregnant, airbag \Rightarrow & severe\,injury\ 25\% \\
driver, pregnant \Rightarrow & severe\,injury\ 30\% \\
driver, pregnant, airbag \Rightarrow & severe\,injury\ 40\% \\
driver, pregnant, less\_than\_3\_month\_pregnancy \Rightarrow & severe\,injury\ 19\% \\
driver, pregnant, less\_than\_3\_month\_pregnancy, airbag \Rightarrow & severe\,injury\ 12\% \\
wear\ glasses, pregnant \Rightarrow & severe\,injury\ 35\%
\end{array}
$$

[2] To simplify the example we suppose that the previous rules still hold with the same confidence.

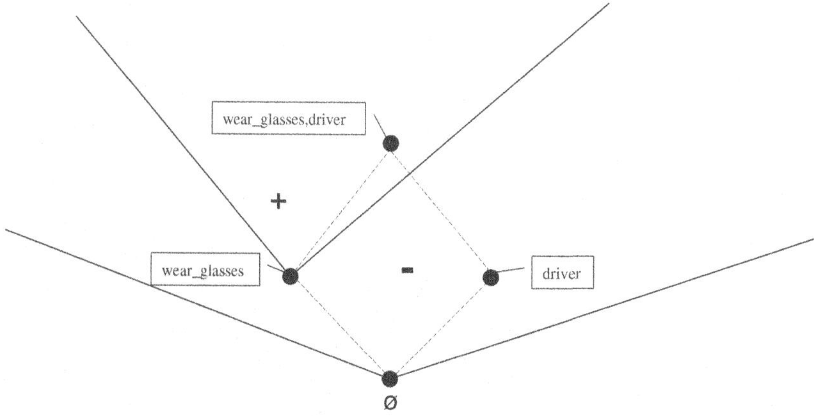

Fig. 1. Association map IC_1

$wear\ glasses, pregnant, airbag \Rightarrow \quad severe\ injury\ 40\%$
$wear\ glasses, driver, pregnant \Rightarrow \quad severe\ injury\ 35\%$
$wear\ glasses, driver, pregnant, airbag \Rightarrow \quad severe\ injury\ 42\%$
$wear\ glasses, driver, pregnant, less\_than\_3\_month\_pregnancy \Rightarrow \quad severe\ injury\ 23\%$
$wear\ glasses, driver, pregnant, airbag, less\_than\_3\_month\_pregnancy \Rightarrow \quad severe\ injury\ 28\%$

We recall that in the context of an IDB these rules are not necessarily extracted and materialized after the insertion of the new data, but from the user point of view they can be used/retrieved at any time.

If we have a close look at these rules, we notice that some of these patterns no longer satisfy IC_1. This can be seen more clearly by drawing the association map corresponding to the whole new set of association rules (for the consequent *severe injury*, the variation criterion *airbag*, and the same support threshold). This map is depicted on Figure 2, where, for readability reasons, the underlying lattice has not been represented. Testing if IC_1 is satisfied or not can then be performed by comparing this map to the map corresponding to IC_1. This is done on Figure 2, where the area of patterns that violate IC_1 is highlighted in grey.

In practice, the use of association maps as integrity constraints in an inductive database can be made as follows. First, the user or the database designer gives a collection of association maps to specify the authorized confidence variations in terms of known effects for specific variation criteria and rule consequents. After an update (or sequence of updates) of the data, the maps describing these effects are computed (from the data) by the inductive database system. Then, they are compared automatically to the ones that have been specified. If a difference is found, the system rejects the update(s) and presents this difference to the user (using eventually a graphical representation with highlighted areas as the one of Figure 2). The user can then assess whether the difference comes from a

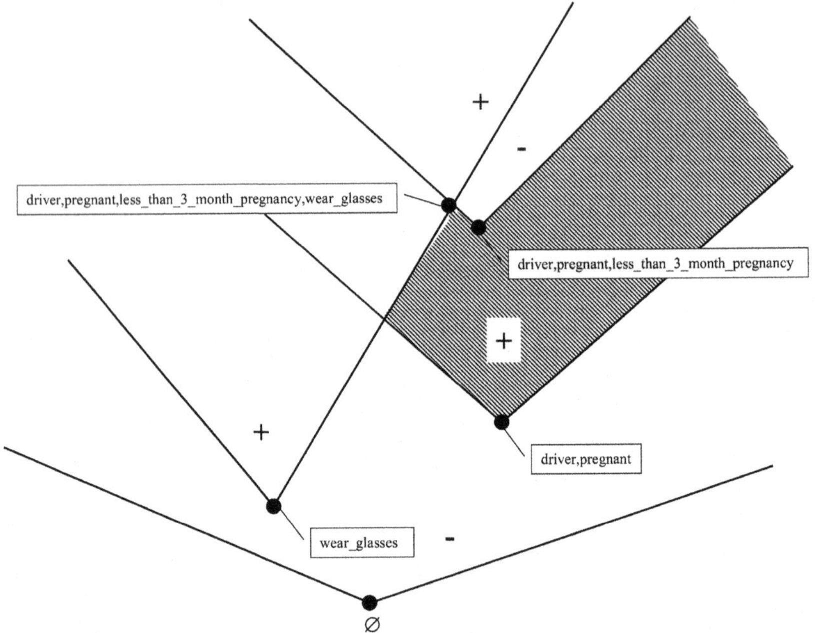

Fig. 2. Patterns that do not satisfy IC_1

corruption of the data or is due to effects that are not correctly specified by the integrity constraints. In the later case, this can leads to the modification of the integrity constraints and be a clue to find an unknown phenomenon.

2.1 Refinement of Maps

The notion of association map presented informally can lead on real data sets to many regions that are not appropriated. We introduce in this section two thresholds used to avoid such situations.

Discarding Extra Regions Using Strong Dependencies. Many exact or nearly exact association rules hold in real data sets and this phenomenon has been used recently to condense huge collections of itemsets [16,7].

Let us consider that we generate a map for consequent C and variation criterion H, and that we have between items A and B the association $A \Rightarrow B$ with a confidence of 100 % (such a rule can be due, for example, to a functional dependency holding in the data). Then $A \Rightarrow C$ and $A, B \Rightarrow C$ have the same confidence, and this is also true for rules $A, H \Rightarrow C$ and $A, B, H \Rightarrow C$. Thus, the effect of H on confidence is the same for antecedents $\{A\}$ and $\{A, B\}$. Moreover, the same holds for any pair of antecedents X and $X \cup \{B\}$, where X is a superset of $\{A\}$. This means that the portion of map generated for supersets of X is redundant with the part constructed for supersets of $X \cup \{B\}$.

Now, suppose that $\{B\}$ is the base of a region, and that the effect of H for $\{A, B\}$ is different than the effect of H on $\{B\}$. In this case, $\{A, B\}$ turns out to be in the border of the region based on $\{B\}$ and results in the generation of an extra region based on $\{A, B\}$. Since we know that the part of the map corresponding to supersets of $\{A, B\}$, is redundant with the one for supersets of $\{A\}$ we can avoid the construction of the extra region based on $\{A, B\}$. For any region based on a Y superset of $\{B\}$ we can make the same simplification when the region based on $Y \cup \{A\}$ has a different effect.

So, we can discard any base X if there exists $Y \subset X$ and $A \in X \setminus Y$ such that $Y \Rightarrow \{A\}$ with a 100% confidence (i.e., if there is an exact rule between items in X).

It should be noticed that exact rules are not likely to be found in noisy data sets or in presence of missing values. In these cases, they appear under the form of rules having a few number of exceptions. So, in the definitions given in Section 3.1 we discard a base X if there is a nearly exact rule between items in X. These nearly exact rules called δ-strong rules (rules with at-most δ exceptions) have been used previously in a different context [7] to condense collections of itemsets and mine frequent patterns more efficiently.

For association map extraction, δ will be a threshold called *freeness.*

Avoiding Regions Created as Artefacts

A confidence is the ratio s_1/s_2 of two integer support values. So, it cannot change in a continuous way, but only by discrete steps.

Let σ be the absolute support threshold used to generate the rules. The greatest discrete step variation due to a single row is encountered when confidence jumps from $(\sigma+1)/(\sigma+1)$ to $\sigma/(\sigma+1)$. Thus, a confidence variation lesser than $1 - \sigma/(\sigma + 1)$ cannot be considered as really significative.

So, we use another threshold τ called *tolerance* to indicate what we consider as a clear confidence variation. When we add the item used as variation criterion to the antecedent of a rule, the variation of confidence must be strictly greater than τ (resp. strictly lower than $-\tau$) to be interpreted as a positive (resp. negative) effect. Otherwise the effect is said to be neutral.

The bases of regions are restricted to be such that their effects must be either positive or negative, except for the first base (the empty set) where the effect is also allowed to be neutral. Thus, in a region, when we encounter a neutral effect we consider that we are still in the same region, and it is only when we find a different and significative positive or negative effect that we generate a border element.

3 Computing Association Maps

In this section, we give more formal definitions and present a way to compute association maps.

3.1 Definitions

Preliminary

Definition 1 (Binary Database). Let R be a set of symbols called items. An *itemset* is a subset of R. A *binary database* r *over* R is a multiset of *rows*, where a row is an itemset. We use the notation $t \in r$ to denote that a particular row t belongs to r.

In this section, we assume that the data set is a binary database r over a set of items R.

Definition 2 (Itemset Support). We denote $\mathcal{M}(r, X) = \{t \in r | X \subseteq t\}$ the multiset of rows in r matched by the itemset X and $Sup(r, X) = |\mathcal{M}(r, X)|$ the *support* of X in r, i.e., the number of rows matched by X.

Definition 3 (Association Rule [2]). Let $Y \subseteq R$ be an itemset. An *association rule over* Y is an expression of the form $X \Rightarrow C$, where $C \in Y$ and $X \subseteq Y \setminus \{C\}$. The support of a rule in r is denoted $Sup(r, X \Rightarrow C)$ and is defined by $Sup(r, X \Rightarrow C) = Sup(r, X \cup \{C\})$. Its confidence is $Conf(r, X \Rightarrow C) = Sup(r, X \cup \{C\})/Sup(r, X)$.

We consider $\sigma \in (0, |r|]$ a support threshold. It should be noticed that it corresponds to an absolute number of rows. However to facilitate the reading of some examples we also use a relative support threshold, that simply corresponds to $\sigma/|r|$.

Definition 4 (Frequent Association Rules). A *frequent association rule* over R w.r.t. σ and r is an association rule $X \Rightarrow C$ over R, such that $Sup(r, X \Rightarrow C) \geq \sigma$. We denote $FreqRules(r, \sigma)$ the set of all frequent rules over R w.r.t. σ and r.

We also recall the definitions of δ-strong rules and $\delta - free$ sets, needed to define association maps. These two notions have been introduced in a different context[3] in [7,6].

Definition 5 (δ-Strong Rule). A *δ-strong rule*[4] in a binary database r is an association rule $X \Rightarrow C$ over R such that $Sup(r, X) - Sup(r, X \cup \{C\}) \leq \delta$, i.e., the rule is violated in no more than δ rows.

In this definition, δ is supposed to have a small value, so a δ-strong rule is intended to be a rule with very few exceptions.

Definition 6 (δ-Free Set). $X \subseteq R$ is a *δ-free set* w.r.t. r if and only if there is no δ-strong rule over X in r. The set of all δ-free sets w.r.t. r is noted $Free(r, \delta)$.

---

[3] Originally $\delta - free$ have been proposed as a condensed representation that can be extracted very efficiently and that can be used to closely approximate the support of all itemsets that are frequent w.r.t. a given support threshold.

[4] Stemming from the notion of *strong rule* of [17].

Since δ is supposed to be rather small, informally, a δ-free set is a set of items such that these items are not related by any very strong positive dependency.

Effect Regions and Association Maps. As presented in Section 2, an association map is defined w.r.t. two items (the consequent and the variation criterion) and three thresholds (support, tolerance and freeness). In this section, we denote respectively $C \in R$ the consequent and $H \in R$ the variation criterion, and we use $\sigma \in (0, |r|]$, $\tau \in [0, 1]$ and an integer δ to represent respectively the support, the tolerance and the freeness threshold.

Definition 7 (Local Effect). Let $X \subseteq R \setminus \{C, H\}$ be an itemset. The *local effect* of H on rule $X \Rightarrow C$, denoted $LocEffect(r, \tau, X, C, H)$ is defined as:

$$LocEffect(r, \tau, X, C, H) = \begin{cases} 1 \;\; if \;\; Conf(r, X \cup \{H\} \Rightarrow C) - \\ \qquad\qquad Conf(r, X \Rightarrow C) > \tau \\ -1 \; if \;\; Conf(r, X \cup \{H\} \Rightarrow C) - \\ \qquad\qquad Conf(r, X \Rightarrow C) < -\tau \\ 0 \;\; otherwise \end{cases}$$

According to its value, the effect is respectively called *positive*, *negative* or *neutral*.

We now define the antecedents that are significative to generate the maps.

Definition 8 (Significant Antecedent). $SigAnte(r, \sigma, \tau, \delta, C, H)$ is the collection of significant antecedents and is defined by $SigAnte(r, \sigma, \tau, \delta, C, H) = \{X \subseteq R | (X \cup \{H\} \Rightarrow C) \in FreqRules(r, \sigma) \wedge X \in Free(r, \delta) \wedge LocEffect(r, \tau, X, C, H) \in \{-1, 1\}\}$.

These antecedents are itemsets that form with H the antecedent of a frequent rule. Moreover, they must be made of items that are not strongly dependent (i.e, they are δ-free) and where the local effect is clearly positive or negative (not neutral). Then, for an itemset X we define the border effect, which is the collection of the minimal supersets of X that are significant antecedents and where the local effect changes strongly (from positive to negative or from negative to positive).

Definition 9 (Effect Border). Let X be an itemset such that $X \subseteq R \setminus \{C, H\}$. The effect border for X is $Border(r, \sigma, \tau, \delta, X, C, H) = \{Y \in R | X \subset Y \wedge Y \in SigAnte(r, \sigma, \tau, \delta, C, H) \wedge LocEffect(r, \tau, X, C, H) \neq LocEffect(r, \tau, Y, C, H) \wedge (\forall Z, X \subset Z \subset Y \Rightarrow LocEffect(r, \tau, Z, C, H) \in \{0, LocEffect(r, \tau, X, C, H)\})\}$.

Now we consider regions of homogeneous effect, i.e., a set of rule antecedents having a common subset and a common local effect. We first define their lower bounds (w.r.t. set inclusion), called effect bases as follows. The empty set is an effect base. A significant antecedent which is in the effect border of an effect base of smaller size is also an effect base. This notion is expressed more formally by the next definition.

Definition 10 (Effect Base). The collection of effect bases is defined inductively as follows.
$Base_0 = \{\emptyset\}$
$Base_i = \{X \subseteq R| \; |X| = i \wedge X \in SigAnte(r, \sigma, \tau, \delta, C, H) \wedge (\exists Y \in \bigcup_{j<i} Base_j,$
$X \in Border(r, \sigma, \tau, \delta, Y, C, H))\}$

$Base(r, \sigma, \tau, \delta, C, H) = \bigcup_i Base_i.$

Then an association map is simply the collection of all effect bases together with their borders.

Definition 11 (Association Map). An association map for a binary database r w.r.t. items C, H and thresholds σ, τ, δ is defined by $AMap(r, \sigma, \tau, \delta, C, H) = \{\langle X, \mathcal{B}\rangle | X \in Base(r, \sigma, \tau, \delta, C, H) \wedge \mathcal{B} = Border(r, \sigma, \tau, \delta, X, C, H)\}$.

Each tuple in an association map corresponds to the lower and upper bounds (w.r.t. set inclusion) of a region where the local effect does not significatively change.

Note that two different effect regions may overlap. This overlapping may occur even when their respective effect bases have opposite local effects, in this case the itemsets that belong to both regions have a neutral local effect.

3.2 Algorithm

We present a generic algorithm called *GenMap* to produce the map for a consequent C, a variation criterion H, and thresholds δ, σ and τ corresponding respectively to freeness, support and tolerance thresholds.

The algorithm calls three functions: $CandAnte, Signif$ and $Effect$.

The algorithm is presented using in its input a set $S = FreqRules(r, \sigma)$ of all frequent association rules along with their supports.

Algorithm 1 (*GenMap*)

Input: *C, H items, n the size of the largest candidate antecedent, set S, thresholds δ, σ and τ.*

Used subprograms: *$CandAnte(S, i, C, H)$ establishes the set of itemsets of size i, not containing C or H, that are candidates for being significant antecedents. $Signif(S, X, C, H, \tau, \delta, \sigma)$, which finds out whether X is a significant antecedent or not. And the function $Effect(S, X, C, H)$ is used to compute the local effect of H for X.*

Output: *a set of tuples containing all effect bases, and their corresponding effects and borders.*

1. **let** $E_\emptyset := Effect(S, \emptyset, C, H), Map := \{\langle \emptyset, E_\emptyset, \emptyset \rangle\}$;
2. **for all** $i \in \{1, \ldots, n\}$ **do**
3.   **for all** $X \in CandAnte(S, i, C, H)$ **do**
4.   **if** $Signif(S, X, C, H, \tau, \delta, \sigma)$ **then**
5.   **let** $E_X := Effect(S, X, C, H)$;
6.   **let** $MaxSubBases_X := \{\langle Y, E_Y, B_Y \rangle \in Map |$
        $Y \subset X \wedge E_Y \neq E_X \wedge \forall W \in B_Y, W \not\subset X\}$;
7.   **if** $MaxSubBases_X \neq \emptyset$ **then**
8.   **let** $Map := Map \cup \{\langle X, E_X, \emptyset \rangle\}$;
9.   **for all** $\langle Z, E_Z, B_Z \rangle \in MaxSubBases_X$ **do**
10.   **let** $Map := (Map \setminus \{\langle Z, E_Z, B_Z \rangle\}) \cup$
            $\{\langle Z, E_Z, B_Z \cup \{X\} \rangle\}$;
11.   **od**
12.   **fi**
13.   **fi**
14.   **od**
15. **od**
16. **output** Map

In line 1, $GenMap$ considers the empty itemset, which is always an effect base according to Definition 10. In line 2, the algorithm enters a loop corresponding to increasing sizes of candidate antecedents.

For each candidate antecedent X the algorithm checks if it is significant (line 4), and if so, $GenMap$ computes in line 6 the set $MaxSubBases_X$ of all bases of regions that contain X in their border.

If at least one of such region exists (line 7), then X is also an effect base, and the corresponding tuple is created in line 8. X is then stored in the borders of all regions having their bases in $MaxSubBases_X$ (lines 9–11).

Theorem 1 (Correctness of $GenMap$). *The algorithm $GenMap$ outputs the effect bases (along with the corresponding border elements and effects) of the association map defined for a consequent C, a variation criterion H, and thresholds δ, σ and τ.*

Proof. *The proof is made by induction on the size of the bases. Note that the effect base $\emptyset$ is included in Map by the first line of the algorithm.*

Hypothesis. *Suppose that for every effect base X of size less or equal to i the algorithm $GenMap$ correctly reported X as a base and as border element in Map.*

Consider an effect base $X \neq \emptyset$ of size $i + 1$. We are going to show that X is correctly reported as a base and as border element in Map.

X is an effect base implies that X is returned by $CandAnte(S, i, C, H)$ (line 3) and not filtered out by $Signif(S, X, C, H, \tau, \delta, \sigma)$ (line 4). Therefore, it will be considered in lines 5–12. By Definition 10, there is at least one effect base $Y \subset X$ such that X is in the border of the region of base Y. Assuming that the induction hypothesis holds, we find all such bases in line 6. Then, X is added as a base to Map in line 8, and correctly reported as border element in lines 9–11.

So, the algorithm correctly reported all effect bases and border elements in Map. The soundness of every update of Map is immediate. ◇

3.3 Computing Association Maps from Association Rules

GenMap can compute the association maps using as input S, the collection of all frequent rule, and running the functions $CandAnte(S, i, C, H)$, $Signif(S, X, C, H, \tau, \delta, \sigma)$ and $Effect(S, X, C, H)$ defined in the following manner.

$CandAnte(S, i, C, H)$ selects from S the rules having an antecedent of size i and consequent C, but skips the rules containing H in their antecedents. Then, it returns the collection of all antecedents of these rules.

$Signif(S, X, C, H, \tau, \delta, \sigma)$ checks if $X \cup \{H\} \Rightarrow C$ is in S (i.e., if the rule is frequent), and if it is the case it tests the local effect of H. To do so, it finds in S the rule $X \Rightarrow C$, and compares the confidences of the two rules. If the absolute value of their difference is less or equal to τ, the function exits returning $false$ (the local effect is neutral). Otherwise the δ-freeness of X is tested by simply checking that for every $A \in X$ the difference between the support of $X \setminus \{A\}$ and X is strictly greater than δ. It should be noticed that the supports of $X \setminus \{A\}$ and X can be obtained using S as follows. Let us consider that we need the support of a frequent itemset Z. Let B be any item such that $B \in Z$, then the rule $Z \setminus \{B\} \Rightarrow B$ is frequent and is in S. By definition 3 the support of Z is equal to the support of this rule.

If X is δ-free, $Signif(S, X, C, H, \tau, \delta, \sigma)$ returns $true$, and $false$ otherwise.

$Effect(S, X, C, H)$ finds the confidences of the rules $X \Rightarrow C$ and $X \cup \{H\} \Rightarrow C$ in S, and then returns the local effect of H according to the difference between the confidences of the two rules.

One can generate association maps using the generic algorithm and the collection of all frequent association rules. Unfortunately, this input collection may be very large. Moreover, for some data sets (e.g. highly correlated census-like data sets), it is an intractable process to mine all frequent association rules at interesting support thresholds.

In the next section, we show that one can avoid extracting all frequent rules, by using more elaborated input collections.

3.4 Computing Association Maps Directly

Let us now consider that S consists of all tuples $\langle Z \setminus \{H, C\}, Z \setminus \{H\}, Z \setminus \{C\}, Z \rangle$ such that Z is a frequent itemset (w.r.t. threshold σ) containing both C and H, and such that $Z \setminus \{H, C\}$ is δ-free. We also consider that we have at hand the supports of the itemsets in the tuples in S.

The main practical advantage of this new input S is that it remains in general many much more smaller than the set of all frequent association rules.

S can be used to generate the association map for consequent C, variation criterion H, thresholds δ, σ, τ, using algorithm *GenMap* when the functions $CandAnte$, $Signif$ and $Effect$ are defined as follows.

$CandAnte(S, i, C, H)$ selects from all tuples in S the ones having a first element of size i and outputs these first elements. By grouping the tuples in S_i

according to the size of the first element at the time we construct S, we can compute the result of $CandAnte(S, i, C, H)$ in a very efficient manner.

$Signif(S, X, C, H, \tau, \delta, \sigma)$ looks for in S $\langle X, X \cup \{C\}, X \cup \{H\}, X \cup \{H, C\}\rangle$. If such a tuple exists in S then, by construction of S, X is δ-free and $X \cup \{H, C\}$ is frequent. Then, to verify that the local effect is not neutral for X, it compares the absolute value of the difference between $Sup(X \cup \{C\})/Sup(X)$ and $Sup(X \cup \{H, C\})/Sup(X \cup \{H\})$ to the tolerance threshold τ.

$Effect(S, X, C, H)$ used S to compute the local effect for X in the same way as function $Signif$. In fact, in an implementation of algorithm $GenMap$ the value of $Effect(S, X, C, H)$ is simply obtained during the computation of $Signif(S, X, C, H, \tau, \delta, \sigma)$.

The generation of S itself can be made using the algorithms presented in [7,6] to mine δ-free sets. In our prototype we choose to generate S using the technique proposed in [9,10] to mine frequent patterns efficiently even in presence of difficult dense data sets. The prototype extracted first a representation called *disjunction-bordered condensation* using the algorithm VLinEx proposed in [9,10] and then generates S from this representation. Finally, it produces the map itself using $GenMap$.

4 Experiments

To check the satisfaction of association map constraints we propose to first extract the corresponding association maps from the data of the IDB, and then to compare these maps with the association map constraints given by the designer of the IDB. We consider that the association map constraints are rather small, thus we neglect the computing cost of the second step and take only the first one into account. In this section, we report experiments showing that the first step (computation of association maps over the IDB) can be done efficiently even in difficult cases.

Conditions of Experiments. We choose *Pumsb*, a very challenging census data set, containing 7117 items, 49046 rows, each with 74 items set to *true*. The particularity of the selected data set is that it is very dense and the combinatorial explosion of the number of frequent itemsets makes the mining of all association rules intractable for low support thresholds [5]. This data set has been preprocessed by researchers from IBM Almaden Research Center[5].

We run experiments on a 1 GHz PC with 512 Mb of RAM and Linux operating system.

To produce difficult conditions for the association map extraction, we choose a value of δ equal to 0 (avoiding only regions due to exact dependencies), and $\tau = 10^{-4}$ (a tolerance close to the minimal tolerance threshold defined in Section 2.1). We also used a heuristic to select ten *hard* pairs consequent/variation criterion, i.e., pairs such that regions in maps tend to be large or numerous.

[5] http://www.almaden.ibm.com/cs/quest/data/ long_patterns.bin.tar

We present this heuristic and the results of the experiments in the following sections.

Selecting Consequents and Variation Criteria. We define a function to associate a score to each pair of consequent C and variation criterion H.

This score is computed from a collection of itemsets denoted $\mathcal{E}_{\sigma,\delta}$ and containing all δ-free itemsets having a support exceeding σ. Let $\mathcal{F}_{C,H}$ be the collection of itemsets in $\mathcal{E}_{\sigma,\delta}$ containing items C and H. Let nb_{neg} (resp. nb_{pos}) be the number of elements in $\mathcal{E}_{\sigma,\delta}$ having a negative (resp. positive) local effect for C, H with tolerance $\tau = 0$. Let M_{neg} (resp. M_{pos}) be the mean value of the confidence variation for all negative (resp. positive) local effects for C,H and $\tau = 0$.

We defined $score(C, H) = T_{CH} * N_{CH} * P_{CH} * abs(M_{neg}) * M_{pos}$.

The factor T_{CH} is $|\mathcal{F}_{C,H}|/|\mathcal{E}_{\sigma,\delta}|$, and represents the ratio of itemsets in $\mathcal{E}_{\sigma,\delta}$ that are candidates to be a base of a region.

The factor N_{CH} is $nb_{neg}/|\mathcal{F}_{C,H}|$ and corresponds to the ratio of negative local effects among all possible local effects. P_{CH} is $nb_{pos}/|\mathcal{F}_{C,H}|$ and corresponds to the same ratio for positive effects. $N_{CH} * P_{CH}$ is maximal when $N_{CH} = P_{CH} = 1/2$, i.e., when the amount of significant antecedents with negative and positive local effects for a given C and H are the same and there is no neutral-effect antecedents. High values of $N_{CH} * P_{CH}$ indicate that the map is likely to contain many changes of effects and thus many regions.

Finally, the factor $abs(M_{neg}) * M_{pos}$ takes into account the amplitude of the changes of the confidence. A higher value implies potential effect bases with clear positive or negative effects, and thus an important number of bases even at high values of the tolerance threshold.

A pair C, H having a high $score(C, H)$ offers a good potentiality of generating maps containing large and numerous regions.

Results. Figure 3 summarizes the results. For various support thresholds, we report the highest (MAX), the lowest (MIN) and the mean (MEAN) extraction time over the ten pairs consequent/variation criterion having the highest $score(C, H)$ values.

The experiments show that on this difficult data set, for support thresholds of 80% to 100%, the extraction of the maps from the data and thus the test of satisfaction of the association map constraints can be done in practice on-line (i.e., during interactive data manipulation sessions). For lower thresholds, the integrity check can be performed reasonably off-line even at a 50% support threshold, which represents very hard conditions on this dense data set[6].

In practice, a large amount of the map extraction time is spent to compute the association rules (or in our prototype, to generate the intermediate representation as presented in Section 3.4). It should be noticed that the computation

[6] Such conditions can be considered as much more difficult than lower support thresholds (e.g., 1% or even less) on many sparse data sets (e.g., basket data, logs of alarms).

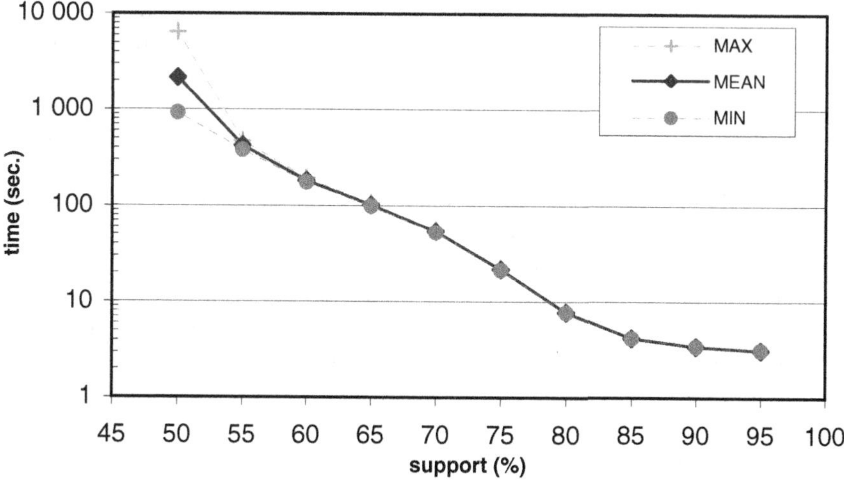

Fig. 3. Extraction times of association maps

of the association rules (or of the intermediate representation) is common to all maps for a given support threshold in a given data set. So, in most cases, when a map has been extracted for a pair consequent/variation, the maps for the other pairs involved in association map constraints can be obtained at a marginal extra cost.

5 Conclusion and Related Work

To our knowledge the notion of integrity constraint in IDB has not been previously explicitly investigated. In this paper we advocated that common data mining tasks such as the detection of corrupted data or of patterns that contradict the expert beliefs [19] can be integrated in a clean way under the concept of integrity constraints for IDB.

We illustrated this possibility by proposing a form of integrity constraints called association map constraints. Such a constraint is a specification of the sign of the variation of association rule confidences when a given attribute is added in the antecedent of the rule. These maps have a simple intuitive meaning and can concisely constrain all association rules. Thus, they allow to express clear and understandable specifications. Moreover, we have shown by means of experiments that the satisfaction of association map constraints can be checked in practice in a reasonably efficient way.

The use of confidence variation has been investigated previously in [4,14] to prune and summarize collection of association rules. As for association map these variations are considered w.r.t. a fixed rule consequent. [4] proposed to select rules $\alpha \Rightarrow C$ showing an increase (or eventually a limited decrease) of confidence with respect to all rules $\beta \Rightarrow C$ where $\beta \subset \alpha$ (i.e., more general rules). If we adapt this idea in the context of integrity constraints for IDB, it leads to specify that some particular rules must have a confidence higher than any of their more general rules, and then to use the algorithm described in [4] to check if this specification is satisfied in the database. In [14] the authors proposed to select rules that are statistically more significant w.r.t. the more general rules and then to summarize this collection of selected rules. Similarly to [4] this approach can be adapted as integrity constraints for IDB.

Compared to these works, association maps are complementary. On one hand, they are more specific in the sense that an association map focuses on the effect of the absence/presence of a particular attribute H (the variation criterion) in the antecedent of the rules. However, it is possible to specify several association maps, each for a different attribute H. On the other hand, an association map is a cartography of all association rules (areas of decrease/increase of confidence w.r.t. the presence of H in the antecedent) and thus give a more general view than the approaches of [4] and [14] that concentrate on rules *better* (in some sense) than the more general ones.

With respect to the association map constraints proposed in this paper, an interesting issue to investigate, is to determine how and in which cases the check of the constraints can be performed incrementally with respect to the updates of the databases.

A more general direction of future work is to investigate how the concepts and techniques proposed previously in the data mining literature, can be adapted and used to specify and check the data and pattern consistency in the context of IDB.

Acknowledgments

We would like to thank the anonymous referees for their helpful comments and suggestions.

References

1. S. Abiteboul, R. Hull, and V. Vianu. *Foundations of Databases.* Addison-Wesley, 1995.
2. R. Agrawal, T. Imielinski, and A. N. Swami. Mining association rules between sets of items in large databases. In *Proceedings of the 1993 ACM SIGMOD International Conference on Management of Data*, pages 207–216, Washington, D.C., USA, May 1993. ACM Press.
3. R. Agrawal, H. Mannila, R. Srikant, H. Toivonen, and A. I. Verkamo. Fast discovery of association rules. In *Advances in Knowledge Discovery and Data Mining*, pages 307–328. AAAI Press, 1996.
4. R. Bayardo, R. Agrawal, and D. Gunopulos. Constraint-based rule mining in large, dense databases. In *Proceedings ICDE'99*, pages 188–197, Sydney, Australia, March 1999.
5. R. J. Bayardo. Efficiently mining long patterns from databases. In *Proceedings of the 1998 ACM SIGMOD International Conference on Management of Data*, pages 85–93. ACM Press, 1998.

6. J.-F. Boulicaut, A. Bykowski, and C. Rigotti. Approximation of frequency queries by mean of free-sets. In *Proc. of the 4th European Conf. on Principles and Practice of Knowledge Discovery in Databases (PKDD'00)*, pages 75–85, Lyon, France, September 2000.
7. J.-F. Boulicaut, A. Bykowski, and C. Rigotti. Free-sets : a condensed representation of boolean data for the approximation of frequency queries. *Journal of Data Mining and Knowledge Discovery*, 7(1):5–22, 2003.
8. J.-F. Boulicaut, M. Klemettinen, and H. Mannila. Querying inductive databases: A case study on the MINE RULE operator. In *Proc. PKDD'98*, volume 1510 of *LNAI*, pages 194–202, Nantes, F, 1998. Springer-Verlag.
9. A. Bykowski and C. Rigotti. A condensed representation to find frequent patterns. In *Proc. of the Twentieth ACM SIGACT-SIGMOD-SIGART Symposium on Principles of Database Systems (PODS'01)*, pages 267–273, Santa Barbara, CA, USA, May 2001. ACM.
10. A. Bykowski and C. Rigotti. Disjunction-bordered condensed representation of frequent patterns. *Information Systems*, To appear.
11. L. Dehaspe and H. Toivonen. Discovery of frequent datalog patterns. *Journal of Data Mining and Knowledge Discovery*, 3(1):7–36, 1999.
12. J. Hipp, U. Güntzer, and G. Nakhaeizadeh. Algorithms for association rule mining – a general survey and comparison. *SIGKDD Explorations*, 2(1):58–64, July 2000.
13. T. Imielinski and H. Mannila. A database perspective on knowledge discovery. *Communications of the ACM*, 39(11):58–64, Nov. 1996.
14. B. Liu, W. Hsu, and Y. Ma. Pruning and summarizing the discovered associations. In *Proc. of the Fifth Int. Conference on Knowledge Discovery and Data Mining (KDD'99)*, pages 125–134, San Diego, CA, USA, August 1999.
15. H. Mannila. Inductive databases and condensed representations for data mining. In *Proc. ILPS'97*, pages 21–30, Port Jefferson, USA, 1997. MIT Press.
16. N. Pasquier, Y. Bastide, R. Taouil, and L. Lakhal. Efficient mining of association rules using closed itemset lattices. *Information Systems*, 24(1):25–46, 1999.
17. G. Piatetsky-Shapiro. Discovery, analysis, and presentation of strong rules. In *Knowledge Discovery in Databases*, pages 229–248. AAAI Press, Menlo Park, CA, 1991.
18. A. Savasere, E. Omiecinski, and S. B. Navathe. An efficient algorithm for mining association rules in large databases. In *Proc. VLDB'95*, pages 432–444, Zurich, Switzerland, September 1995. Morgan Kaufmann.
19. A. Silberschatz and A. Tuzhilin. On subjective measures of interestingness in knowledge discovery. In *Proc. of the First Int. Conference on Knowledge Discovery and Data Mining (KDD'95)*, pages 275–281, Montreal, Canada, August 1995.
20. R. Srikant and R. Agrawal. Mining quantitative association rules in large relational tables. In *Proc. ACM SIGMOD'96*, pages 1–12, Montreal, Quebec, Canada, June 1996. ACM Press.
21. J. Ullman. *Database and Knowledge-Base Systems, vol. II.* Computer Science Press, Rockville, MD, 1989.

Author Index

Lecture Notes in Artificial Intelligence (LNAI)

Vol. 3139: F. Iida, R. Pfeifer, L. Steels, Y. Kuniyoshi (Eds.), Embodied Artificial Intelligence. IX, 331 pages. 2004.

Vol. 3131: V. Torra, Y. Narukawa (Eds.), Modeling Decisions for Artificial Intelligence. XI, 327 pages. 2004.

Vol. 3127: K.E. Wolff, H.D. Pfeiffer, H.S. Delugach (Eds.), Conceptual Structures at Work. XI, 403 pages. 2004.

Vol. 3123: A. Belz, R. Evans, P. Piwek (Eds.), Natural Language Generation. X, 219 pages. 2004.

Vol. 3120: J. Shawe-Taylor, Y. Singer (Eds.), Learning Theory. X, 648 pages. 2004.

Vol. 3097: D. Basin, M. Rusinowitch (Eds.), Automated Reasoning. XII, 493 pages. 2004.

Vol. 3071: A. Omicini, P. Petta, J. Pitt (Eds.), Engineering Societies in the Agents World. XIII, 409 pages. 2004.

Vol. 3070: L. Rutkowski, J. Siekmann, R. Tadeusiewicz, L.A. Zadeh (Eds.), Artificial Intelligence and Soft Computing - ICAISC 2004. XXV, 1208 pages. 2004.

Vol. 3068: E. André, L. Dybkjær, W. Minker, P. Heisterkamp (Eds.), Affective Dialogue Systems. XII, 324 pages. 2004.

Vol. 3067: M. Dastani, J. Dix, A. El Fallah-Seghrouchni (Eds.), Programming Multi-Agent Systems. X, 221 pages. 2004.

Vol. 3066: S. Tsumoto, R. Słowiński, J. Komorowski, J.W. Grzymała-Busse (Eds.), Rough Sets and Current Trends in Computing. XX, 853 pages. 2004.

Vol. 3065: A. Lomuscio, D. Nute (Eds.), Deontic Logic in Computer Science. X, 275 pages. 2004.

Vol. 3060: A.Y. Tawfik, S.D. Goodwin (Eds.), Advances in Artificial Intelligence. XIII, 582 pages. 2004.

Vol. 3056: H. Dai, R. Srikant, C. Zhang (Eds.), Advances in Knowledge Discovery and Data Mining. XIX, 713 pages. 2004.

Vol. 3055: H. Christiansen, M.-S. Hacid, T. Andreasen, H.L. Larsen (Eds.), Flexible Query Answering Systems. X, 500 pages. 2004.

Vol. 3040: R. Conejo, M. Urretavizcaya, J.-L. Pérez-de-la-Cruz (Eds.), Current Topics in Artificial Intelligence. XIV, 689 pages. 2004.

Vol. 3035: M.A. Wimmer (Ed.), Knowledge Management in Electronic Government. XII, 326 pages. 2004.

Vol. 3034: J. Favela, E. Menasalvas, E. Chávez (Eds.), Advances in Web Intelligence. XIII, 227 pages. 2004.

Vol. 3030: P. Giorgini, B. Henderson-Sellers, M. Winikoff (Eds.), Agent-Oriented Information Systems. XIV, 207 pages. 2004.

Vol. 3029: B. Orchard, C. Yang, M. Ali (Eds.), Innovations in Applied Artificial Intelligence. XXI, 1272 pages. 2004.

Vol. 3025: G.A. Vouros, T. Panayiotopoulos (Eds.), Methods and Applications of Artificial Intelligence. XV, 546 pages. 2004.

Vol. 3020: D. Polani, B. Browning, A. Bonarini, K. Yoshida (Eds.), RoboCup 2003: Robot Soccer World Cup VII. XVI, 767 pages. 2004.

Vol. 3012: K. Kurumatani, S.-H. Chen, A. Ohuchi (Eds.), Multi-Agnets for Mass User Support. X, 217 pages. 2004.

Vol. 3010: K.R. Apt, F. Fages, F. Rossi, P. Szeredi, J. Váncza (Eds.), Recent Advances in Constraints. VIII, 285 pages. 2004.

Vol. 2990: J. Leite, A. Omicini, L. Sterling, P. Torroni (Eds.), Declarative Agent Languages and Technologies. XII, 281 pages. 2004.

Vol. 2980: A. Blackwell, K. Marriott, A. Shimojima (Eds.), Diagrammatic Representation and Inference. XV, 448 pages. 2004.

Vol. 2977: G. Di Marzo Serugendo, A. Karageorgos, O.F. Rana, F. Zambonelli (Eds.), Engineering Self-Organising Systems. X, 299 pages. 2004.

Vol. 2972: R. Monroy, G. Arroyo-Figueroa, L.E. Sucar, H. Sossa (Eds.), MICAI 2004: Advances in Artificial Intelligence. XVII, 923 pages. 2004.

Vol. 2961: P. Eklund (Ed.), Concept Lattices. IX, 411 pages. 2004.

Vol. 2953: K. Konrad, Model Generation for Natural Language Interpretation and Analysis. XIII, 166 pages. 2004.

Vol. 2934: G. Lindemann, D. Moldt, M. Paolucci (Eds.), Regulated Agent-Based Social Systems. X, 301 pages. 2004.

Vol. 2930: F. Winkler (Ed.), Automated Deduction in Geometry. VII, 231 pages. 2004.

Vol. 2926: L. van Elst, V. Dignum, A. Abecker (Eds.), Agent-Mediated Knowledge Management. XI, 428 pages. 2004.

Vol. 2923: V. Lifschitz, I. Niemelä (Eds.), Logic Programming and Nonmonotonic Reasoning. IX, 365 pages. 2004.

Vol. 2915: A. Camurri, G. Volpe (Eds.), Gesture-Based Communication in Human-Computer Interaction. XIII, 558 pages. 2004.

Vol. 2913: T.M. Pinkston, V.K. Prasanna (Eds.), High Performance Computing - HiPC 2003. XX, 512 pages. 2003.

Vol. 2903: T.D. Gedeon, L.C.C. Fung (Eds.), AI 2003: Advances in Artificial Intelligence. XVI, 1075 pages. 2003.

Vol. 2902: F.M. Pires, S.P. Abreu (Eds.), Progress in Artificial Intelligence. XV, 504 pages. 2003.

Vol. 2892: F. Dau, The Logic System of Concept Graphs with Negation. XI, 213 pages. 2003.

Vol. 2891: J. Lee, M. Barley (Eds.), Intelligent Agents and Multi-Agent Systems. X, 215 pages. 2003.

Vol. 2882: D. Veit, Matchmaking in Electronic Markets. XV, 180 pages. 2003.

Vol. 2871: N. Zhong, Z.W. Raś, S. Tsumoto, E. Suzuki (Eds.), Foundations of Intelligent Systems. XV, 697 pages. 2003.

Vol. 2854: J. Hoffmann, Utilizing Problem Structure in Planing. XIII, 251 pages. 2003.

Vol. 2843: G. Grieser, Y. Tanaka, A. Yamamoto (Eds.), Discovery Science. XII, 504 pages. 2003.

Vol. 2842: R. Gavaldá, K.P. Jantke, E. Takimoto (Eds.), Algorithmic Learning Theory. XI, 313 pages. 2003.

Vol. 2838: N. Lavrač, D. Gamberger, L. Todorovski, H. Blockeel (Eds.), Knowledge Discovery in Databases: PKDD 2003. XVI, 508 pages. 2003.

Vol. 2837: N. Lavrač, D. Gamberger, L. Todorovski, H. Blockeel (Eds.), Machine Learning: ECML 2003. XVI, 504 pages. 2003.

Vol. 2835: T. Horváth, A. Yamamoto (Eds.), Inductive Logic Programming. X, 401 pages. 2003.

Vol. 2821: A. Günter, R. Kruse, B. Neumann (Eds.), KI 2003: Advances in Artificial Intelligence. XII, 662 pages. 2003.

Vol. 2807: V. Matoušek, P. Mautner (Eds.), Text, Speech and Dialogue. XIII, 426 pages. 2003.

Vol. 2801: W. Banzhaf, J. Ziegler, T. Christaller, P. Dittrich, J.T. Kim (Eds.), Advances in Artificial Life. XVI, 905 pages. 2003.

Vol. 2797: O.R. Zaïane, S.J. Simoff, C. Djeraba (Eds.), Mining Multimedia and Complex Data. XII, 281 pages. 2003.

Vol. 2792: T. Rist, R.S. Aylett, D. Ballin, J. Rickel (Eds.), Intelligent Virtual Agents. XV, 364 pages. 2003.

Vol. 2782: M. Klusch, A. Omicini, S. Ossowski, H. Laamanen (Eds.), Cooperative Information Agents VII. XI, 345 pages. 2003.

Vol. 2780: M. Dojat, E. Keravnou, P. Barahona (Eds.), Artificial Intelligence in Medicine. XIII, 388 pages. 2003.

Vol. 2777: B. Schölkopf, M.K. Warmuth (Eds.), Learning Theory and Kernel Machines. XIV, 746 pages. 2003.

Vol. 2752: G.A. Kaminka, P.U. Lima, R. Rojas (Eds.), RoboCup 2002: Robot Soccer World Cup VI. XVI, 498 pages. 2003.

Vol. 2741: F. Baader (Ed.), Automated Deduction – CADE-19. XII, 503 pages. 2003.

Vol. 2705: S. Renals, G. Grefenstette (Eds.), Text- and Speech-Triggered Information Access. VII, 197 pages. 2003.

Vol. 2703: O.R. Zaïane, J. Srivastava, M. Spiliopoulou, B. Masand (Eds.), WEBKDD 2002 - MiningWeb Data for Discovering Usage Patterns and Profiles. IX, 181 pages. 2003.

Vol. 2700: M.T. Pazienza (Ed.), Extraction in the Web Era. XIII, 163 pages. 2003.

Vol. 2699: M.G. Hinchey, J.L. Rash, W.F. Truszkowski, C.A. Rouff, D.F. Gordon-Spears (Eds.), Formal Approaches to Agent-Based Systems. IX, 297 pages. 2002.

Vol. 2691: V. Mařík, J.P. Müller, M. Pechoucek (Eds.), Multi-Agent Systems and Applications III. XIV, 660 pages. 2003.

Vol. 2684: M.V. Butz, O. Sigaud, P. Gérard (Eds.), Anticipatory Behavior in Adaptive Learning Systems. X, 303 pages. 2003.

Vol. 2682: R. Meo, P.L. Lanzi, M. Klemettinen (Eds.), Database Support for Data Mining Applications. XII, 325 pages. 2004.

Vol. 2671: Y. Xiang, B. Chaib-draa (Eds.), Advances in Artificial Intelligence. XIV, 642 pages. 2003.

Vol. 2663: E. Menasalvas, J. Segovia, P.S. Szczepaniak (Eds.), Advances in Web Intelligence. XII, 350 pages. 2003.

Vol. 2661: P.L. Lanzi, W. Stolzmann, S.W. Wilson (Eds.), Learning Classifier Systems. VII, 231 pages. 2003.

Vol. 2654: U. Schmid, Inductive Synthesis of Functional Programs. XXII, 398 pages. 2003.

Vol. 2650: M.-P. Huget (Ed.), Communications in Multiagent Systems. VIII, 323 pages. 2003.

Vol. 2645: M.A. Wimmer (Ed.), Knowledge Management in Electronic Government. XI, 320 pages. 2003.

Vol. 2639: G. Wang, Q. Liu, Y. Yao, A. Skowron (Eds.), Rough Sets, Fuzzy Sets, Data Mining, and Granular Computing. XVII, 741 pages. 2003.

Vol. 2637: K.-Y. Whang, J. Jeon, K. Shim, J. Srivastava, Advances in Knowledge Discovery and Data Mining. XVIII, 610 pages. 2003.

Vol. 2636: E. Alonso, D. Kudenko, D. Kazakov (Eds.), Adaptive Agents and Multi-Agent Systems. XIV, 323 pages. 2003.

Vol. 2627: B. O'Sullivan (Ed.), Recent Advances in Constraints. X, 201 pages. 2003.

Vol. 2600: S. Mendelson, A.J. Smola (Eds.), Advanced Lectures on Machine Learning. IX, 259 pages. 2003.

Vol. 2592: R. Kowalczyk, J.P. Müller, H. Tianfield, R. Unland (Eds.), Agent Technologies, Infrastructures, Tools, and Applications for E-Services. XVII, 371 pages. 2003.

Vol. 2586: M. Klusch, S. Bergamaschi, P. Edwards, P. Petta (Eds.), Intelligent Information Agents. VI, 275 pages. 2003.

Vol. 2583: S. Matwin, C. Sammut (Eds.), Inductive Logic Programming. X, 351 pages. 2003.

Vol. 2581: J.S. Sichman, F. Bousquet, P. Davidsson (Eds.), Multi-Agent-Based Simulation. X, 195 pages. 2003.

Vol. 2577: P. Petta, R. Tolksdorf, F. Zambonelli (Eds.), Engineering Societies in the Agents World III. X, 285 pages. 2003.

Vol. 2569: D. Karagiannis, U. Reimer (Eds.), Practical Aspects of Knowledge Management. XIII, 648 pages. 2002.

Vol. 2560: S. Goronzy, Robust Adaptation to Non-Native Accents in Automatic Speech Recognition. XI, 144 pages. 2002.

Vol. 2557: B. McKay, J. Slaney (Eds.), AI 2002: Advances in Artificial Intelligence. XV, 730 pages. 2002.

Vol. 2554: M. Beetz, Plan-Based Control of Robotic Agents. XI, 191 pages. 2002.